Advances in Experimental Medicine and Biology

Volume 1362

Advances in Experimental Medicine and Biology provides a platform for scientific contributions in the main disciplines of the biomedicine and the life sciences. This series publishes thematic volumes on contemporary research in the areas of microbiology, immunology, neurosciences, biochemistry, biomedical engineering, genetics, physiology, and cancer research. Covering emerging topics and techniques in basic and clinical science, it brings together clinicians and researchers from various fields.

Advances in Experimental Medicine and Biology has been publishing exceptional works in the field for over 40 years, and is indexed in SCOPUS, Medline (PubMed), Journal Citation Reports/Science Edition, Science Citation Index Expanded (SciSearch, Web of Science), EMBASE, BIOSIS, Reaxys, EMBiology, the Chemical Abstracts Service (CAS), and Pathway Studio.

2020 Impact Factor: 2.622

More information about this series at https://link.springer.com/bookseries/5584

Mohammed S. Razzaque
Editor

Phosphate Metabolism

From Physiology to Toxicity

Springer

Editor
Mohammed S. Razzaque, MBBS, PhD
Professor of Pathology
Lake Erie College of Osteopathic Medicine
Erie, PA, USA

ISSN 0065-2598 ISSN 2214-8019 (electronic)
Advances in Experimental Medicine and Biology
ISBN 978-3-030-91621-3 ISBN 978-3-030-91623-7 (eBook)
https://doi.org/10.1007/978-3-030-91623-7

This Springer imprint is published by the registered company Springer Nature Switzerland AG
The registered company address is: Gewerbestrasse 11, 6330 Cham, Switzerland

Preface

I am pleased to present this book, entitled *Phosphate Metabolism: From Physiology to Toxicity*. Phosphate is a widely distributed mineral in the human body. It is an integral component of bone and is also involved in cell signaling, energy metabolism, and nucleic acid synthesis. Phosphate is present in virtually every human system, and its optimal balance is essential for the biological activities of the cells. This book volume contains the works of the authors who are actively involved in research in determining various aspects of phosphate regulation in health and diseases. All chapters present the current state of knowledge, ranging from physiologic regulation of phosphate homeostasis to adverse effects of phosphate toxicity, and discuss future clinical perspectives. The chapters in this book fall broadly into three groups based on their focus: (1) factors regulating phosphate homeostasis, (2) mode of cytotoxic effects of phosphate, and (3) organ damage induced by phosphate toxicity. Despite widening our understanding of in vivo regulation of phosphate, there has not been much in-depth research conducted to determine the underlying mechanisms of organ-specific phosphate toxicity and its long-term cumulative effects on human health. I expect that this book will encourage more interdisciplinary collaboration to enhance our understanding of phosphate toxicity and identify possible interventions to delay or reduce the eventual debilitating health consequences associated with the dietary phosphate burden. My time and efforts of writing, editing, and organizing this book will be worthwhile if the content inspires young physicians and scientists to take on the challenges of finding innovative ways of reducing the amount of phosphate-based preservatives in our processed foods and drinks. As editor, I hope that the broad portfolio of its contents offered to the readership will help foster a more enlightened insight into phosphate regulation in the human body.

I would like to take this opportunity to express my thanks and gratitude to each of the contributors of this book volume for sharing their knowledge and expertise. Finally, I acknowledge my family's kind support and encouragement (Rafi, Yuki, Lisa, Newaz, Zahid, Muhit, Shahed, and my mother, Nilufar Begum) that helped me complete this book.

Erie, PA, USA Mohammed S. Razzaque

Contents

Phosphate Metabolism: From Physiology to Toxicity

1

Mohammed S. Razzaque

Abstract

Systemic phosphate homeostasis is tightly controlled by the delicate cross-organ talk among intestine, kidney, bone, and parathyroid glands. The endocrine regulation of phosphate homeostasis is primarily mediated by fibroblast growth factor 23 (FGF23), vitamin D, and parathyroid hormone (PTH). Bone-derived FGF23 acts on the proximal tubular epithelial cells of the kidney to partly maintain the homeostatic balance of the phosphate. FGF23, through binding with its cell surface receptors in the presence of klotho, can activate downstream signaling kinases to reduce the functionality of the sodium-phosphate (NaPi) co-transporters of the kidney to influence the systemic phosphate homeostasis. Given the complexity of molecular regulation of phosphate homeostasis, providing information on all aspects of its homeostatic control in a single volume of a book is an overwhelming task. As the Editor, I have organized the chapters that I believe will provide necessary information on the physiologic regulation and pathologic dysregulation of phosphate in health and diseases. Readers will be able to use this volume as a quick reference for updated information on phosphate metabolism without prior acquaintance with the field.

Keywords

FGF23 · Klotho · Vitamin D · PTH · EMT · Cytotoxicity

1.1 Phosphate Homeostasis

Phosphate exerts essential biological functions in humans; around 85% of it is present in the bone [1]. Low phosphate status is linked to musculoskeletal deformities in humans, including rickets or osteomalacia and skeletal myopathy [2–4]. Contrary, high phosphate status is linked to cardiovascular calcification, commonly observed in patients with chronic kidney disease (CKD) on hemodialysis [5–8]. Fibroblast growth factor 23 (FGF23) is the master regulator of systemic phosphate homeostasis [2, 4]. Several other humoral factors, including parathyroid hormone (PTH) and 1,25-dihydroxyvitamin D [1,25 $(OH)_2D$], also play crucial roles in fine-tuning phosphate balance [9, 10]. Of relevance, phosphate, 1,25$(OH)_2D$, and PTH have feedback regulation on FGF23, while calcium can also induce FGF23 (Fig. 1.1). The osteoblasts and osteocytes mainly produce FGF23; it can specifically bind to the FGF receptors in klotho expressing organs, including kidney and parathyroid glands [11–13]. In the kidney FGF23 suppresses the

M. S. Razzaque (✉)
Department of Pathology, Lake Erie College of Osteopathic Medicine, Erie, PA, USA
e-mail: mrazzaque@lecom.edu; msr.nagasaki@gmail.com

M. S. Razzaque (ed.), *Phosphate Metabolism*, Advances in Experimental Medicine and Biology 1362,
https://doi.org/10.1007/978-3-030-91623-7_1

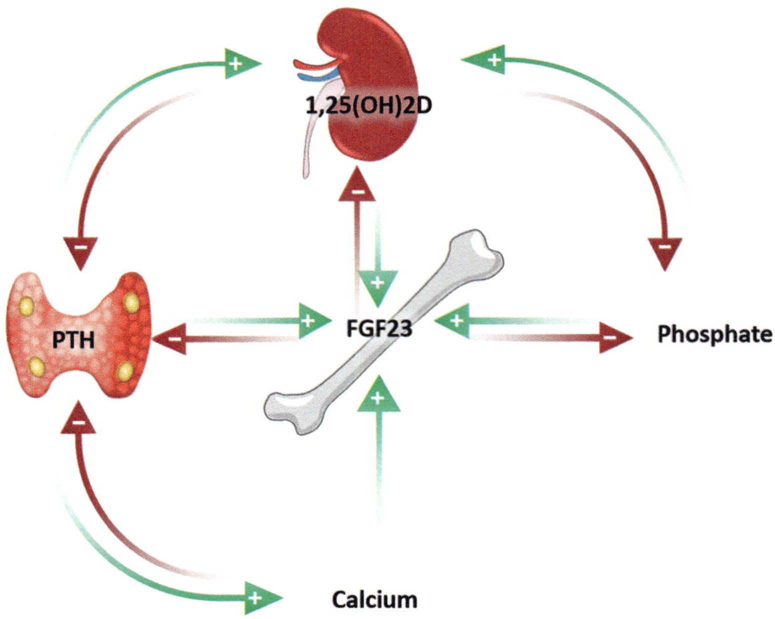

Fig. 1.1 Simplified diagram showing feed-back regulation of FGF23 by phosphate, 1,25(OH)2D, and PTH. Calcium can also positively influence FGF23 synthesis

expression of sodium-phosphate (NaPi) cotransporters (type IIa and type IIc) in the proximal renal tubules to reduce renal phosphate reabsorption [14–19]. FGF23 also reduces the generation of the active form of vitamin D [1,25 $(OH)_2D$] by suppressing the renal expression of 1α-hydroxylase, which converts inactive 25-hydroxyvitamin D [$25(OH)_2D$] to active 1,25 $(OH)_2D$. Multiple steps of vitamin D metabolism are illustrated in our earlier publications [20–25]. Of importance, $1,25(OH)_2D$ can increase intestinal phosphate absorption. FGF23, therefore, can reduce both intestinal phosphate absorption (indirectly by reducing vitamin D activities) and renal phosphate reabsorption (directly by suppressing NaPi co-transporter activities) to reduce the overall phosphate content of the body, which might reflect as a low serum level of phosphate. Bone cell-derived full-length FGF23 is biologically active [26, 27], while proteases like Furin can cleave the full-length FGF23 into inactive smaller fragments [28]. A phosphate-rich diet and $1,25(OH)_2D_3$ can increase the serum level of full-length FGF23 in experimental animals [29]. Furthermore, transforming growth factor β (TGFβ), erythropoietin (EPO), inflammation, iron deficiency, and hypoxia are the positive regulator of FGF23, whereas insulin is shown to be a negative regulator [30–33] (Fig. 1.2).

1.2 Book Chapters

The chapters in this book fall broadly into three groups based on their focus: (1) factors regulating phosphate homeostasis, (2) mode of cytotoxic effects of phosphate, and (3) organ damage induced by phosphate toxicity. The organ cross-talk during physiologic and pathologic phosphate regulations by various factors is elaborated in the chapter contributed by Akimbekov et al. [34]. The chapter is adequately illustrated to present simplistic views of the roles and regulation of FGF23, vitamin D, and PTH in controlling phosphate homeostasis [34]. In a follow-up chapter, Nakatani et al. has discussed the effects of FGF23 on vitamin D metabolism and explained how increased serum FGF23 might be an important indicator of adverse clinical outcomes of patients with kidney diseases [35]. Existing evidence suggests that even in individuals with normal renal function, FGF23 plays an important role in vitamin D metabolism; an increased level of FGF23 and a decreased level of vitamin D are

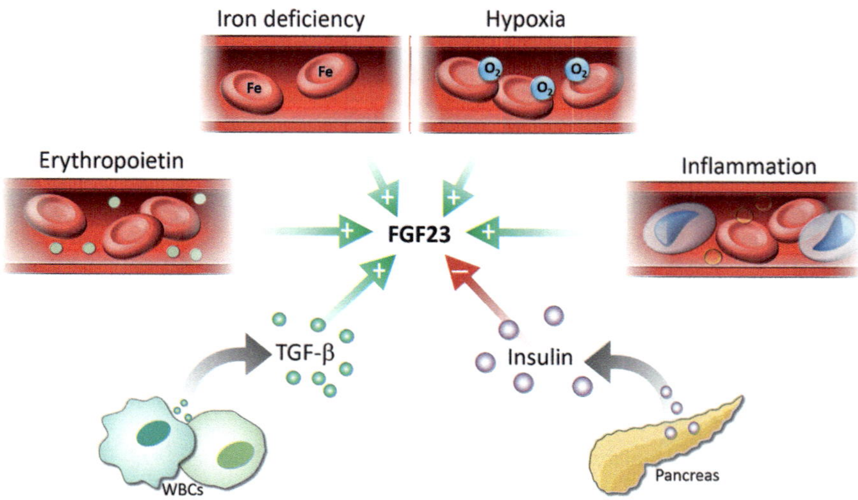

Fig. 1.2 Factors influencing production of FGF23. Transforming growth factor β (TGFβ), erythropoietin (EPO), inflammation, iron deficiency, and hypoxia are the positive regulator of FGF23, whereas insulin is a negative regulator

associated with commonly encountered adverse events in patients with CKD [36].

One of the unresolved areas of phosphate homeostasis is how the body senses the alteration of phosphate status to trigger the release of phosphate-regulatory factors. The chapter contributed by Takashi and Fukumoto detailed about phosphate sensing, and explained how FGFR1c has the potential to be a phosphate sensor [37]. Studies have also proposed PiT1, PiT2, and calcium-sensing receptor (CaSR) as phosphate-sensors, although detail mechanisms are not yet clear [37]. In a separate chapter, Abbasian and colleagues have elucidated how endothelial cells can sense the elevated inorganic phosphate concentration to generate subcellular signals to induce various vascular responses, ranging from angiogenesis to endothelial-mesenchymal transition (EndoMT) [38]. Phosphate toxicity can cause cytotoxicity (cell stress, senescence, apoptosis, and necrosis) and induce epithelial to mesenchymal transition (EMT) [39]. Potential subcellular regulations of phosphate-induced TGF-β-dependent and -independent EMT are elaborated by Lewis and colleagues [40]. Hu and Moe, in their chapter, described the underlying mechanisms of phosphate-induced cellular senescence and detailed the potential roles of klotho and plasminogen activator inhibitor-1 in senescence [41]. The role of phosphate-induced inflammation in tissue /organ damage and tumorigenesis is an evolving area of research and is briefly deliberated in another chapter (Fig. 1.3) [42]. A separate chapter by Michigami et al. discussed how phosphate could generate abnormal cell signaling and oxidative stress to induce cytotoxicity and inflammatory events [43].

Processed foods often contain high phosphate-rich additives that are more readily absorbed into the body than organic phosphate sources. The Food and Drug Administration (FDA) does not mandate food manufacturers or retailers to report per serving phosphate amounts on food labels [44]. A chapter of this book volume is devoted to elaborating the common dietary sources of natural and artificial phosphate-containing foods in the U.S. and Japan [45]. Cardiovascular anomalies are the major consequence of phosphate toxicity [5, 12, 24, 44, 46]. How phosphate can act as one of the cardiovascular toxins, is discussed in the chapter contributed by Leifheit-Nestler et al. [46]. Studies have shown that magnesium-based phosphate binders can reduce phosphate-induced cardiovascular calcification [47–49]. Bruna et al. reviewed how bacteria sense and respond to intracellular changes in phosphate and magnesium concentrations. The

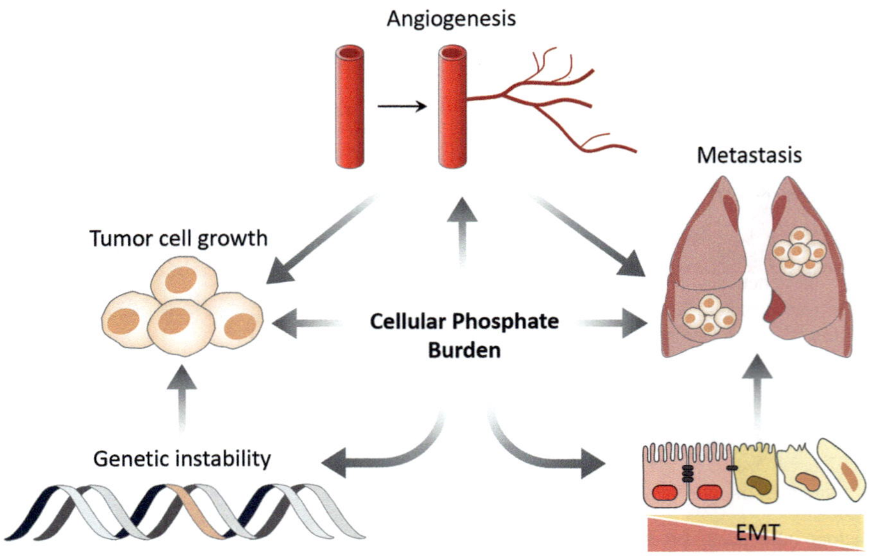

Fig. 1.3 Cellular phosphate burden can stimulate tumorigenesis, possibly by exerting mitogenic effects on tumor cells, promoting angiogenesis, inducing chromosome instability, and facilitating metastasis

authors provided experimental evidence to explain how these two minerals are functionally linked, and that excessive cytoplasmic phosphate microenvironment mimics conditions resulting from insufficient cytoplasmic magnesium and contrariwise [50].

Further understandings into the molecular details of phosphate-medicated cytotoxicity, phenotypic alteration of cell behaviors, and inflammatory events are likely to offer novel therapeutic targets for minimizing phosphate toxicity-induced tissue and organ damages, with far-reaching impacts on tumorigenesis, vasculogenesis, fibrogenesis and neuronal toxicity [51–53].

1.3 Conclusion

All the chapters in this book present the current state of understanding of physiologic and pathologic regulations of phosphate homeostasis. The wide range of topics that are covered will provide the reader with a fundamental understanding of phosphate regulation during health and diseases. The goal is to encourage more interdisciplinary collaboration to enhance our understanding of organ-specific phosphate toxicity and identify possible approaches to reduce the amount of phosphate-based preservatives used in the processed foods and drinks. The sincere expectation will be that among the readers, a few will be encouraged to take up the challenge and reap the rewards for themselves to enhance further understanding of pathomechanisms of phosphate toxicity that will lead to better patient care.

Acknowledgement I want to express my sincere gratitude to Dr. Nuraly Akimbekov (Al-Farabi Kazakh National University, Kazakhstan) for his help in drawing the illustrations. I also wish to thank Dr. Margo Wolfe for reading the manuscript and providing useful suggestions.

References

1. Osuka S, Razzaque MS (2012) Can features of phosphate toxicity appear in normophosphatemia? J Bone Miner Metab 30:10–18
2. Razzaque MS (2009) The FGF23-Klotho axis: endocrine regulation of phosphate homeostasis. Nat Rev Endocrinol 5:611–619
3. Kinoshita Y, Fukumoto S (2018) X-linked hypophosphatemia and FGF23-related

Hypophosphatemic diseases: prospect for new treatment. Endocr Rev 39:274–291
4. Razzaque MS (2009) FGF23-mediated regulation of systemic phosphate homeostasis: is Klotho an essential player? Am J Physiol Renal Physiol 296:F470–F476
5. Razzaque MS (2013) Phosphate toxicity and vascular mineralization. Contrib Nephrol 180:74–85
6. Razzaque MS (2011) The dualistic role of vitamin D in vascular calcifications. Kidney Int 79:708–714
7. Ohnishi M, Nakatani T, Lanske B, Razzaque MS (2009) In vivo genetic evidence for suppressing vascular and soft-tissue calcification through the reduction of serum phosphate levels, even in the presence of high serum calcium and 1,25-dihydroxyvitamin d levels. Circ Cardiovasc Genet 2:583–590
8. Memon F, El-Abbadi M, Nakatani T, Taguchi T, Lanske B, Razzaque MS (2008) Does Fgf23-klotho activity influence vascular and soft tissue calcification through regulating mineral ion metabolism? Kidney Int 74:566–570
9. Florenzano P, Cipriani C, Roszko KL, Fukumoto S, Collins MT, Minisola S, Pepe J (2020) Approach to patients with hypophosphataemia. Lancet Diabetes Endocrinol 8:163–174
10. Lanske B, Razzaque MS (2014) Molecular interactions of FGF23 and PTH in phosphate regulation. Kidney Int 86:1072–1074
11. Urakawa I, Yamazaki Y, Shimada T, Iijima K, Hasegawa H, Okawa K, Fujita T, Fukumoto S, Yamashita T (2006) Klotho converts canonical FGF receptor into a specific receptor for FGF23. Nature 444:770–774
12. Razzaque MS (2011) Phosphate toxicity: new insights into an old problem. Clin Sci (Lond) 120:91–97
13. Razzaque MS (2012) The role of Klotho in energy metabolism. Nat Rev Endocrinol 8:579–587
14. Nakatani T, Ohnishi M, Razzaque MS (2009) Inactivation of klotho function induces hyperphosphatemia even in presence of high serum fibroblast growth factor 23 levels in a genetically engineered hypophosphatemic (Hyp) mouse model. FASEB J 23:3702–3711
15. Nakatani T, Sarraj B, Ohnishi M, Densmore MJ, Taguchi T, Goetz R, Mohammadi M, Lanske B, Razzaque MS (2009) In vivo genetic evidence for klotho-dependent, fibroblast growth factor 23 (Fgf23) -mediated regulation of systemic phosphate homeostasis. FASEB J 23:433–441
16. Ohnishi M, Nakatani T, Lanske B, Razzaque MS (2009) Reversal of mineral ion homeostasis and soft-tissue calcification of klotho knockout mice by deletion of vitamin D 1alpha-hydroxylase. Kidney Int 75: 1166–1172
17. Breusegem SY, Takahashi H, Giral-Arnal H, Wang X, Jiang T, Verlander JW, Wilson P, Miyazaki-Anzai S, Sutherland E, Caldas Y, Blaine JT, Segawa H, Miyamoto K, Barry NP, Levi M (2009) Differential regulation of the renal sodium-phosphate cotransporters NaPi-IIa, NaPi-IIc, and PiT-2 in dietary potassium deficiency. Am J Physiol Renal Physiol 297:F350–F361
18. Fujii T, Shiozaki Y, Segawa H, Nishiguchi S, Hanazaki A, Noguchi M, Kirino R, Sasaki S, Tanifuji K, Koike M, Yokoyama M, Arima Y, Kaneko I, Tatsumi S, Ito M, Miyamoto KI (2019) Analysis of opossum kidney NaPi-IIc sodium-dependent phosphate transporter to understand Pi handling in human kidney. Clin Exp Nephrol 23:313–324
19. Shiozaki Y, Segawa H, Ohnishi S, Ohi A, Ito M, Kaneko I, Kido S, Tatsumi S, Miyamoto K (2015) Relationship between sodium-dependent phosphate transporter (NaPi-IIc) function and cellular vacuole formation in opossum kidney cells. J Med Investig 62:209–218
20. Uwitonze AM, Razzaque MS (2018) Role of magnesium in vitamin D activation and function. J Am Osteopath Assoc 118:181–189
21. Uwitonze AM, Rahman S, Ojeh N, Grant WB, Kaur H, Haq A, Razzaque MS (2020) Oral manifestations of magnesium and vitamin D inadequacy. J Steroid Biochem Mol Biol 200:105636
22. Lanske B, Razzaque MS (2007) Vitamin D and aging: old concepts and new insights. J Nutr Biochem 18: 771–777
23. Erem S, Atfi A, Razzaque MS (2019) Anabolic effects of vitamin D and magnesium in aging bone. J Steroid Biochem Mol Biol 193:105400
24. Brown RB, Haq A, Stanford CF, Razzaque MS (2015) Vitamin D, phosphate, and vasculotoxicity. Can J Physiol Pharmacol 93:1077–1082
25. Akimbekov NS, Digel I, Sherelkhan DK, Lutfor AB, Razzaque MS (2020) Vitamin D and the host-gut microbiome: a brief overview. Acta Histochem Cytochem 53:33–42
26. Fukumoto S, Yamashita T (2007) FGF23 is a hormone-regulating phosphate metabolism--unique biological characteristics of FGF23. Bone 40:1190–1195
27. Yamazaki Y, Tamada T, Kasai N, Urakawa I, Aono Y, Hasegawa H, Fujita T, Kuroki R, Yamashita T, Fukumoto S, Shimada T (2008) Anti-FGF23 neutralizing antibodies show the physiological role and structural features of FGF23. J Bone Miner Res 23:1509–1518
28. Tagliabracci VS, Engel JL, Wiley SE, Xiao J, Gonzalez DJ, Nidumanda Appaiah H, Koller A, Nizet V, White KE, Dixon JE (2014) Dynamic regulation of FGF23 by Fam20C phosphorylation, GalNAc-T3 glycosylation, and furin proteolysis. Proc Natl Acad Sci U S A 111:5520–5525
29. Saito H, Maeda A, Ohtomo S, Hirata M, Kusano K, Kato S, Ogata E, Segawa H, Miyamoto K, Fukushima N (2005) Circulating FGF-23 is regulated by 1alpha,25-dihydroxyvitamin D3 and phosphorus in vivo. J Biol Chem 280:2543–2549
30. Bar L, Feger M, Fajol A, Klotz LO, Zeng S, Lang F, Hocher B, Foller M (2018) Insulin suppresses the

production of fibroblast growth factor 23 (FGF23). Proc Natl Acad Sci U S A 115:5804–5809
31. Bar L, Stournaras C, Lang F, Foller M (2019) Regulation of fibroblast growth factor 23 (FGF23) in health and disease. FEBS Lett 593:1879–1900
32. Daryadel A, Bettoni C, Haider T, Imenez Silva PH, Schnitzbauer U, Pastor-Arroyo EM, Wenger RH, Gassmann M, Wagner CA (2018) Erythropoietin stimulates fibroblast growth factor 23 (FGF23) in mice and men. Pflugers Arch 470:1569–1582
33. Farrow EG, Yu X, Summers LJ, Davis SI, Fleet JC, Allen MR, Robling AG, Stayrook KR, Jideonwo V, Magers MJ, Garringer HJ, Vidal R, Chan RJ, Goodwin CB, Hui SL, Peacock M, White KE (2011) Iron deficiency drives an autosomal dominant hypophosphatemic rickets (ADHR) phenotype in fibroblast growth factor-23 (Fgf23) knock-in mice. Proc Natl Acad Sci U S A 108:E1146–E1155
34. Akimbekov NS, Digel I, Sherelkhan DK, Razzaque MS (2022) Vitamin D and phosphate interactions in health and disease. In: Razzaque MS (ed) Phosphate metabolism, Advances in experimental medicine and biology 1362. Springer, Cham, pp 37–46
35. Nakatani S, Nakatani A, Mori K, Emoto M, Inaba M, Razzaque MS (2022) Fibroblast growth factor 23 as regulator of vitamin D metabolism. In: Razzaque MS (ed) Phosphate metabolism, Advances in experimental medicine and biology 1362. Springer, Cham, pp 47–54
36. Razzaque MS (2009) Does FGF23 toxicity influence the outcome of chronic kidney disease? Nephrol Dial Transplant 24:4–7
37. Takashi Y, Fukumoto S (2022) Phosphate-sensing.- In: Razzaque MS (ed) Phosphate metabolism, Advances in experimental medicine and biology 1362. Springer, Cham, pp 27–36
38. Abbasian N, Bevington A, Dylan Burger D (2022) Phosphate & endothelial function: how sensing of elevated inorganic phosphate concentration generates signals in endothelial cells. In: Razzaque MS (ed) Phosphate metabolism, Advances in experimental medicine and biology 1362. Springer, Cham, pp 85–98
39. He P, Mann-Collura O, Fling J, Edara N, Hetz R, Razzaque MS (2021) High phosphate actively induces cytotoxicity by rewiring pro-survival and pro-apoptotic signaling networks in HEK293 and HeLa cells. FASEB J 35:e20997
40. Lewis E, Seltun F, Razzaque MS, He P (2022) Excessive phosphate-mediated EMT in the context of phosphate toxicity. In: Razzaque MS (ed) Phosphate metabolism, Advances in experimental medicine and biology 1362. Springer, Cham, pp 73–84
41. Hu MC, Moe OW (2022) Phosphate and cellular senescence. Adv Exp Med Biol. In: Razzaque MS (ed) Phosphate metabolism, Advances in experimental medicine and biology 1362. Springer, Cham, pp 55–72
42. Erem S, Osuka S, Razzaque MS (2022) Phosphate burden and inflammation. In: Razzaque MS (ed) Phosphate metabolism, Advances in experimental medicine and biology 1362. Springer, Cham, pp 7–14
43. Michigami T, Yamazaki M, Razzaque MS (2022) Extracellular phosphate, inflammation and cytotoxicity. In: Razzaque MS (ed) Phosphate metabolism, Advances in experimental medicine and biology 1362. Springer, Cham, pp 15–26
44. Erem S, Razzaque MS (2018) Dietary phosphate toxicity: an emerging global health concern. Histochem Cell Biol 150:711–719
45. Miyamoto K, Oh J, Razzaque MS (2022) Common dietary sources of natural and artificial phosphorus in food. In: Razzaque MS (ed) Phosphate metabolism, Advances in experimental medicine and biology 1362. Springer, Cham, pp 99–106
46. Leifheit-Nestler M, Vogt I, Haffner D, Richter B (2022) Phosphate is a cardiovascular toxin. In: Razzaque MS (ed) Phosphate metabolism, Advances in experimental medicine and biology 1362. Springer, Cham, pp 107–134
47. Guillot AP, Hood VL, Runge CF, Gennari FJ (1982) The use of magnesium-containing phosphate binders in patients with end-stage renal disease on maintenance hemodialysis. Nephron 30:114–117
48. Neven E, De Schutter TM, Dams G, Gundlach K, Steppan S, Buchel J, Passlick-Deetjen J, D'Haese PC, Behets GJ (2014) A magnesium based phosphate binder reduces vascular calcification without affecting bone in chronic renal failure rats. PLoS One 9: e107067
49. De Schutter TM, Behets GJ, Geryl H, Peter ME, Steppan S, Gundlach K, Passlick-Deetjen J, D'Haese PC, Neven E (2013) Effect of a magnesium-based phosphate binder on medial calcification in a rat model of uremia. Kidney Int 83:1109–1117
50. Bruna RE, Kendra CG, Pontes MH (2022) Coordination of phosphate and magnesium metabolism in bacteria. In: Razzaque MS (ed) Phosphate metabolism, Advances in experimental medicine and biology 1362. Springer, Cham, pp 135–150
51. Razzaque MS (2022) Salivary phosphate as a biomarker for human diseases. FASEB Bioadv 4: 102–108. https://doi.org/10.1096/fba.2021-00104
52. Hetz R, Beeler E, Janoczkin A, Kiers S, Li L, Willard BB, Razzaque MS, He P (2021) Excessive inorganic-phosphate burden perturbed intracellular signaling: quantitative proteomics and phosphoproteomic-sanalyses. Front Nutr 8:765391. https://doi.org/10.3389/fnut.2021.765391
53. Acquaviva J, Abdelhady HG, Razzaque MS (2022) Phosphate dysregulation and neurocognitive sequelae. In: Razzaque MS (ed) Phosphate metabolism, Advances in experimental medicine and biology 1362. Springer, Cham, pp 151–160

Phosphate Burden and Inflammation 2

Anna Sarah Erem, Satoko Osuka, and Mohammed S. Razzaque

Abstract

Phosphate is an essential macromineral often introduced to the body through dietary intake. The mechanisms for maintaining phosphate levels are tightly controlled via hormonal interactions and excretion via the kidneys. However, western diets consist of high levels of inorganic phosphate, which can overwhelm the regulatory mechanisms in place for maintaining homeostasis. Recent studies have found that phosphate burden can lead to activation of inflammatory signaling in various parts of the body. In addition, individuals with impaired kidney function may also experience exacerbated symptoms of phosphate overload due to decreased filtration and elimination. Many disease states can arise as a result of phosphate burden and subsequent inflammatory signaling, including cardiovascular diseases, tumorigenesis, depression, and neuronal disorders. While the pathophysiological causes of these diseases have been elucidated, there remains a need to address the clinical impacts of excessive dietary phosphate intake and to clarify potential drug candidates that may help alleviate these conditions. This brief chapter looks to explain the overall connection between phosphate burden and inflammation in various diseases.

Keywords

Inflammation · Phosphate burden · FGF23 · Cytokines · IL-1 · Tumorigenesis

2.1 Phosphate in the Human Body

Phosphate was first discovered by Hennig Brand in 1669 using a urine preparation; since then, extensive research has been performed investigating its essential role in living organisms. This mineral is a crucial component in the structure of nucleic acids and phospholipid membranes; in addition, it is involved in several biological processes, including the phosphorylation of proteins, and the formation of cyclic AMP and ATP [19]. Phosphate is a relatively abundant mineral found in the human body, measuring between 500 and 800 grams (g) [13]. The normal plasma concentration of inorganic phosphate in adults is 2.5–4.5 mg/dL; men typically have a

A. S. Erem
Department of Pathology and Laboratory Medicine, Emory University School of Medicine, Atlanta, GA, USA
e-mail: a.erem@saba.edu

S. Osuka
Department of Obstetrics and Gynecology, Nagoya University Graduate School of Medicine, Nagoya, Japan
e-mail: satokoosuka@med.nagoya-u.ac.jp

M. S. Razzaque (✉)
Department of Pathology, Lake Erie College of Osteopathic Medicine, Erie, PA, USA
e-mail: mrazzaque@lecom.edu; msr.nagasaki@gmail.com

M. S. Razzaque (ed.), *Phosphate Metabolism*, Advances in Experimental Medicine and Biology 1362,
https://doi.org/10.1007/978-3-030-91623-7_2

slightly higher concentration than women [13]. Around 85% of phosphate is found in the bones and teeth, 10–15% in soft tissues, and approximately 1% in blood. Serum phosphate levels, therefore, might not always reflect the total amount and distribution of phosphate. Blood is constantly being filtered by the kidneys, and phosphate is partly reabsorbed there. In individuals with a normal phosphate intake, between 80 and 90% of filtered phosphate is reabsorbed [17].

Fibroblastic growth factor 23 (FGF23), calcitriol, and parathyroid hormone (PTH) levels are responsible for regulating phosphate concentrations in the body by regulating the amount of urinary excretion [31, 32]. Intake of dietary phosphate stimulates PTH secretion which in turn increases bone FGF23 synthesis and release, and is key in triggering the synthesis of calcitriol by the kidneys. FGF23 suppresses both PTH and calcitriol levels, and calcitriol inhibits PTH synthesis and secretion, while stimulating FGF23 release. The regulatory triangle formed between these three substrates creates a fine balance for the maintenance of serum phosphate levels. Since filtration and urinary excretion rely heavily on the kidneys, individuals with reduced kidney function are susceptible to high levels of phosphate in their blood. This disruption in homeostasis can lead to the development of hyperphosphatemia.

Hyperphosphatemia is a condition defined by high plasma concentrations of phosphate (>4.5 mg/dL) [19]. It is often caused by a decrease in renal processing and excretion of phosphate; in other cases, thyroid diseases such as hypoparathyroidism or pseudohypothyroidism may be to blame [19]. While most individuals with hyperphosphatemia are asymptomatic, other conditions, such as hypocalcemia may exacerbate symptoms (e.g., tetany) [13]. Hyperphosphatemia is sometimes associated with concomitant cardiovascular disease in patients with chronic kidney disease (CKD). In those with normal kidney function, elevated phosphate levels have been associated with cardiovascular events and vascular calcification.

2.2 Inflammation Associated with Phosphate Toxicity

Inflammation is a biological process generated in response to pathogens and tissue injury. It is a feed-forward system that, once activated, continues to recruit inflammatory cells and drive inflammation at the cellular level. While it is beneficial in most cases, chronic inflammation can increase the risk of developing diseases such as cancer, CKD, and cardiovascular disease (CVD). Traditionally, identified mediators of inflammation are cytokines and chemokines; however, lipid mediators involving phosphate may also play a role in inflammation, and tumorigenesis [13].

Phosphate is a relatively abundant mineral found in almost all the places in the human body. While it has important physiologic roles in maintaining musculoskeletal functions, it can also become pathogenic when it precipitates with calcium in extra-skeletal systems. This ectopic precipitation can lead to cell damage and initiate inflammatory responses. In healthy individuals, calcium-phosphate crystals are adsorbed by the serum protein fetuin-A; this interaction prevents the crystals from forming large, pathogenic precipitates. However, these formations can also break off into nanoparticles, known as calciprotein particles (CPPs), which circulate in the blood. Over time, serum CPP levels increase due to aging and elevated phosphate. CPPs have been implicated in chronic non-infectious inflammation and vascular stiffness [17].

The kidneys are vital in maintaining phosphate homeostasis, as excess phosphate is excreted from the body in urine. Elevated levels of plasma FGF23 have been linked to expression of inflammatory cytokines (IL-6, TNF-α), C-reactive protein (CRP), fibrinogen, and flares in autoimmune diseases in patients with CKD [39]. Increases in serum phosphate have also been found to promote inflammation in individuals with CKD via induction of inflammatory gene programs in the liver [9]. In rat models of CKD, fluctuations in dietary phosphate intake were found to influence inflammation and vascular calcification [37].

The role of dietary phosphorus in inflammation and oral diseases has also been studied in human diseases. Of relevance, Western diets contain high levels of refined carbohydrates, fat, sodium, and phosphorus, contributing to the growing number of cardiovascular and metabolic diseases. For instance, inorganic phosphate salts are used as a flavor enhancer and preservative [17]. Ingested phosphate is absorbed through the small intestine after it is cleaved by alkaline phosphatases in the cells lining the intestine [17]. Higher phosphorus intake significantly correlated with the presence of gingivitis and elevated levels of pro-inflammatory cytokine IL-1 beta, and inversely correlated with the anti-inflammatory cytokine IL-4 in saliva [12].

As mentioned, phosphate toxicity is associated with the activation of cellular stress response systems and inflammation. Cortisol, which is released by the hypothalamic-pituitary-adrenal axis response to stress and inflammation, is speculated to be associated with phosphate toxicity and depression [6]. In turn, phosphate toxicity may negatively impact adrenal gland function, potentially leading to adrenal insufficiency and increased depression. Furthermore, Alzheimer's disease is associated with hyperphosphorylated tau protein which self-assembles into neurofibrillary tangles, perhaps utilizing excessive amounts of phosphate in the brain and central nervous system (CNS) [6].

2.3 Mechanisms of Phosphate-Induced Inflammation

Exposure to inorganic phosphate activates inflammatory signaling pathways, particularly through activation of the transcription factor nuclear factor κ-light-chain-enhancer of activated B cells (NK- κB) [45]. This has been studied extensively in vascular smooth muscle cells (VSMCs) in regard to inflammatory signaling and its effects on calcification in cardiovascular disease. NF-κB affects multiple signaling pathways to mediate inflammatory signaling and calcification. For example, it inhibits IκB kinase (IKK), which is known to have an anti-calcific effect [1, 41]. NF-κ B can also be activated by the pro-inflammatory mediator TNF and TNF-related weak inducer of apoptosis (TWEAK) [16]. VSMCs that have been exposed to high levels of inorganic phosphate *in vitro* have been shown to release the inflammatory cytokine IL-6 [44]. IL-6 induces expression of CBFA1 through activation of the transcription factor STAT3 [18]. In addition, IL-6 plays a role in the inorganic phosphate-induced senescence of VSMCs. This is linked to oxidative stress, which activates a DNA damage response and can further induce inflammation [10, 43]. Exposure to inorganic phosphate can also result in upregulation of IL-1β and the NALP3 inflammasome, which can promote senescence [15, 42].

2.4 Phosphate Toxicity and Cardiovascular Disease

Phosphate toxicity is associated with an increased risk of cardiovascular disease [11, 28, 34]. This includes heart failure, chronic obstructive pulmonary disease (COPD), and cardiovascular mortality caused by stroke or noncoronary heart diseases [39]. While it has been shown that patients with CKD are more susceptible to CVD, circulating excess phosphate can also play a role in vascular calcification and cardiovascular events in individuals with normal kidney function [8, 39].

Research now shows that the connection between hyperphosphatemia and cardiovascular events is linked to pro-inflammatory signaling and effects [3, 38, 40]. Circulating phosphate levels can be associated with inflammatory mediators such as IL-6 and CRP [39]. Phosphate exposure activates pro-inflammatory cellular signaling via the expression of the transcription factor NF-κB in VSMCs. In VMSCs, exposure to phosphate can initiate oxidative stress; however, when phosphate burden is pharmacologically blocked, VSMC-calcification is inhibited

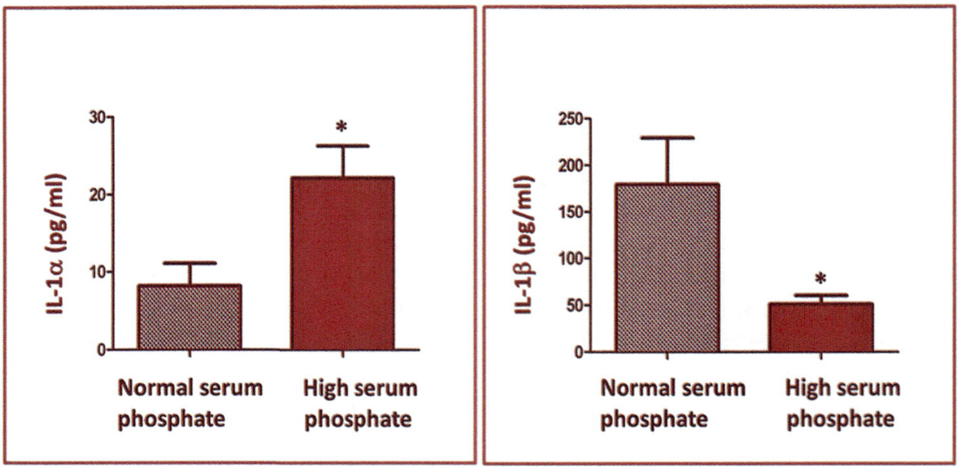

Fig. 2.1 Serum IL-1α and IL-1β levels in hyperphosphatemic klotho knockout mice ($n = 5$) and age-matched (6-weeks-old) normophosphatemic wild-type mice ($n = 4$). Note that compared to control wild-type mice (with normal serum phosphate), klotho knockout mice (with high serum phosphate) show elevated IL-1α and reduced IL-1β levels (*: $p < 0.05$)

[2, 20]. Oxidative stress causes DNA damage, which then activates a DNA damage response. This process can lead to more inflammation and vascular calcification [36].

In CKD patients, high phosphate levels are associated with cardiovascular disease. Multiple *in vitro* studies have investigated the effects of phosphate concentrations on human aortic smooth muscle cells. Exposure to phosphate induces the expression of pro-inflammatory mediators such as interleukins (IL-1β, IL-6, IL-8), and TNF-α [21, 44, 46]. In addition, an increase in reactive oxygen/nitrogen species (ROS/RNS) production, can be detected, further indicating an inflammatory response [21]. Animal models of CKD have also demonstrated that high-phosphate diets induce inflammation both locally in the arteries and systemically, which may also play a role in the pathogenesis of the vascular calcification associated with hyperphosphatemia [29]. It is worth noting, klotho knockout mice are a well-studied model for phosphate toxicity that can lead to cardiovascular calcification and premature aging [22–26, 33, 35]. Dysregulated inflammatory cytokines were detected in serum collected from hyperphosphatemic klotho knockout mice compared to their age-matched normophosphatemic wild-type mice (Fig. 2.1). Such molecular signature of inflammation in klotho knockout mice with phosphate toxicity was associated with significantly reduced longevity. More importantly, reducing phosphate toxicity in klotho knockout mice could reduce cardiovascular calcification and increase longevity [23, 27].

2.5 Phosphate Toxicity and Tumorigenesis

Phosphate toxicity leading to cellular phosphate burden has been identified as a cause of cancer cell growth. Tumor cells are unable to regulate phosphate homeostasis due to their higher expression of phosphate cotransporters and ability to store more phosphate than healthy cells. Animal models have demonstrated that high dietary intake of phosphate leads to the growth of skin and lung tumors [39].

A potential interaction between the excess of blood vessels supplying a tumor (inflammatory hyperemia) and hyperphosphatemia has been speculated in the formation and progress of cancer [5]. Hyperemia increases the blood flow rate and volume to tumors. This combined with the inflammatory nature of hyperphosphatemia leads to an increase in inorganic phosphate circulating within the tumor microenvironment [5]. Elevated intracellular phosphorus levels may trigger the synthesis of ribosomal RNA (rRNA), which leads to increased protein synthesis, driving tumor growth [5]. Research has also been performed investigating the role of sphingosine-1-phosphate (S1P) and ceramide-1-phosphate (C1P) in inflammation and cancer [7, 14]. These pro-inflammatory metabolites have been found to participate in and promote tumorigenesis [4, 30].

2.6 Conclusions

Phosphate toxicity can induce a wide range of tissue injuries, including cardiovascular damage and premature aging (Fig. 2.2) [26, 27]. Recent investigations have highlighted the inflammatory effects of excess phosphate and various disease states. An elevated level of circulating phosphate is partly the result of an increase in dietary phosphate intake, insufficient kidney function, and abnormal FGF23 and PTH levels. Combined, these factors can lead to tumorigenesis, cardiovascular events, and mental health conditions. These facts raise the point that phosphate intake should be monitored, particularly in susceptible groups such as those with chronic kidney disease. Further work is needed in elucidating the exact mechanisms and treatments available to prevent inflammation from worsening any of these conditions.

Acknowledgement We want to express our sincere gratitude to Dr. Nuraly Akimbekov (Al-Farabi Kazakh National University, Kazakhstan) for his help in drawing the illustration.

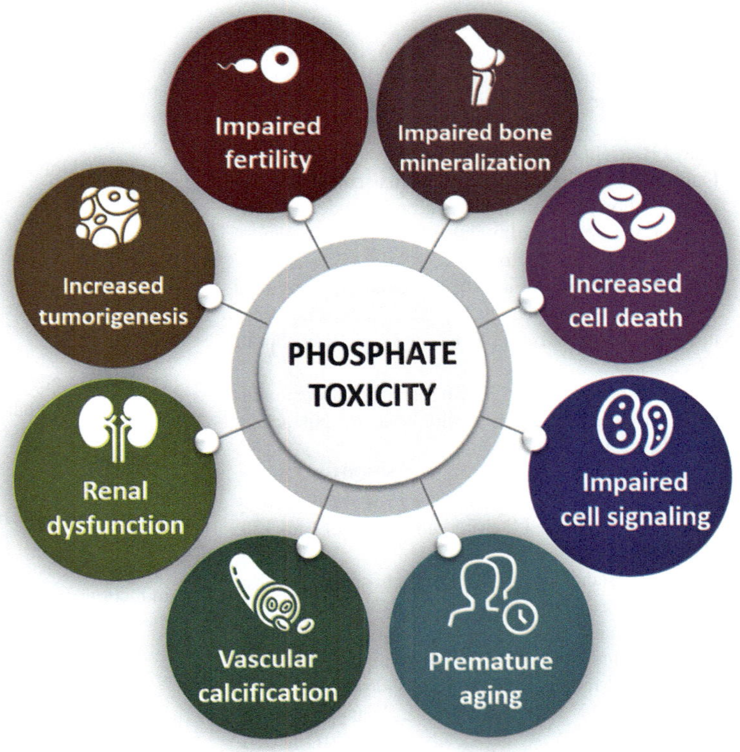

Fig. 2.2 Partial list of various pathologies induced by phosphate toxicity [34]

References

1. Al-Huseini I, Ashida N, Kimura T (2018) Deletion of ikappaB-kinase beta in smooth muscle cells induces vascular calcification through beta-catenin-runt-related transcription factor 2 signaling. J Am Heart Assoc 7
2. Alesutan I, Moritz F, Haider T, Shouxuan S, Gollmann-Tepekoylu C, Holfeld J, Pieske B, Lang F, Eckardt KU, Heinzmann SS, Voelkl J (2020) Impact of beta-glycerophosphate on the bioenergetic profile of vascular smooth muscle cells. J Mol Med (Berl) 98: 985–997
3. Benz K, Hilgers KF, Daniel C, Amann K (2018) Vascular calcification in chronic kidney disease: the role of inflammation. Int J Nephrol 2018:4310379
4. Berwick ML, Dudley BA, Maus K, Chalfant CE (2019) The role of ceramide 1-phosphate in inflammation, cellular proliferation, and wound healing. Adv Exp Med Biol 1159:65–77
5. Brown RB (2019) Potential interaction of inflammatory hyperemia and hyperphosphatemia in tumorigenesis. Future Oncol 15:3909–3916
6. Brown RB (2020) Stress, inflammation, depression, and dementia associated with phosphate toxicity. Mol Biol Rep 47:9921–9929
7. Brown RB, Razzaque MS (2018) Phosphate toxicity and tumorigenesis. Biochim Biophys Acta Rev Cancer 1869:303–309
8. Cozzolino M, Ciceri P, Galassi A, Mangano M, Carugo S, Capelli I, Cianciolo G (2019) The key role of phosphate on vascular calcification. Toxins (Basel) 11
9. Czaya B, Richter B, Yanucil C, Campos I, Heitman K, Faul C (2019) Hyperphosphatemia contributes to inflammation and iron dysregulation in models of Normal and impaired renal function. American Society of Hematology, Washington, DC
10. Dai L, Qureshi AR, Witasp A, Lindholm B, Stenvinkel P (2019) Early vascular ageing and cellular senescence in chronic kidney disease. Comput Struct Biotechnol J 17:721–729
11. Erem S, Razzaque MS (2018) Dietary phosphate toxicity: an emerging global health concern. Histochem Cell Biol 150:711–719
12. Goodson JM, Shi P, Razzaque MS (2019) Dietary phosphorus enhances inflammatory response: a study of human gingivitis. J Steroid Biochem Mol Biol 188: 166–171
13. Goyal R, Jialal I (2020) Hyperphosphatemia. In: StatPearls. Treasure Island
14. Hait NC, Maiti A (2017) The role of sphingosine-1-phosphate and ceramide-1-phosphate in inflammation and cancer. Mediat Inflamm 2017:4806541
15. Han L, Zhang Y, Zhang M, Guo L, Wang J, Zeng F, Xu D, Yin Z, Xu Y, Wang D, Zhou H (2020) Interleukin-1beta-induced senescence promotes osteoblastic transition of vascular smooth muscle cells. Kidney Blood Press Res 45:314–330
16. Henaut L, Sanz AB, Martin-Sanchez D, Carrasco S, Villa-Bellosta R, Aldamiz-Echevarria G, Massy ZA, Sanchez-Nino MD, Ortiz A (2016) TWEAK favors phosphate-induced calcification of vascular smooth muscle cells through canonical and non-canonical activation of NFkappaB. Cell Death Dis 7:e2305
17. Kuro OM (2020) Phosphate as a pathogen of arteriosclerosis and aging. J Atheroscler Thromb
18. Kurozumi A, Nakano K, Yamagata K, Okada Y, Nakayamada S, Tanaka Y (2019) IL-6 and sIL-6R induces STAT3-dependent differentiation of human VSMCs into osteoblast-like cells through JMJD2B-mediated histone demethylation of RUNX2. Bone 124:53–61
19. Lewis Iii JL (2020) Hyperphosphatemia [Online]. Merck Manual. Available: https://www.merckmanuals.com/professional/endocrine-and-metabolic-disorders/electrolyte-disorders/hyperphosphatemia [Accessed]
20. Luong TTD, Schelski N, Boehme B, Makridakis M, Vlahou A, Lang F, Pieske B, Alesutan I, Voelkl J (2018) Fibulin-3 attenuates phosphate-induced vascular smooth muscle cell calcification by inhibition of oxidative stress. Cell Physiol Biochem 46:1305–1316
21. Martinez-Moreno JM, Herencia C, De Oca AM, Diaz-Tocados JM, Vergara N, Gomez-Luna MJ, Lopez-Arguello SD, Camargo A, Peralbo-Santaella E, Rodriguez-Ortiz ME, Canalejo A, Rodriguez M, Munoz-Castaneda JR, Almaden Y (2017) High phosphate induces a pro-inflammatory response by vascular smooth muscle cells and modulation by vitamin D derivatives. Clin Sci (Lond) 131:1449–1463
22. Nakatani T, Ohnishi M, Razzaque MS (2009a) Inactivation of klotho function induces hyperphosphatemia even in presence of high serum fibroblast growth factor 23 levels in a genetically engineered hypophosphatemic (Hyp) mouse model. FASEB J 23:3702–3711
23. Nakatani T, Sarraj B, Ohnishi M, Densmore MJ, Taguchi T, Goetz R, Mohammadi M, Lanske B, Razzaque MS (2009b) In vivo genetic evidence for klotho-dependent, fibroblast growth factor 23 (Fgf23) -mediated regulation of systemic phosphate homeostasis. FASEB J 23:433–441
24. Ohnishi M, Kato S, Akiyoshi J, Atfi A, Razzaque MS (2011a) Dietary and genetic evidence for enhancing glucose metabolism and reducing obesity by inhibiting klotho functions. FASEB J 25:2031–2039
25. Ohnishi M, Kato S, Razzaque MS (2011b) Genetic induction of phosphate toxicity significantly reduces the survival of hypercholesterolemic obese mice. Biochem Biophys Res Commun 415:434–438
26. Ohnishi M, Nakatani T, Lanske B, Razzaque MS (2009) In vivo genetic evidence for suppressing vascular and soft-tissue calcification through the reduction of serum phosphate levels, even in the presence of high serum calcium and 1,25-dihydroxyvitamin d levels. Circ Cardiovasc Genet 2:583–590

27. Ohnishi M, Razzaque MS (2010) Dietary and genetic evidence for phosphate toxicity accelerating mammalian aging. FASEB J 24:3562–3571
28. Osuka S, Razzaque MS (2012) Can features of phosphate toxicity appear in normophosphatemia? J Bone Miner Metab 30:10–18
29. Peri-Okonny P, Baskin KK, Iwamoto G, Mitchell JH, Smith SA, Kim HK, Szweda LI, Bassel-Duby R, Fujikawa T, Castorena CM, Richardson J, Shelton JM, Ayers C, Berry JD, Malladi VS, Hu MC, Moe OW, Scherer PE, Vongpatanasin W (2019) High-phosphate diet induces exercise intolerance and impairs fatty acid metabolism in mice. Circulation 139:1422–1434
30. Presa N, Gomez-Larrauri A, Rivera IG, Ordonez M, Trueba M, Gomez-Munoz A (2016) Regulation of cell migration and inflammation by ceramide 1-phosphate. Biochim Biophys Acta 1861:402–409
31. Razzaque MS (2009a) The FGF23-Klotho axis: endocrine regulation of phosphate homeostasis. Nat Rev Endocrinol 5:611–619
32. Razzaque MS (2009b) FGF23-mediated regulation of systemic phosphate homeostasis: is Klotho an essential player? Am J Physiol Renal Physiol 296:F470–F476
33. Razzaque MS (2011a) Osteo-renal regulation of systemic phosphate metabolism. IUBMB Life 63:240–247
34. Razzaque MS (2011b) Phosphate toxicity: new insights into an old problem. Clin Sci (Lond) 120: 91–97
35. Razzaque MS (2014) Bone-kidney axis in systemic phosphate turnover. Arch Biochem Biophys 561: 154–158
36. Sanchis P, Ho CY, Liu Y, Beltran LE, Ahmad S, Jacob AP, Furmanik M, Laycock J, Long DA, Shroff R, Shanahan CM (2019) Arterial "inflammaging" drives vascular calcification in children on dialysis. Kidney Int 95:958–972
37. Tani M, Tanaka S, Takamiya K, Sakaue M, Ito M (2020) Effects of repetitive diet-induced fluctuations in plasma phosphorus on vascular calcification and inflammation in rats with early-stage chronic kidney disease. J Clin Biochem Nutr 66:139–145
38. Voelkl J, Cejka D, Alesutan I (2019a) An overview of the mechanisms in vascular calcification during chronic kidney disease. Curr Opin Nephrol Hypertens 28:289–296
39. Voelkl J, Egli-Spichtig D, Alesutan I, Wagner CA (2021) Inflammation: a putative link between phosphate metabolism and cardiovascular disease. Clin Sci (Lond) 135:201–227
40. Voelkl J, Lang F, Eckardt KU, Amann K, Kuro OM, Pasch A, Pieske B, Alesutan I (2019b) Signaling pathways involved in vascular smooth muscle cell calcification during hyperphosphatemia. Cell Mol Life Sci 76:2077–2091
41. Voelkl J, Luong TT, Tuffaha R, Musculus K, Auer T, Lian X, Daniel C, Zickler D, Boehme B, Sacherer M, Metzler B, Kuhl D, Gollasch M, Amann K, Muller DN, Pieske B, Lang F, Alesutan I (2018) SGK1 induces vascular smooth muscle cell calcification through NF-kappaB signaling. J Clin Invest 128: 3024–3040
42. Wen C, Yang X, Yan Z, Zhao M, Yue X, Cheng X, Zheng Z, Guan K, Dou J, Xu T, Zhang Y, Song T, Wei C, Zhong H (2013) Nalp3 inflammasome is activated and required for vascular smooth muscle cell calcification. Int J Cardiol 168:2242–2247
43. Xu D, Zeng F, Han L, Wang J, Yin Z, Lv L, Guo L, Wang D, Xu Y, Zhou H (2019) The synergistic action of phosphate and interleukin-6 enhances senescence-associated calcification in vascular smooth muscle cells depending on p53. Mech Ageing Dev 182: 111124
44. Zhang D, Bi X, Liu Y, Huang Y, Xiong J, Xu X, Xiao T, Yu Y, Jiang W, Huang Y, Zhang J, Zhang B, Zhao J (2017) High phosphate-induced calcification of vascular smooth muscle cells is associated with the TLR4/NF-kappab signaling pathway. Kidney Blood Press Res 42:1205–1215
45. Zhao MM, Xu MJ, Cai Y, Zhao G, Guan Y, Kong W, Tang C, Wang X (2011) Mitochondrial reactive oxygen species promote p65 nuclear translocation mediating high-phosphate-induced vascular calcification in vitro and in vivo. Kidney Int 79:1071–1079
46. Zickler D, Luecht C, Willy K, Chen L, Witowski J, Girndt M, Fiedler R, Storr M, Kamhieh-Milz J, Schoon J, Geissler S, Ringden O, Schindler R, Moll G, Dragun D, Catar R (2018) Tumour necrosis factor-alpha in uraemic serum promotes osteoblastic transition and calcification of vascular smooth muscle cells via extracellular signal-regulated kinases and activator protein 1/c-FOS-mediated induction of interleukin 6 expression. Nephrol Dial Transplant 33:574–585

Extracellular Phosphate, Inflammation and Cytotoxicity

3

Toshimi Michigami, Miwa Yamazaki, and Mohammed S. Razzaque

Abstract

Phosphorus is an essential nutrient that plays a crucial role in various biological processes, including cell membrane integrity, synthesis of nucleic acids, energy metabolism, intracellular signaling, and hard tissue mineralization. Therefore, the control of phosphorus balance is critical in all living organisms, and the fibroblast growth factor 23 (FGF23)-αKlotho system is central to maintain phosphate homeostasis in mammals. Although phosphate is indispensable for basic cellular functions, its excessive retention is toxic and can affect almost all organ systems' functionality. In human patients, hyperphosphatemia has been implicated in an increase in morbidity and mortality. Also, mouse models with hyperphosphatemia generated by disruption of the FGF23-αKlotho system exhibit extensive tissue damage, premature aging, and a short lifespan. Experimental studies using cell and animal models suggest that cytotoxic and inflammatory effects of elevated phosphate are partly mediated by abnormal cell signaling and oxidative stress. This review provides an overview of our current understanding regarding the toxicity of phosphate.

Keywords

Phosphate toxicity · Fibroblast growth factor 23 (FGF23) · αKlotho · Phosphate-induced signaling · Oxidative stress

3.1 Introduction

Phosphorus is an essential nutrient for all organisms, being involved in various biological processes that include cellular membrane composition, synthesis of deoxyribonucleic acid (DNA) and ribonucleic acid (RNA), intracellular signaling and energy metabolism, as well as skeletal mineralization in vertebrates. Phosphorus ingested through food usually exists as both organic and inorganic forms of phosphate in the body. In human adults, ~90% of the total phosphorus is distributed in the bone as hydroxyapatite (calcium-phosphate) crystals, and the remainder is mostly present in soft tissues. The extracellular fluid contains <1% of phosphorus [54, 55]. Phosphorus in serum exists mostly as inorganic phosphate (Pi) in the form of free ions HPO_4^{2-} and $H_2PO_4^{-}$, and HPO_4^{2-} is dominant in physiological pH. Intracellular phosphate mostly exists as a bound form or inorganic

T. Michigami (✉) · M. Yamazaki
Department of Bone and Mineral Research, Research Institute, Osaka Women's and Children's Hospital, Izumi, Osaka, Japan
e-mail: michigami@wch.opho.jp; miwayama@wch.opho.jp

M. S. Razzaque
Department of Pathology, Lake Erie College of Osteopathic Medicine, Erie, PA, USA
e-mail: mrazzaque@lecom.edu

M. S. Razzaque (ed.), *Phosphate Metabolism*, Advances in Experimental Medicine and Biology 1362,
https://doi.org/10.1007/978-3-030-91623-7_3

phosphate esters, phospholipids in the cell membrane, or phosphorylated intermediate molecules involved in various biochemical processes, which include the generation, storage, and transport of cellular energy through the formation of adenosine 5′-triphosphate (ATP) by oxidative phosphorylation [54, 55].

Mammalian cells take up Pi from extracellular fluid mainly through membrane transporters that function dependently on sodium (Na^+) gradient across the plasma membrane. These Na^+-dependent Pi transporters (Na^+/Pi cotransporters) have been classified into three families in mammals [87]. Type I transporter that belongs to the solute carrier family 17 (SLC17) is involved in the transport of organic ions in addition to Pi, and human type 1 transporter NPT1 has been suggested to function as a urate transporter [16]. Among type II transporters, type IIa and IIc encoded by the *SLC34A1* and *SLC34A3* genes in human, respectively, are predominantly expressed in brush border membrane of proximal tubules in the kidney and responsible for renal reabsorption of Pi, while type IIb transporter encoded by *SLC34A2* is expressed in various tissues and accountable for active transcellular absorption of Pi in the small intestine. Type III Na^+/Pi cotransporters PiT1 and PiT2 encoded by *SLC20A1* and SLC20A2, respectively, are widely expressed with different expression patterns, and their main function is likely to supply Pi to individual cells [32, 87].

Renal reabsorption and intestinal absorption of Pi are critical for Pi homeostasis in mammals. Renal Pi reabsorption by type IIa and IIc Na^+/Pi cotransporters (NaPi2a and NaPi2c) is suppressed by several humoral factors, including parathyroid hormone (PTH) and fibroblast growth factor 23 (FGF23), which increases urinary Pi excretion [66, 67]. Intestinal Pi absorption by type IIb Na^+/Pi cotransporter (NaPi2b) is increased by both 1,25-dihydroxyvitamin D [1,25(OH)$_2$D] and low dietary phosphate [30]. It was reported that mice deficient for NaPi2b absorbed approximately 50% less phosphate than wild-type animals, confirming the major contribution of this transporter in Pi homeostasis [76]. However, in humans, the daily need for phosphate is covered by intestinal absorption from ingested food, so serum Pi level is virtually maintained by renal phosphate excretion.

3.2 FGF23-αKlotho System and Phosphate Metabolism

Mounting evidence has established the central roles of FGF23 in Pi metabolism. FGF23 is a secreted protein of 32 kDa produced by bone and exerts its effects on the distant target organs, including the kidney, in an endocrine fashion. In the kidney, FGF23 increases Pi excretion by suppressing the expression of NaPi2a and NaPi2c. Also, FGF23 decreases the production of 1,25(OH)$_2$D by suppressing the expression of 25-hydroxyvitamin D-1α-hydroxylase and inducing that of 25-hydroxyvitamin D-24-hydroxylase [55, 66, 67]. To evoke its signal through FGF receptor (FGFR), FGF23 requires αKlotho [46, 86]. αKlotho is a 130-kDa type I membrane protein, which was initially identified as an aging-related factor [45]. In addition to the transmembrane form, soluble form of αKlotho is also detectable in serum and CSF, although its physiological roles are not clarified yet [37, 93, 94]. Because of the predominant expression of a transmembrane form of αKlotho in the kidney, parathyroid gland, and the choroid plexus, these organs have been considered to be the physiological targets for FGF23 action [45]. In the parathyroid gland, it has been shown that FGF23 suppresses the gene expression and secretion of PTH [8]. In addition, we previously reported that the placenta also expresses αKlotho and that a high level of maternal FGF23 up-regulates the placental expression of the *Cyp24a1* gene encoding 25-hydroxyvitamin D-24-hydroxylase and affects fetal vitamin D metabolism in a mouse model of human X-linked hypophosphatemic rickets [61, 62].

Disruption of FGF23-αKlotho system by loss-of-function mutations in FGF23, αKlotho, or GalNAc-transferase 3 that is an enzyme responsible for *O*-glycosylation of FGF23, causes hyperphosphatemic familial tumoral calcinosis [3, 35, 36]. Similarly, mice deficient for *Fgf23* or *αKlotho* also exhibit hyperphosphatemia and increased level of 1,25$(OH)_2$D [70, 77]. Thus, FGF23-αKlotho system is central in Pi and vitamin D metabolism, and its impairment leads to excessive retention of Pi in the body.

On the other hand, the excess action of FGF23 has been implicated in the pathogenesis of various hypophosphatemic diseases, including some heritable rickets/osteomalacia [26]. Mutations in FGF23 that make the protein resistant to cleavage are responsible for autosomal dominant hypophosphatemic rickets (ADHR), which is characterized by renal Pi wasting, hypophosphatemia, and inappropriately low levels of serum 1,25$(OH)_2$D [2]. In addition, loss-of-function mutations in several genes such as the p*hosphate-regulating gene homologous to endopeptidases on X chromosome* (*PHEX*), *dentin matrix protein 1* (*DMP1*) and *family with sequence similarity 20, member C* (*FAM20C*) cause increased levels of FGF23 and hypophosphatemic rickets [25, 34, 49, 84], and overproduction of FGF23 by tumors also causes hypophosphatemic osteomalacia [78]. Recently, Burosumab, a monoclonal antibody against FGF23, has been developed as a new drug to treat FGF23-related hypophosphatemic rickets/osteomalacia [13, 26].

Although the physiological actions of FGF23 require αKlotho, recent studies have suggested that massively elevated levels of circulating FGF23 as found in patients with chronic kidney disease (CKD) might result in pathological changes in cells and tissues lacking αKlotho [74].

3.3 Phosphate and Morbidity and Mortality

Although phosphorus is an essential nutrient for all living organisms, excessive retention of Pi in the body is toxic and causes various cellular and tissue injuries [24, 38, 68, 69]. Mice deficient for *Fgf23* or *αKlotho*, which exhibit markedly elevated levels of serum Pi and 1,25$(OH)_2$D, suffer from premature aging, vascular calcification, and early mortality. They also manifest hypogonadism, infertility, emphysema, and generalized tissue atrophy [45, 56]. Interestingly, it was shown that a low-phosphate diet corrected hyperphosphatemia, prevented vascular calcification, and prolonged survival in *Fgf23*-deficient mice, despite the persistent elevation in 1,25$(OH)_2$D levels [82]. Moreover, genetic inactivation of NaPi2a in *αKlotho*-deficient mice restored severe hyperphosphatemia and reduced vascular and soft tissue calcification, even in the presence of extremely high levels of serum calcium and 1,25$(OH)_2$D. These mice lacking both *NaPi2a* and *αKlotho* also recovered body weight, regained reproductive ability, reduced their tissue atrophy, and exhibited the longer survival, compared to the mice deficient for *αKlotho* alone [63, 64]. However, when fed with a high-phosphate diet, the *NaPi2a*/*αKlotho* double-deficient mice again suffered from premature aging and shortened lifespan [64]. These observations in mouse models clearly indicate that excessive Pi can be toxic to multiple organs and accelerates the aging process.

Human studies also have suggested the toxic effects of Pi. In CKD (CKD), elevated serum Pi has been implicated as a risk factor for cardiovascular diseases and higher mortality in both dialysis and non-dialysis patients [9, 23, 27, 43, 65, 81]. Hyperphosphatemia and elevated calcium-phosphate product levels are associated with calcification of soft tissues and blood vessel walls in CKD patients and are predictive of high morbidity and mortality [20, 27]. In addition to the passive precipitation of calcium-phosphate in soft tissues, substantial evidence suggests that high extracellular Pi induces the expression of osteoblastic genes in vascular smooth muscle cells, which contributes to calcification [40]. It has also been reported that higher serum Pi is associated with high mortality even in individuals with preserved renal function. Tonelli, et al. performed a post hoc analysis of the data from the Cholesterol And Recurrent Events (CARE) study and found a graded independent relation between higher serum Pi and the risk of death and cardiovascular events in 4127 subjects, most of whom had serum Pi levels within the normal range [85]. Li, et al. performed a meta-analysis study using 24 clinical trials with a total of

147,634 patients without CKD and revealed a positive association between serum Pi level and mortality [47]. A prospective study demonstrated that a higher serum Pi predicted mortality in renal transplantation recipients [18].

Phosphate overload can also be associated with a diet rich in phosphate additives and treatment with phosphate-containing laxatives or enema [7]. It was reported that the high-phosphate diet containing much food additives increased the serum and urinary Pi levels as well as the excretion of hydroxyproline and cyclic adenosine monophosphate (cAMP), while decreased serum and urinary calcium levels [7]. A randomized controlled trial demonstrated that avoiding phosphate-containing food additives resulted in a modest improvement in hyperphosphatemia among patients with end-stage renal disease [83]. The administration of phosphate-containing enemas also caused serum Pi elevation and decreased serum calcium [29]. Several case reports describe the complications such as tetany, hypocalcemic coma, and brain damage, induced by administration of phosphate-containing enema [22, 50, 80].

3.4 Phosphate and Pathological Calcification

Pathological calcification is one of the primary mechanisms for harmful effects induced by phosphate overload. Hyperphosphatemia and the increased calcium-phosphate products may facilitate the ectopic calcification of soft tissue and blood vessels [72, 73]. In addition to the accelerated formation of calcium-phosphate crystals, *in vitro* studies using cultured vascular smooth muscle cells (VSMCs) demonstrated that elevated extracellular Pi caused osteoblastic transdifferentiation [17, 40]. The treatment of VSMCs with high Pi induced the expression of *RUNX2*, a master transcription factor required for osteoblastic differentiation, and *BGLAP* encoding osteocalcin, an osteoblast-specific matrix protein [40]. Since it was reported that ablation of Runx2 prevented vascular calcification in mice [48], the Pi-induced up-regulation of Runx2 might substantially contribute to the ectopic calcification caused by phosphate overload.

3.5 Phosphate and Inflammation

Inflammation is a common feature of advanced renal disease [60]. It was reported that serum Pi and calcium-phosphate product levels directly correlated with serum levels of C-reactive protein (CRP) and interleukin-6 (IL-6) in 133 patients with CKD not on dialysis and not receiving calcium supplements, phosphate binders, or vitamin D, whereas HDL-cholesterol and estimated glomerular filtration rate (eGFR) inversely correlated with the levels of inflammatory state [57]. The authors of this study suggested that serum Pi was an independent risk factor for the presence of an inflammatory state, based on their logistic regression analysis. In an animal study, Yamada, et al. reported that dietary phosphate overload resulted in an increase in serum and tissue levels of TNF-α and developed malnutrition in adenine-induced CKD rat model [91]. Emerging evidence has suggested that pro-inflammatory response might mediate the development of phosphate overload-related vascular calcification [88].

Duchenne muscular dystrophy is a lethal inherited disease caused by dystrophin deficiency and characterized by progressive muscle degeneration, increased macrophage infiltration, and ectopic calcification. It was reported that dietary phosphorus overload dramatically aggravated the dystrophic phenotype in the dystrophin-deficient *mdx* mouse, a model of Duchenne muscular dystrophy, by increasing the number of necrotic muscle fibers and the degree of inflammation associated with infiltration of M1 macrophages [89]. More recently, a human study has demonstrated that dietary phosphorus enhances inflammatory response in gingivitis [28]. Thus, phosphate overload may induce inflammation in various conditions.

3.6 Phosphate-Induced Signaling and Cytotoxicity

Although the underlying mechanisms are still not fully understood, the direct inflammatory and cytotoxic effects of extracellular Pi are likely to be involved. It has been shown that elevated extracellular Pi itself triggers signaling to regulate gene expression and cellular functions in some cell types [6, 41, 44, 55, 93, 94]. Especially, Pi-induced signaling has been extensively investigated in bone cells. In an osteoblastic cell line MC3T3-E1, elevated extracellular Pi induced the expression of several genes, including that for osteopontin [4–6, 19, 59, 71]. In early chondrocytes of proliferating stage, extracellular Pi up-regulated the *cyclin D1* expression [44]. In mature chondrocytes, elevated Pi induced the expression of matrix Gla protein (MGP) that is related to mineralization [41]. These studies revealed that gene regulation by the elevated extracellular Pi involved the type III Na^{+}/Pi cotransporter Pit-1 and the activation of MEK/ERK pathway [4, 41, 44, 55]. It is also reported that signaling pathways activated by elevated Pi include FGF receptor signaling, G-protein signaling, N-ras signaling [11] and Akt/mTORC signaling [31, 42]. In addition to the effects on gene expression, elevated extracellular Pi facilitates apoptosis in terminally differentiated chondrocytes [51, 52, 75]. Thus, extracellular Pi is physiologically an important regulator of proliferation, differentiation and function of bone cells. However, as mentioned above, a pathologically elevated concentration of extracellular Pi induced the expression of several osteoblast-specific genes such as *Runx2* and *osteocalcin* in vascular smooth muscle cells in a Pit-1-dependent manner, leading to a phenotypical change to predispose calcification [40]. By expressing wild-type or various mutant Pit-1 proteins in Pit-1-deficient vascular smooth muscle cells, Chavkin, et al. demonstrated that both Pi-uptake-dependent and –independent functions of Pit-1 are involved in vascular calcification [14].

As described above, an elevation in extracellular Pi also triggers a signal for apoptosis in hypertrophic mature chondrocytes [52, 75]. In pathological conditions, elevated levels of extracellular Pi may accelerate apoptosis in other cell types as well. Indeed, the apoptosis rate was increased in various organs in the hyperphosphatemic *αKlotho*-deficient mice, which was restored in the *NaPi2a*/*αKlotho* double-deficient mice with lowered serum Pi [64]. ISO-HAS is an endothelial cell line originally established from human hemangiosarcoma [53]. Using this cell line (obtained from Cell Resource Center for Biomedical Research Institute for Development, Aging and Cancer, Tohoku University, Japan), elevated Pi-induced apoptosis was associated with the cleavage of caspase-3 (Fig. 3.1).

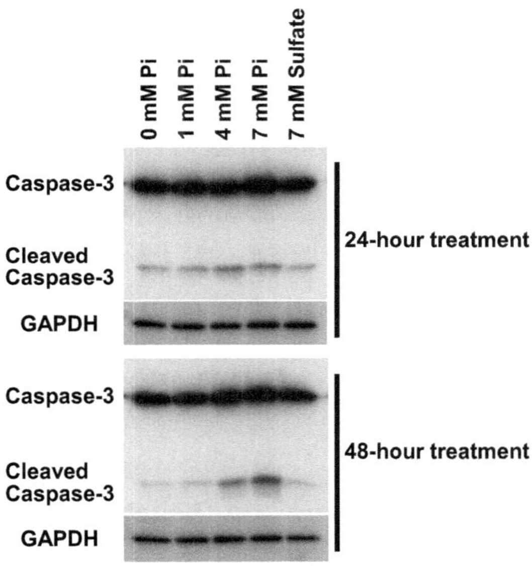

Fig. 3.1 Increased extracellular phosphate (Pi) accelerated the activation of caspase-3, a crucial mediator of apoptosis, in an endothelial cell line, ISO-HAS. The cells were starved in the Pi-free medium for 24 h and then treated for 24 or 48 h with the indicated concentration of Pi or sulfate as a negative control. Media containing Pi or sulfate were prepared by addition of sodium salts to Pi-free medium. The experiments were performed in the presence of 0.1% fetal bovine serum, which provided additional 0.01 mM Pi. Whole cell lysates were harvested and subjected to Western blotting using antibodies against caspase-3 and glyceraldehyde 3-phosphate dehydrogenase (GAPDH). Treatment with high Pi caused an increase in the amount of cleaved caspase-3, suggesting apoptosis

Fig. 3.2 Simplified diagram illustrating various toxic effects of the elevated extracellular phosphate

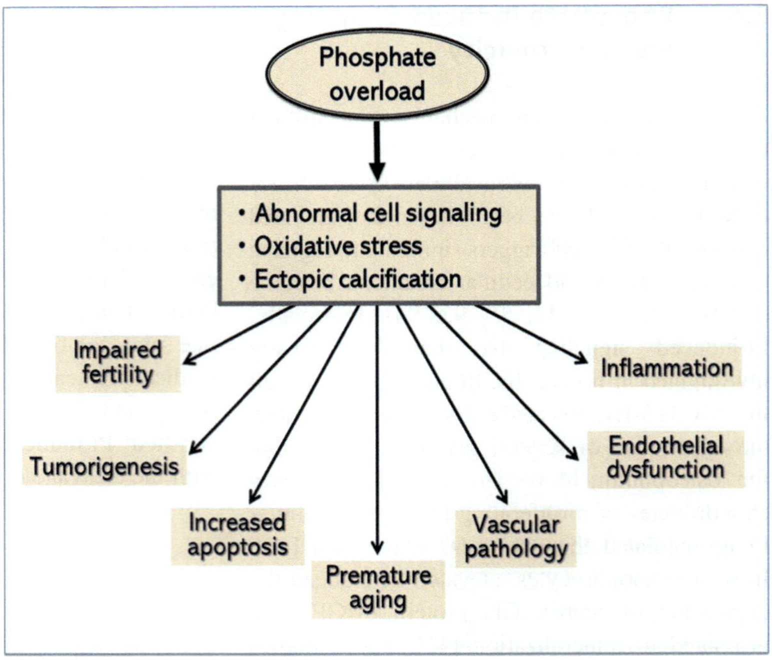

A recent study using HEK293 cells and HeLa cells suggested that abnormally elevated extracellular Pi might rewire interwoven network of signaling pathways such as Akt pathway, ERK pathway and JNK pathway, leading to a wide variety of cytotoxic effects including abberant proliferation, endoplasmic reticulum (ER) stress, epithelial-mesenchymal transition (EMT) and cell death [31] (Fig. 3.2).

3.7 Phosphate and Oxidative Stress

A line of evidence has revealed the involvement of oxidative stress in the harmful effects by elevated extracellular Pi. Zhao, et al. reported that high extracellular Pi induced the generation of mitochondrial reactive oxygen species (ROS), which promoted the nuclear translocation of nuclear factor κB p65 and accelerated the osteogenic transdifferentiation of the cultured smooth muscle cells [95]. Nguyen, et al. demonstrated that high extracellular Pi induced defective insulin secretion and cytotoxicity in pancreatic β-cells, which was also mediated by oxidative stress and hyperpolarization of mitochondria [58]. It has been demonstrated that oxidative-stress induces DNA damage and senescence in human fibroblasts [15]. In *αklotho*-deficient mice characterized by tissue atrophy and premature aging, urinary 8-hydeoxy-2′-deoxyguanosine (8-OHdG), a marker for oxidative stress-induced DNA damage, is increased [42, 92]. In brown adipose tissue of *αklotho*-KO mice, we demonstrated that hyperphosphatemia-induced activation of Akt/mTORC1 signaling pathway might be involved in suppressing antioxidant genes and oxidative damage [42], thus, the cytotoxic effects of Pi overload appear to be partly mediated by increased oxidative stress. It has also been reported that elevated Pi induces autophagy, which counteracts vascular calcification by reducing matrix vesicle release [21].

It is reported that hyperphosphatemia impairs endothelial cell function [79]. Dialysis patients have an increased concentration of circulating procoagulant endothelial microparticles, contributing to cardiovascular occlusive events [10]. Abbasian, et al. reported that treatment of cultured human endothelial cells with elevated extracellular Pi led to a rise in intracellular Pi concentration and a marked increase in cellular tropomyosin-3, plasma membrane blebbing, and release of microparticles. This effect of Pi was suggested to be independent of oxidative stress or apoptosis [1].

3.8 Effects of Phosphate on Tumorigenesis

The toxicity of phosphate is also implicated in tumorigenesis. It was shown that a high Pi diet increased the development of lung and skin cancers in animal models [12, 39]. In a mouse model of lung cancer, high dietary Pi stimulated pulmonary Akt activity and increased lung tumorigenesis [39], and knockdown of type IIb Na^+/Pi co-transporter in the lung suppressed the lung tumorigenesis [33]. In humans, a population-based prospective study investigated the relationship between serum Pi and risk of cancer in human, and showed a positive link between Pi quartiles and the risk of cancer of the pancreas, lung, thyroid gland, and bone in men, and cancer of the esophagus, lung, and nonmelanoma skin cancer in women [90]. Interestingly, the risks of breast, endometrial and other endocrine cancers were lower in the population of higher Pi levels in both men and women, probably due to the effects of hormonal factors [90].

3.9 Conclusions

Although phosphate is an essential nutrient involved in various biological processes, its overload in the body can be toxic and cause damage to multiple organs. The pathological effects of Pi include premature aging, calcification and blood vessels and soft tissues, impaired fertility, increased inflammation, accelerated cell death, and tumorigenesis, leading to a short lifespan. Abnormal cell signaling and oxidative stress have been implicated in the direct cytotoxicity of the elevated extracellular Pi, although the underlying molecular mechanisms are still not fully understood. Excess phosphate retention can be associated with various conditions, including renal dysfunction, administration of phosphate-containing laxatives and enemas, and ingestion of diet rich in phosphate food additives. Since phosphate overload can induce severe medical complications, phosphate balance in the diet should be considered for health promotion.

References

1. Abbasian N, Burton JO, Herbert KE, Tregunna BE, Brown JR, Ghaderi-Najafabadi M, Brunskill NJ, Goodall AH, Bevington A (2015) Hyperphosphatemia, phosphoprotein phosphatases, and microparticle release in vascular endothelial cells. J Am Soc Nephrol 26(9):2152–2162
2. ADHR-CONSORTIUM (2000) Autosomal dominant hypophosphataemic rickets is associated with mutations in FGF23. Nat Genet 26(3):345–348
3. Araya K, Fukumoto S, Backenroth R, Takeuchi Y, Nakayama K, Ito N, Yoshii N, Yamazaki Y, Yamashita T, Silver J, Igarashi T, Fujita T (2005) A novel mutation in fibroblast growth factor 23 gene as a cause of tumoral calcinosis. J Clin Endocrinol Metab 90(10):5523–5527
4. Beck GR Jr, Knecht N (2003) Osteopontin regulation by inorganic phosphate is ERK1/2-, protein kinase C-, and proteasome-dependent. J Biol Chem 278(43): 41921–41929
5. Beck GR Jr, Moran E, Knecht N (2003) Inorganic phosphate regulates multiple genes during osteoblast differentiation, including Nrf2. Exp Cell Res 288(2): 288–300
6. Beck GR Jr, Zerler B, Moran E (2000) Phosphate is a specific signal for induction of osteopontin gene expression. Proc Natl Acad Sci USA 97(15): 8352–8357
7. Bell RR, Draper HH, Tzeng DY, Shin HK, Schmidt GR (1977) Physiological responses of human adults to foods containing phosphate additives. J Nutr 107(1): 42–50
8. Ben-Dov IZ, Galitzer H, Lavi-Moshayoff V, Goetz R, Kuro-o M, Mohammadi M, Sirkis R, Naveh-Many T, Silver J (2007) The parathyroid is a target organ for FGF23 in rats. J Clin Invest 117(12):4003–4008
9. Block GA, Hulbert-Shearon TE, Levin NW, Port FK (1998) Association of serum phosphorus and calcium x phosphate product with mortality risk in chronic hemodialysis patients: a national study. Am J Kidney Dis 31(4):607–617
10. Burton JO, Hamali HA, Singh R, Abbasian N, Parsons R, Patel AK, Goodall AH, Brunskill NJ (2013) Elevated levels of procoagulant plasma microvesicles in dialysis patients. PLoS One 8(8): e72663
11. Camalier CE, Yi M, Yu LR, Hood BL, Conrads KA, Lee YJ, Lin Y, Garneys LM, Bouloux GF, Young MR,

Veenstra TD, Stephens RM, Colburn NH, Conrads TP, Beck GR Jr (2013) An integrated understanding of the physiological response to elevated extracellular phosphate. J Cell Physiol 228(7):1536–1550
12. Camalier CE, Young MR, Bobe G, Perella CM, Colburn NH, Beck GR Jr (2010) Elevated phosphate activates N-ras and promotes cell transformation and skin tumorigenesis. Cancer Prev Res (Phila) 3(3): 359–370
13. Carpenter TO, Whyte MP, Imel EA, Boot AM, Hogler W, Linglart A, Padidela R, Van't Hoff W, Mao M, Chen CY, Skrinar A, Kakkis E, San Martin J, Portale AA (2018) Burosumab therapy in children with X-linked hypophosphatemia. N Engl J Med 378(21):1987–1998
14. Chavkin NW, Chia JJ, Crouthamel MH, Giachelli CM (2015) Phosphate uptake-independent signaling functions of the type III sodium-dependent phosphate transporter, PiT-1, in vascular smooth muscle cells. Exp Cell Res 333(1):39–48
15. Chen JH, Stoeber K, Kingsbury S, Ozanne SE, Williams GH, Hales CN (2004) Loss of proliferative capacity and induction of senescence in oxidatively stressed human fibroblasts. J Biol Chem 279(47): 49439–49446
16. Chiba T, Matsuo H, Kawamura Y, Nagamori S, Nishiyama T, Wei L, Nakayama A, Nakamura T, Sakiyama M, Takada T, Taketani Y, Suma S, Naito M, Oda T, Kumagai H, Moriyama Y, Ichida K, Shimizu T, Kanai Y, Shinomiya N (2015) NPT1/ SLC17A1 is a renal urate exporter in humans and its common gain-of-function variant decreases the risk of renal underexcretion gout. Arthritis Rheumatol 67(1): 281–287
17. Ciceri P, Elli F, Cappelletti L, Tosi D, Braidotti P, Bulfamante G, Cozzolino M (2015) A new in vitro model to delay high phosphate-induced vascular calcification progression. Mol Cell Biochem 410(1–2): 197–206
18. Connolly GM, Cunningham R, McNamee PT, Young IS, Maxwell AP (2009) Elevated serum phosphate predicts mortality in renal transplant recipients. Transplantation 87(7):1040–1044
19. Conrads KA, Yi M, Simpson KA, Lucas DA, Camalier CE, Yu LR, Veenstra TD, Stephens RM, Conrads TP, Beck GR Jr (2005) A combined proteome and microarray investigation of inorganic phosphate-induced pre-osteoblast cells. Mol Cell Proteomics 4(9):1284–1296
20. Cozzolino M, Dusso AS, Slatopolsky E (2001) Role of calcium-phosphate product and bone-associated proteins on vascular calcification in renal failure. J Am Soc Nephrol 12(11):2511–2516
21. Dai XY, Zhao MM, Cai Y, Guan QC, Zhao Y, Guan Y, Kong W, Zhu WG, Xu MJ, Wang X (2013) Phosphate-induced autophagy counteracts vascular calcification by reducing matrix vesicle release. Kidney Int 83(6):1042–1051
22. Domico MB, Huynh V, Anand SK, Mink R (2006) Severe hyperphosphatemia and hypocalcemic tetany after oral laxative administration in a 3-month-old infant. Pediatrics 118(5):e1580–e1583
23. Eddington H, Hoefield R, Sinha S, Chrysochou C, Lane B, Foley RN, Hegarty J, New J, O'Donoghue DJ, Middleton RJ, Kalra PA (2010) Serum phosphate and mortality in patients with chronic kidney disease. Clin J Am Soc Nephrol 5(12):2251–2257
24. Erem S, Razzaque MS (2018) Dietary phosphate toxicity: an emerging global health concern. Histochem Cell Biol 150(6):711–719
25. Feng JQ, Ward LM, Liu S, Lu Y, Xie Y, Yuan B, Yu X, Rauch F, Davis SI, Zhang S, Rios H, Drezner MK, Quarles LD, Bonewald LF, White KE (2006) Loss of DMP1 causes rickets and osteomalacia and identifies a role for osteocytes in mineral metabolism. Nat Genet 38(11):1310–1315
26. Fukumoto S (2021) FGF23-related hypophosphatemic rickets/osteomalacia: diagnosis and new treatment. J Mol Endocrinol 66(2):R57–R65
27. Ganesh SK, Stack AG, Levin NW, Hulbert-Shearon T, Port FK (2001) Association of elevated serum PO(4), Ca x PO(4) product, and parathyroid hormone with cardiac mortality risk in chronic hemodialysis patients. J Am Soc Nephrol 12(10):2131–2138
28. Goodson JM, Shi P, Razzaque MS (2019) Dietary phosphorus enhances inflammatory response: a study of human gingivitis. J Steroid Biochem Mol Biol 188: 166–171
29. Grosskopf I, Graff E, Charach G, Binyamin G, Spinrad S, Blum I (1991) Hyperphosphataemia and hypocalcaemia induced by hypertonic phosphate enema—an experimental study and review of the literature. Hum Exp Toxicol 10(5):351–355
30. Hattenhauer O, Traebert M, Murer H, Biber J (1999) Regulation of small intestinal Na-P(i) type IIb cotransporter by dietary phosphate intake. Am J Phys 277(4 Pt 1):G756–G762
31. He P, Mann-Collura O, Fling J, Edara N, Hetz R, Razzaque MS (2021) High phosphate actively induces cytotoxicity by rewiring pro-survival and pro-apoptotic signaling networks in HEK293 and HeLa cells. FASEB J 35(1):e20997
32. Hernando N, Gagnon K, Lederer E (2021) Phosphate transport in epithelial and nonepithelial tissue. Physiol Rev 101(1):1–35
33. Hong SH, Minai-Tehrani A, Chang SH, Jiang HL, Lee S, Lee AY, Seo HW, Chae C, Beck GR Jr, Cho MH (2013) Knockdown of the sodium-dependent phosphate co-transporter 2b (NPT2b) suppresses lung tumorigenesis. PLoS One 8(10):e77121
34. HYP-CONSORTIUM (1995) A gene (PEX) with homologies to endopeptidases is mutated in patients with X-linked hypophosphatemic rickets. The HYP Consortium. Nat Genet 11(2):130–136
35. Ichikawa S, Imel EA, Kreiter ML, Yu X, Mackenzie DS, Sorenson AH, Goetz R, Mohammadi M, White

KE, Econs MJ (2007) A homozygous missense mutation in human KLOTHO causes severe tumoral calcinosis. J Clin Invest 117(9):2684–2691
36. Ichikawa S, Sorenson AH, Austin AM, Mackenzie DS, Fritz TA, Moh A, Hui SL, Econs MJ (2009) Ablation of the Galnt3 gene leads to low-circulating intact fibroblast growth factor 23 (Fgf23) concentrations and hyperphosphatemia despite increased Fgf23 expression. Endocrinology 150(6): 2543–2550
37. Imura A, Iwano A, Tohyama O, Tsuji Y, Nozaki K, Hashimoto N, Fujimori T, Nabeshima Y (2004) Secreted Klotho protein in sera and CSF: implication for post-translational cleavage in release of Klotho protein from cell membrane. FEBS Lett 565(1–3): 143–147
38. Jain N, Elsayed EF (2013) Dietary phosphate: what do we know about its toxicity. J Nephrol 26(5):856–864
39. Jin H, Chang SH, Xu CX, Shin JY, Chung YS, Park SJ, Lee YS, An GH, Lee KH, Cho MH (2007) High dietary inorganic phosphate affects lung through altering protein translation, cell cycle, and angiogenesis in developing mice. Toxicol Sci 100(1):215–223
40. Jono S, McKee MD, Murry CE, Shioi A, Nishizawa Y, Mori K, Morii H, Giachelli CM (2000) Phosphate regulation of vascular smooth muscle cell calcification. Circ Res 87(7):E10–E17
41. Julien M, Magne D, Masson M, Rolli-Derkinderen M, Chassande O, Cario-Toumaniantz C, Cherel Y, Weiss P, Guicheux J (2007) Phosphate stimulates matrix Gla protein expression in chondrocytes through the extracellular signal regulated kinase signaling pathway. Endocrinology 148(2):530–537
42. Kawai M, Kinoshita S, Ozono K, Michigami T (2016) Inorganic phosphate activates the AKT/mTORC1 pathway and shortens the life span of an alphaKlotho-deficient model. J Am Soc Nephrol 27(9):2810–2824
43. Kestenbaum B, Sampson JN, Rudser KD, Patterson DJ, Seliger SL, Young B, Sherrard DJ, Andress DL (2005) Serum phosphate levels and mortality risk among people with chronic kidney disease. J Am Soc Nephrol 16(2):520–528
44. Kimata M, Michigami T, Tachikawa K, Okada T, Koshimizu T, Yamazaki M, Kogo M, Ozono K (2010) Signaling of extracellular inorganic phosphate up-regulates cyclin D1 expression in proliferating chondrocytes via the Na+/Pi cotransporter Pit-1 and Raf/MEK/ERK pathway. Bone 47(5):938–947
45. Kuro-o M, Matsumura Y, Aizawa H, Kawaguchi H, Suga T, Utsugi T, Ohyama Y, Kurabayashi M, Kaname T, Kume E, Iwasaki H, Iida A, Shiraki-Iida T, Nishikawa S, Nagai R, Nabeshima YI (1997) Mutation of the mouse klotho gene leads to a syndrome resembling ageing. Nature 390(6655):45–51
46. Kurosu H, Ogawa Y, Miyoshi M, Yamamoto M, Nandi A, Rosenblatt KP, Baum MG, Schiavi S, Hu MC, Moe OW, Kuro-o M (2006) Regulation of fibroblast growth factor-23 signaling by klotho. J Biol Chem 281(10):6120–6123
47. Li JW, Xu C, Fan Y, Wang Y, Xiao YB (2014) Can serum levels of alkaline phosphatase and phosphate predict cardiovascular diseases and total mortality in individuals with preserved renal function? A systemic review and meta-analysis. PLoS One 9(7):e102276
48. Lin ME, Chen T, Leaf EM, Speer MY, Giachelli CM (2015) Runx2 expression in smooth muscle cells is required for arterial medial calcification in mice. Am J Pathol 185(7):1958–1969
49. Lorenz-Depiereux B, Bastepe M, Benet-Pages A, Amyere M, Wagenstaller J, Muller-Barth U, Badenhoop K, Kaiser SM, Rittmaster RS, Shlossberg AH, Olivares JL, Loris C, Ramos FJ, Glorieux F, Vikkula M, Juppner H, Strom TM (2006) DMP1 mutations in autosomal recessive hypophosphatemia implicate a bone matrix protein in the regulation of phosphate homeostasis. Nat Genet 38(11):1248–1250
50. Loughnan P, Mullins GC (1977) Brain damage following a hypertonic phosphate enema. Am J Dis Child 131(9):1032
51. Magne D, Bluteau G, Faucheux C, Palmer G, Vignes-Colombeix C, Pilet P, Rouillon T, Caverzasio J, Weiss P, Daculsi G, Guicheux J (2003) Phosphate is a specific signal for ATDC5 chondrocyte maturation and apoptosis-associated mineralization: possible implication of apoptosis in the regulation of endochondral ossification. J Bone Miner Res 18(8):1430–1442
52. Mansfield K, Rajpurohit R, Shapiro IM (1999) Extracellular phosphate ions cause apoptosis of terminally differentiated epiphyseal chondrocytes. J Cell Physiol 179(3):276–286
53. Masuzawa M, Fujimura T, Hamada Y, Fujita Y, Hara H, Nishiyama S, Katsuoka K, Tamauchi H, Sakurai Y (1999) Establishment of a human hemangiosarcoma cell line (ISO-HAS). Int J Cancer 81(2):305–308
54. Michigami T (2013) Extracellular phosphate as a signaling molecule. Contrib Nephrol 180:14–24
55. Michigami T, Kawai M, Yamazaki M, Ozono K (2018) Phosphate as a signaling molecule and its sensing mechanism. Physiol Rev 98(4):2317–2348
56. Nakatani T, Sarraj B, Ohnishi M, Densmore MJ, Taguchi T, Goetz R, Mohammadi M, Lanske B, Razzaque MS (2009) In vivo genetic evidence for klotho-dependent, fibroblast growth factor 23 (Fgf23) -mediated regulation of systemic phosphate homeostasis. FASEB J 23(2):433–441
57. Navarro-Gonzalez JF, Mora-Fernandez C, Muros M, Herrera H, Garcia J (2009) Mineral metabolism and inflammation in chronic kidney disease patients: a cross-sectional study. Clin J Am Soc Nephrol 4(10): 1646–1654
58. Nguyen TT, Quan X, Hwang KH, Xu S, Das R, Choi SK, Wiederkehr A, Wollheim CB, Cha SK, Park KS (2015) Mitochondrial oxidative stress mediates high phosphate-induced secretory defects and apoptosis in

insulin-secreting cells. Am J Physiol Endocrinol Metab: ajpendo 00009:02015
59. Nishino J, Yamazaki M, Kawai M, Tachikawa K, Yamamoto K, Miyagawa K, Kogo M, Ozono K, Michigami T (2017) Extracellular phosphate induces the expression of dentin matrix protein 1 through the FGF receptor in osteoblasts. J Cell Biochem 118(5): 1151–1163
60. Oberg BP, McMenamin E, Lucas FL, McMonagle E, Morrow J, Ikizler TA, Himmelfarb J (2004) Increased prevalence of oxidant stress and inflammation in patients with moderate to severe chronic kidney disease. Kidney Int 65(3):1009–1016
61. Ohata Y, Arahori H, Namba N, Kitaoka T, Hirai H, Wada K, Nakayama M, Michigami T, Imura A, Nabeshima Y, Yamazaki Y, Ozono K (2011) Circulating levels of soluble alpha-Klotho are markedly elevated in human umbilical cord blood. J Clin Endocrinol Metab 96(6):E943–E947
62. Ohata Y, Yamazaki M, Kawai M, Tsugawa N, Tachikawa K, Koinuma T, Miyagawa K, Kimoto A, Nakayama M, Namba N, Yamamoto H, Okano T, Ozono K, Michigami T (2014) Elevated fibroblast growth factor 23 exerts its effects on placenta and regulates vitamin D metabolism in pregnancy of Hyp mice. J Bone Miner Res 29(7):1627–1638
63. Ohnishi M, Nakatani T, Lanske B, Razzaque MS (2009) In vivo genetic evidence for suppressing vascular and soft-tissue calcification through the reduction of serum phosphate levels, even in the presence of high serum calcium and 1,25-dihydroxyvitamin d levels. Circ Cardiovasc Genet 2(6):583–590
64. Ohnishi M, Razzaque MS (2010) Dietary and genetic evidence for phosphate toxicity accelerating mammalian aging. FASEB J 24(9):3562–3571
65. Palmer SC, Hayen A, Macaskill P, Pellegrini F, Craig JC, Elder GJ, Strippoli GF (2011) Serum levels of phosphorus, parathyroid hormone, and calcium and risks of death and cardiovascular disease in individuals with chronic kidney disease: a systematic review and meta-analysis. JAMA 305(11):1119–1127
66. Quarles LD (2008) Endocrine functions of bone in mineral metabolism regulation. J Clin Invest 118(12): 3820–3828
67. Razzaque MS (2009) The FGF23-Klotho axis: endocrine regulation of phosphate homeostasis. Nat Rev Endocrinol 5(11):611–619
68. Razzaque MS (2011) Phosphate toxicity: new insights into an old problem. Clin Sci (Lond) 120(3):91–97
69. Razzaque MS (2013) Phosphate toxicity and vascular mineralization. Contrib Nephrol 180:74–85
70. Razzaque MS, Lanske B (2006) Hypervitaminosis D and premature aging: lessons learned from Fgf23 and Klotho mutant mice. Trends Mol Med 12(7):298–305
71. Rendenbach C, Yorgan TA, Heckt T, Otto B, Baldauf C, Jeschke A, Streichert T, David JP, Amling M, Schinke T (2014) Effects of extracellular phosphate on gene expression in murine osteoblasts. Calcif Tissue Int 94(5):474–483
72. Reynolds JL, Joannides AJ, Skepper JN, McNair R, Schurgers LJ, Proudfoot D, Jahnen-Dechent W, Weissberg PL, Shanahan CM (2004) Human vascular smooth muscle cells undergo vesicle-mediated calcification in response to changes in extracellular calcium and phosphate concentrations: a potential mechanism for accelerated vascular calcification in ESRD. J Am Soc Nephrol 15(11):2857–2867
73. Ribeiro S, Ramos A, Brandao A, Rebelo JR, Guerra A, Resina C, Vila-Lobos A, Carvalho F, Remedio F, Ribeiro F (1998) Cardiac valve calcification in haemodialysis patients: role of calcium-phosphate metabolism. Nephrol Dial Transplant 13(8): 2037–2040
74. Richter B, Faul C (2018) FGF23 actions on target tissues-with and without Klotho. Front Endocrinol (Lausanne) 9:189
75. Sabbagh Y, Carpenter TO, Demay MB (2005) Hypophosphatemia leads to rickets by impairing caspase-mediated apoptosis of hypertrophic chondrocytes. Proc Natl Acad Sci USA 102(27): 9637–9642
76. Sabbagh Y, O'Brien SP, Song W, Boulanger JH, Stockmann A, Arbeeny C, Schiavi SC (2009) Intestinal npt2b plays a major role in phosphate absorption and homeostasis. J Am Soc Nephrol 20(11): 2348–2358
77. Shimada T, Kakitani M, Yamazaki Y, Hasegawa H, Takeuchi Y, Fujita T, Fukumoto S, Tomizuka K, Yamashita T (2004) Targeted ablation of Fgf23 demonstrates an essential physiological role of FGF23 in phosphate and vitamin D metabolism. J Clin Invest 113(4):561–568
78. Shimada T, Mizutani S, Muto T, Yoneya T, Hino R, Takeda S, Takeuchi Y, Fujita T, Fukumoto S, Yamashita T (2001) Cloning and characterization of FGF23 as a causative factor of tumor-induced osteomalacia. Proc Natl Acad Sci USA 98(11):6500–6505
79. Shuto E, Taketani Y, Tanaka R, Harada N, Isshiki M, Sato M, Nashiki K, Amo K, Yamamoto H, Higashi Y, Nakaya Y, Takeda E (2009) Dietary phosphorus acutely impairs endothelial function. J Am Soc Nephrol 20(7):1504–1512
80. Sotos JF, Cutler EA, Finkel MA, Doody D (1977) Hypocalcemic coma following two pediatric phosphate enemas. Pediatrics 60(3):305–307
81. Stevens LA, Djurdjev O, Cardew S, Cameron EC, Levin A (2004) Calcium, phosphate, and parathyroid hormone levels in combination and as a function of dialysis duration predict mortality: evidence for the complexity of the association between mineral metabolism and outcomes. J Am Soc Nephrol 15(3): 770–779
82. Stubbs JR, Liu S, Tang W, Zhou J, Wang Y, Yao X, Quarles LD (2007) Role of hyperphosphatemia and 1,25-dihydroxyvitamin D in vascular calcification and mortality in fibroblastic growth factor 23 null mice. J Am Soc Nephrol 18(7):2116–2124
83. Sullivan C, Sayre SS, Leon JB, Machekano R, Love TE, Porter D, Marbury M, Sehgal AR (2009) Effect of food additives on hyperphosphatemia among patients with end-stage renal disease: a randomized controlled trial. JAMA 301(6):629–635

84. Takeyari S, Yamamoto T, Kinoshita Y, Fukumoto S, Glorieux FH, Michigami T, Hasegawa K, Kitaoka T, Kubota T, Imanishi Y, Shimotsuji T, Ozono K (2014) Hypophosphatemic osteomalacia and bone sclerosis caused by a novel homozygous mutation of the FAM20C gene in an elderly man with a mild variant of Raine syndrome. Bone 67:56–62
85. Tonelli M, Curhan G, Pfeffer M, Sacks F, Thadhani R, Melamed ML, Wiebe N, Muntner P (2009) Relation between alkaline phosphatase, serum phosphate, and all-cause or cardiovascular mortality. Circulation 120(18):1784–1792
86. Urakawa I, Yamazaki Y, Shimada T, Iijima K, Hasegawa H, Okawa K, Fujita T, Fukumoto S, Yamashita T (2006) Klotho converts canonical FGF receptor into a specific receptor for FGF23. Nature 444(7120):770–774
87. Virkki LV, Biber J, Murer H, Forster IC (2007) Phosphate transporters: a tale of two solute carrier families. Am J Physiol Renal Physiol 293(3):F643–F654
88. Voelkl J, Egli-Spichtig D, Alesutan I, Wagner CA (2021) Inflammation: a putative link between phosphate metabolism and cardiovascular disease. Clin Sci (Lond) 135(1):201–227
89. Wada E, Yoshida M, Kojima Y, Nonaka I, Ohashi K, Nagata Y, Shiozuka M, Date M, Higashi T, Nishino I, Matsuda R (2014) Dietary phosphorus overload aggravates the phenotype of the dystrophin-deficient mdx mouse. Am J Pathol 184(11):3094–3104
90. Wulaningsih W, Michaelsson K, Garmo H, Hammar N, Jungner I, Walldius G, Holmberg L, Van Hemelrijck M (2013) Inorganic phosphate and the risk of cancer in the Swedish AMORIS study. BMC Cancer 13:257
91. Yamada S, Tokumoto M, Tatsumoto N, Taniguchi M, Noguchi H, Nakano T, Masutani K, Ooboshi H, Tsuruya K, Kitazono T (2014) Phosphate overload directly induces systemic inflammation and malnutrition as well as vascular calcification in uremia. Am J Physiol Renal Physiol 306(12):F1418–F1428
92. Yamamoto M, Clark JD, Pastor JV, Gurnani P, Nandi A, Kurosu H, Miyoshi M, Ogawa Y, Castrillon DH, Rosenblatt KP, Kuro-o M (2005) Regulation of oxidative stress by the anti-aging hormone klotho. J Biol Chem 280(45):38029–38034
93. Yamazaki M, Ozono K, Okada T, Tachikawa K, Kondou H, Ohata Y, Michigami T (2010a) Both FGF23 and extracellular phosphate activate Raf/MEK/ERK pathway via FGF receptors in HEK293 cells. J Cell Biochem 111(5):1210–1221
94. Yamazaki Y, Imura A, Urakawa I, Shimada T, Murakami J, Aono Y, Hasegawa H, Yamashita T, Nakatani K, Saito Y, Okamoto N, Kurumatani N, Namba N, Kitaoka T, Ozono K, Sakai T, Hataya H, Ichikawa S, Imel EA, Econs MJ, Nabeshima Y (2010b) Establishment of sandwich ELISA for soluble alpha-Klotho measurement: age-dependent change of soluble alpha-Klotho levels in healthy subjects. Biochem Biophys Res Commun 398(3):513–518
95. Zhao MM, Xu MJ, Cai Y, Zhao G, Guan Y, Kong W, Tang C, Wang X (2011) Mitochondrial reactive oxygen species promote p65 nuclear translocation mediating high-phosphate-induced vascular calcification in vitro and in vivo. Kidney Int 79(10):1071–1079

Phosphate-Sensing

4

Yuichi Takashi and Seiji Fukumoto

Abstract

The blood level of phosphate is tightly regulated in a narrow range. Hyperphosphatemia and hypophosphatemia both lead to the development of diseases, such as hyperphosphatemic tumoral calcinosis and rickets/osteomalacia, respectively. Although several humoral factors have been known to affect blood phosphate levels, fibroblast growth factor 23 (FGF23) is the principal hormone involved in the regulation of blood phosphate. This hormone is produced by bone, particularly by osteocytes and osteoblasts, and has the effect of lowering the blood level of phosphate in the renal proximal tubules. Therefore, some phosphate-sensing mechanism should exist, at least in the bone. However, the mechanisms through which bone senses changes in the blood level of phosphate, and through which the bone regulates FGF23 production remain to be fully elucidated. Our recent findings demonstrate that high extracellular phosphate phosphorylates FGF receptor 1c (FGFR1c). Its downstream extracellular signal-regulated kinase (ERK) kinase (MEK)/ERK signaling pathway regulates the expression of several transcription factors and the *GALNT3* gene, which encodes GalNAc-T3, which plays a role in the regulation of post-translational modification of FGF23 protein, which in turn enhances FGF23 production. The FGFR1c-*GALNT3* gene axis is considered to be the most important mechanism for regulating the production of FGF23 in bone in the response to a high phosphate diet. Thus—in the regulation of FGF23 production and blood phosphate levels—FGFR1c may be considered to function as a phosphate-sensing molecule. A feedback mechanism, in which FGFR1c and FGF23 are involved, is present in blood phosphate regulation. In addition, other reports indicate that PiT1 and PiT2 (type III sodium-phosphate cotransporters), and calcium-sensing receptor are also involved in the phosphate-sensing mechanism. In the present chapter, we summarize new insights on phosphate-sensing mechanisms.

Keywords

FGF23 · Phosphate-sensing · PiT1, PiT2 · *GALNT3* · 1,25$(OH)_2$D · FGFR1c

Y. Takashi
Department of Endocrinology and Diabetes Mellitus, Fukuoka University School of Medicine, Fukuoka, Japan
e-mail: y.takashi.si@fukuoka-u.ac.jp

S. Fukumoto (✉)
Department of Molecular Endocrinology, Fujii Memorial Institute of Medical Sciences, Institute of Advanced Medical Sciences, Tokushima University, Tokushima, Japan
e-mail: fukumoto-tky@umin.ac.jp

M. S. Razzaque (ed.), *Phosphate Metabolism*, Advances in Experimental Medicine and Biology 1362,
https://doi.org/10.1007/978-3-030-91623-7_4

4.1 General

Phosphate is an essential mineral for humans and plays many functions in the body. Approximately 85% of phosphate is stored in the bone as a hydroxyapatite [$Ca_{10}(PO_4)_6(OH)_2$] in humans [24]. The skeleton, which is composed of hydroxyapatite crystals and matrix proteins, is physically hard enough to support the weight of the body. Thus, chronic hypophosphatemia induces rickets/osteomalacia and skeletal muscle myopathy [26]. On the other hand, hyperphosphatemia is also harmful to the body. Hyperphosphatemia leads to ectopic calcification, which is typically seen in patients with hyperphosphatemic tumoral calcinosis [26]. Hyperphosphatemia is also well known to induce vascular calcification, resulting in ischemic heart disease and stroke in patients with end-stage renal disease (ESRD) on hemodialysis [45]. Therefore, the blood phosphate level needs to be regulated within a narrow range. It is well known that there are several humoral factors that help to maintain blood phosphate level in an appropriate range [16]. Fibroblast growth factor 23 (FGF23), a bone derived hormone, is the principal hormone in the regulation of blood phosphate [14, 31, 42]. Therefore, it could be considered that a phosphate-sensing mechanism has the function of regulating FGF23 production in bone. However, phosphate-sensing mechanism has been largely unknown.

4.2 Fibroblast Growth Factor 23: FGF23

Several humoral factors maintain the blood level of phosphate in humans, FGF23, parathyroid hormone (PTH) and 1,25-dihydroxyvitamin D [$1,25(OH)_2D$] [16]. Among these humoral factors, FGF23 is the principal hormone in the regulation of blood phosphate levels [14, 31, 42]. This hormone is produced by bone (particularly osteocytes and osteoblasts). The main action of FGF23 is observed in the renal proximal tubules [14, 31, 42]. In the presence of α-Klotho (the expression of which is limited to several tissues, including the kidney and parathyroid glands), FGF23 can bind to the FGF receptor 1c (FGFR1c) [27, 28, 50]. FGF23 suppresses the type IIa and IIc sodium-phosphate cotransporters expression in the renal proximal tubules and inhibits proximal tubular phosphate reabsorption. In addition, FGF23 reduces the synthesis of $1,25(OH)_2D$ by suppressing the expression of *CYP27B1*, which produces 25-hydroxyvitamin D-1α-hydroxylase, and by enhancing the expression of *CYP24A1*, which encodes 25-hydroxyvitamin D-24-hydroxylase. Because $1,25(OH)_2D$ enhances intestinal phosphate absorption, FGF23 reduces the blood phosphate level by inhibiting both proximal tubular phosphate reabsorption and intestinal phosphate absorption via the reduction of $1,25(OH)_2D$ [42].

4.2.1 The Structure of FGF23 Protein

FGF23 protein (a peptide with 251 amino acids) is produced by bone. After cleavage of a signal peptide (24 amino acids), FGF23 protein (227 amino acids) is secreted into circulation [40]. Before or during its secretion, part of this protein is proteolytically cleaved into inactive fragments. FGF23 can be cleaved between arginine (Arg) 179 and serine (Ser) 180 by enzymes, such as a Furin, which recognize the Arg176-X177-X178-Arg179 motif [42]. The important thing is that only full-length FGF23 protein is biologically active [18, 41, 53]. This cleavage was previously demonstrated to be inhibited by *O*-glycosylation of threonine (Thr) 178 in FGF23 protein initiated by UDP-*N*-acetyl-alpha-D-galactosamine: polypeptide *N*-acetylgalactosaminyltransderase 3 (GalNAc-T3)—a gene product of *UDP-N-acetyl-alpha-D-galactosamine: polypeptide N-acetylgalactosaminyltransderase 3 (GALNT3)*— [17]. Thus, *O*-glycosylation of FGF23 protein—which is initiated by GalNAc-T3—works to increase active full-length FGF23 protein. On the other hand, the phosphorylation of Ser180 in FGF23 protein by family with sequence similarity 20, member C (FAM20C) was reported to inhibit *O*-glycosylation of FGF23 protein by GalNAc-T3 and enhance the processing of FGF23 protein (Fig. 4.1) [44].

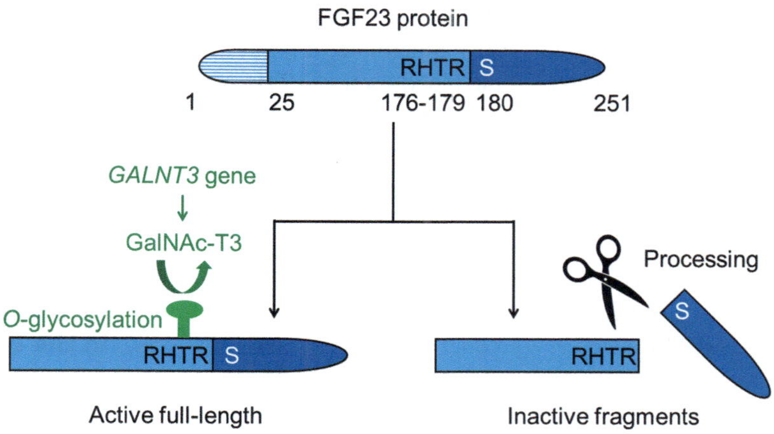

Fig. 4.1 The structure of FGF23 protein and posttranslational modification via a gene product of *GALNT3*

4.2.2 The Posttranslational Modification of FGF23 Protein Via a Phosphate Responsive Gene: *GALNT3*

Inactivating mutations of *GALNT3* cause hyperphosphatemic familial tumoral calcinosis (HFTC) in humans [48]. These patients show that hyperphosphatemia and high blood 1,25 $(OH)_2D$ levels like *Fgf23* knockout mice [48]. It could be considered that FGF23 protein, in which *O*-glycosylation is impaired, is susceptible to this processing, which results in low active full-length FGF23 and impairs the actions of FGF23. *GALNT3* is one of 20 *GALNT* gene families in humans; however, the gene product of *GALNT3* is necessary for the initiation of *O*-glycosylation of Thr178 in FGF23. No other *GALNT* gene products can initiate *O*-glycosylation [5]. According to these facts, it could be considered that the activity of FGF23, namely the level of full-length FGF23 in the blood, is regulated by the transcription and translation of *FGF23*, as well as by the posttranslational modification of FGF23 by the gene product of *GALNT3*. Previously, it was reported that a high phosphate diet increased blood levels of full-length FGF23 in both mice and humans [15, 37]. Recent data demonstrated that a high phosphate diet did not enhance the expression of *Fgf23* in the bone, while a high phosphate diet increased blood full-length FGF23 level in mice [46]. On the contrary, a high phosphate diet increases *Galnt3* levels in bone [46]. Taken together, a high phosphate diet is considered to increase the blood level of full-length FGF23 by increasing the expression of the *Galnt3* gene. Furthermore, when we performed Western blotting in *in vitro* experiments to determine the ratio of full-length FGF23 to cleaved fragments of FGF23 in culture media of the osteoblastic UMR106 cell line, the ratio of full-length FGF23 protein under a high extracellular phosphate condition was higher in comparison to that under a low phosphate condition [46]. *In vitro* experiments demonstrated that high extracellular phosphate levels increased the *Galnt3* gene expression in a dose dependent manner, and enhanced the GalNAc-T3 protein (encoded by *Galnt3*) expression [46]. Taken together, we consider *Galnt3* to be a phosphate responsive gene.

As mentioned above, the posttranslational modification of FGF23 protein was reported to be regulated not only by GalNAc-T3 but also by FAM20C. Both a high phosphate diet and high levels of extracellular phosphate also enhanced the expression of *Fam20c* [46]. From the observation that the blood full-length FGF23 level was increased by a high phosphate diet, the role of GalNAc-T3 in response to high phosphate levels seemed to be dominant over that of FAM20C. In addition, the processing of FGF23 protein is mediated by proteases like Furin [44]. Neither a high phosphate diet nor a high level of extracellular phosphate changed the expression of *Furin* [46]. It was already reported that $1,25(OH)_2D_3$ increases the blood level of full-length FGF23 in murine models [38]. $1,25(OH)_2D_3$ enhanced the expression of *Fgf23* in osteoblastic UMR106 cells, however, $1,25(OH)_2D_3$ had no effect on the expression of *Galnt3* [46]. Furthermore, a high phosphate diet increased the blood level of PTH, while PTH was reported to suppress the expression of *Galnt3* [23]. It is therefore considered unlikely that $1,25(OH)_2D_3$ and PTH are involved in the increase in the expression of *Galnt3* or in the increase in the blood level of full-length FGF23 that is caused by high phosphate levels.

Taken together, posttranslational modification of FGF23 by the gene product of *Galnt3* is considered to be the main mechanism of the response to high dietary phosphate and *Galnt3* is considered to be responsive to phosphate.

4.3 Phosphate-Sensing

4.3.1 Intracellular Signaling Induced by Phosphate: MEK/ERK Pathway

Alternations of extracellular phosphate have been shown transduce signals into the cells to regulate the expression of gene and the behavior of cells [33]. High levels of extracellular phosphate were shown to stimulate the extracellular signal-regulated kinase (ERK) kinase (MEK)/ERK pathway [4]. Furthermore, high levels of extracellular phosphate were demonstrated to induce a number of genes, including *osteopontin* (*Opn*), *dentin matrix protein 1* (*Dmp1*), *cyclin D1*, and *early growth response 1* (*Egr1*) through this same pathway [3, 25, 35, 54].

It was recently reported—based on a DNA microarray analysis using osteoblastic UMR106 cells—that high levels of extracellular phosphate induced the expression of several transcription factors downstream of the MEK/ERK pathway [46]. High extracellular phosphate induced the phosphorylation of ERK1/2 in a dose-dependent manner [46]. In addition, inhibition of the MEK/ERK pathway suppressed the enhancement of *Galnt3* gene, a phosphate responsive gene as mentioned above, by high extracellular phosphate [46]. MEK/ERK pathway activation by high levels of extracellular phosphate is therefore considered necessary for the induction of *Galnt3*. The same study also revealed the necessity of a number of transcription factors (e.g., *Egr1* and *ETS variant 5* [*Etv5*]) for high levels of extracellular phosphate to induce the expression of *Galnt3* [46]. However, *Egr1* and *Etv5* were not sufficient to enhance the *Galnt3* gene expression; thus, other transcriptional activators are needed.

4.3.2 A Potential Candidate Molecule for Phosphate-Sensing in the Bone: FGFR1c

As mentioned above, high extracellular phosphate was reported to activate the MEK/ERK pathway by several research groups [3, 4, 25, 35, 46, 54]. Until today, however, the upstream molecules of the MEK/ERK pathway have been unknown. The MEK/ERK pathway is activated by a number of receptor tyrosine kinases (RTKs), including—but not limited to—epidermal growth factor receptor (EGFR) and FGFRs [32]. An experiment was performed in order to investigate whether high extracellular phosphate levels can activate RTKs, and if so, to determine the RTKs that are activated using proteomics. Osteoblastic UMR106 cells were treated with high levels of extracellular phosphate and an LC-MS/MS analysis was performed after digestion of the cell extract with trypsin and immunoprecipitation with an anti-phosphotyrosine antibody [1]. The proteomic analysis demonstrated that FGFR1 was the only RTK to be phosphorylated by high levels of extracellular phosphate [46].

Alternative splicing produces two types of FGFR1 (FGFR1b and FGFR1c) [36]. Osteoblastic UMR106 cells only express FGFR1c endogenously [46]. Furthermore, this analysis identified, two types of FGFR1 peptides—one with phosphotyrosine 583 and 585, the other with phosphotyrosine 653 and 654—in which high extracellular phosphate levels induced tyrosine (Tyr) phosphorylation [46]. Six Tyr residues, which are located in the cytoplasmic region of FGFR1, Tyr653, 583, 463, 766, 585, and 654, are sequentially phosphorylated in this order following the activation of FGFR1 [19, 29]. Among them, the phosphorylation of the two Tyr residues (Tyr653 and 654) is known to dramatically increase the tyrosine kinase activity of FGFR1

[36]. Parallel-reaction monitoring (PRM) was applied to quantify the amount of this peptide with phosphotyrosine 653 and 654, which revealed that high extracellular phosphate levels increased the amount of this peptide approximately threefold [46]. Moreover, treatment with an FGFR inhibitor and *Fgfr1* silencing aborted both the enhanced *Galnt3* expression and the ERK1/2 phosphorylation induced by high levels of extracellular phosphate [46]. These findings were in line with previous reports that suggested that FGFR1 is involved in the cellular responses to high levels of extracellular phosphate [35, 54]. In addition, these findings are supported by the findings of a previous study reporting that— in patients with osteoglophonic dysplasia—some activating mutations in *FGFR1* gene induce high blood levels of FGF23 as well as hypophosphatemia [52].

4.3.2.1 The Significance of FGFR1c as a Phosphate-Sensing Molecule *In Vivo*

In addition to these *in vitro* data, some *in vivo* experiments to examine the significance of FGFR1c have been reported. First, enhanced phosphorylation of ERK1/2 in the bone by a high phosphate diet was visible in the whole tissue extracts by immunoblotting in intact animals [46]. Second, an FGFR inhibitor, NVP-BGJ398 [22], was administered to mice fed a high phosphate diet. Although the blood phosphate levels of mice fed a high phosphate diet were significantly higher than those of mice fed a control diet, under NVP-BGJ398 treatment, neither the blood level of full-length FGF23 nor *Galnt3* gene expression in the bone showed a clear increase, confirming the proposed function of FGFR in *Galnt3* gene induction [46]. Finally, a selective ablation of *Fgfr1* in the bone by crossing *Osteocalcin-Cre* [55] mice with floxed *Fgfr1* [49] mice abolished the increase of the blood level of full-length FGF23 and the *Galnt3* gene upregulation in bone by a high phosphate diet [46]. These *in vivo* data again support the function of FGFR1c in the regulation of the *Galnt3* gene expression by high phosphate.

4.3.2.2 The Specific FGFR1c Signal Transduction Mediated by Phosphate

Although high levels of extracellular phosphate induced the expression of the *Galnt3* gene through activated FGFR1c, it was unclear whether canonical FGFR ligands (e.g., FGF2) also enhance the expression of *Galnt3*. However, FGF2 did not enhance the expression of *Galnt3* gene [46]. It is therefore possible that a molecular mechanism is involved in the differences of FGFR1c activation caused by high levels of extracellular phosphate and the canonical FGFR ligands. Given that FGFR1 activates the MEK/ERK pathway through FGFR substrate 2α (FRS2α)—a phosphorylation substrate of activated FGFR—and that FRS2α is necessary to activate the downstream MEK/ERK pathway [20, 21], we investigated the phosphorylation of FRS2α. FGF2 was observed to phosphorylate Tyr196 and 436 of FRS2α; however, only Tyr196 was phosphorylated by high levels of extracellular phosphate. Monitoring of the time course of ERK1/2 activation revealed that ERK1/2 activation by high levels of extracellular phosphate was transient; in contrast, activation by FGF2 was sustained [46]. Thus, there are differences in signal transduction by high extracellular phosphate and the canonical FGFR ligands.

Three typical intracellular signaling pathways are known to exist downstream of the FGFRs; these are the MEK/ERK pathway, the phosphoinositide 3-kinase (PI3K)/protein kinase B (Akt) pathway, and the phospholipase C-γ (PLCγ)/calcineurin/Nuclear Factor of Activated T cells (NFAT) pathway [20]. Among these, FRS2α activates the MEK/ERK and PI3K/Akt pathways. In contrast, PLCγ/calcineurin/NFAT activation occurs independently of the phosphorylation of FRS2α [20]. PRM demonstrated that the protein level of ERK2 with phosphorylated Tyr183 and 185 was increased by both high extracellular phosphate and FGF2. However, while FGF2 effectively increased the protein level of phosphorylated PLCγ, high extracellular phosphate did not. While Akt was also phosphorylated by FGF2, Akt was not phosphorylated by high extracellular phosphate. Furthermore, a PI3K inhibitor and a calcineurin inhibitor did not influence the induction of *Galnt3* gene expression by high levels of extracellular phosphate [46]. Taken together, these findings indicate that the PI3K/Akt and PLCγ/calcineurin/NFAT pathways are not involved in the induction of the expression of *Galnt3* by high levels of extracellular phosphate.

4.3.2.3 The Activation Model of Unliganded FGFR1c by Phosphate

FGFR1c is necessary for the induction of the *Galnt3* gene expression by high levels of phosphate. Thus, phosphate is considered to function as the first messenger to regulate the level of full-length FGF23 in the blood and a feedback system—which involves FGFR1c and FGF23—exists in regulation of the blood level of phosphate. However, the exact mechanism through which FGFR1c is activated by high phosphate levels remains to be elucidated. FGFRs can form either homodimers to facilitate the binding of paracrine FGFs (e.g., FGF1 and FGF2) or heterodimers with coreceptors (e.g., Klothos) to bind endocrine FGFs including FGF23 [20]. The phosphorylation and dimerization of FGFRs are known to be coupled with the activation of FGFR by the binding of FGFR ligands, while the strength of FGFR activation differs according to the type of FGFR ligand [39]. Several reports have proposed a dimerization of unliganded FGFR model [12, 30, 39]. For the dimerization of unliganded FGFRs on the cell surface, the transmembrane domain is pivotal, while the extracellular and intracellular tyrosine kinase domains are not [12]. Activating mutations in *FGFR1* that change amino acids in the FGFR1 transmembrane domain have been identified in patients with osteoglophonic dysplasia [52]. Furthermore, FGFR1 with these mutations has been reported to be more prone to induce FGFR1 dimerization in comparison to intact FGFR1 in the absence of canonical FGFR ligands [12]. We hypothesize that the activation of unliganded FGFR1c by high extracellular phosphate levels is mediated by an alternation in the protein structure of FGFR1c (e.g., stabilization of the interaction between the transmembrane domains).

4.3.3 The Involvement of Type III Sodium-Phosphate Cotransporters: PiT1 and PiT2

Several studies indicate the involvement of sodium-phosphate cotransporters in the process of phosphate-sensing [4, 25, 35, 54]. Bacteria and yeast use some types of phosphate transporters as phosphate-sensors [33]. However, such a phosphate "transceptor" model has not been established in mammals. Mammalian sodium-phosphate cotransporters are classified into two distinct families: type II and III sodium-phosphate cotransporters [6]. *SLC20A1* and *SLC20A2* encode the two types of type III sodium-phosphate cotransporters, PiT1 and PiT2, respectively [51]. Treatment with phosphonoformic acid (foscarnet: PFA), which is a sodium-phosphate cotransporter inhibitor, was shown to block the uptake of phosphate by cells as well as the effects of phosphate on the expression of *Opn* [4]. Studies in several types of cultured cells suggested the involvement of PiT1 in the signal transduction and gene regulation triggered by high extracellular phosphate [25, 54]. In addition, several reports have indicated that silencing *Slc20a1*, which encodes PiT1, reduced the activation of FGFR1 and MEK/ERK pathway by high levels of extracellular phosphate [25, 54]. Conversely, PiT2—rather than PiT1—was recently reported to be involved in the regulation of FGF23 secretion high extracellular phosphate [8]. Moreover, these authors reported that extracellular phosphate both induced PiT1-PiT2 heterodimerization and mediated the MEK/ERK pathway activation independently of the uptake of phosphate. They concluded that phosphate binding to Ser128 (in PiT1) and Ser113 (in PiT2) is the key factor in mediating phosphate signaling through the PiT proteins [7]. In contrast, our data showed that a high phosphate diet did not induce the expression of *Slc20a2*, which encodes PiT2, in the bone, and the induction of *Galnt3* gene by high extracellular phosphate was not suppressed by the silencing on *Slc20a2* in osteoblastic UMR106 cells [46]. It remains to be clarified how the transporters solely can activate the MEK/ERK signaling in mammalian cells. Furthermore, whether or not PiT1 and

PiT2 (type III sodium-phosphate cotransporters) play a role in the activation of unliganded FGFR1c by high phosphate remains unclear.

4.3.4 A Molecule for Phosphate-Sensing in the Parathyroid Glands: Calcium-Sensing Receptor

Calcium-sensing receptor (CaSR) is the key molecule for the regulated secretion of PTH, which is the principal hormone to maintain the blood calcium (Ca) level [10]. PTH not only works to increase blood Ca level but also inhibit phosphate reabsorption in the renal proximal tubules [9, 13]. Although increased blood phosphate level stimulates PTH secretion from parathyroid glands, phosphate-sensing mechanism on the parathyroid cells also remained unclear [2, 34, 43, 47]. It was recently reported that extracellular phosphate stimulates PTH secretion through CaSR in parathyroid cells. Phosphate was found to function as a noncompetitive antagonist for CaSR and phosphate-binding sites in the extracellular domain of CaSR [11]. Thus, several molecules may have phosphate-sensing mechanisms that work in a cell- or tissue-specific manner.

4.4 Perspectives

The identification of FGF23 has advanced the understanding of not only the regulatory mechanism of phosphate metabolism but also the pathogenesis of hypophosphatemic and hyperphosphatemic diseases. Recent findings have uncovered an unreported function of FGFR1c and demonstrated a novel molecular basis for phosphate-sensing in regulated production of FGF23 in bone (Fig. 4.2). However, the precise

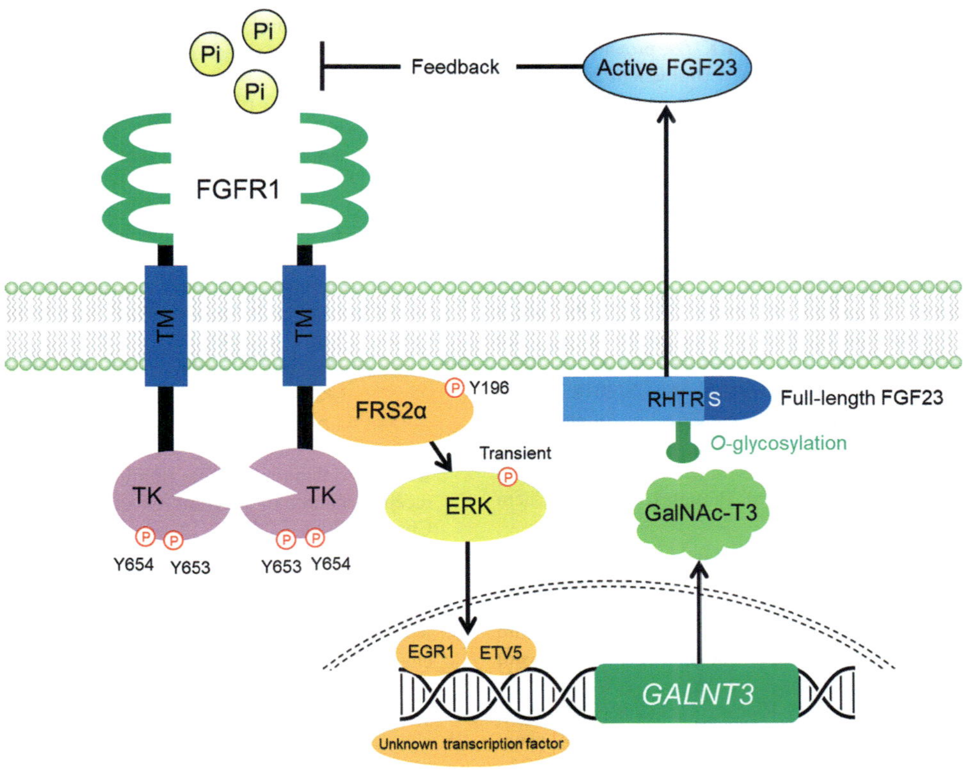

Fig. 4.2 The model of phosphate-sensing in the bone to regulate the production of FGF23 and the blood level of phosphate

mechanism through which FGFR1c is activated by high phosphate remains to be fully elucidated. Furthermore, while PiT1 and PiT2 are also reported to play a role in phosphate-sensing, the precise mechanism of their involvement has not been determined.

As the identification of CaSR has led to the discovery of new drugs for the treatment of primary and secondary hyperparathyroidism, elucidating the mechanism of phosphate-sensing may facilitate the optimization of treatment strategies for abnormal phosphate metabolism.

References

1. Abe Y, Nagano M, Tada A et al (2017) Deep phosphotyrosine proteomics by optimization of phosphotyrosine enrichment and MS/MS parameters. J Proteome Res 16:1077–1086
2. Almaden Y, Canalejo A, Hernandez A et al (1996) Direct effect of phosphorus on PTH secretion from whole rat parathyroid glands in vitro. J Bone Miner Res 11:970–976
3. Beck GR Jr, Knecht N (2003) Osteopontin regulation by inorganic phosphate is ERK1/2-, protein kinase C-, and proteasome-dependent. J Biol Chem 278:41921–41929
4. Beck GR Jr, Zerler B, Moran E (2000) Phosphate is a specific signal for induction of osteopontin gene expression. Proc Natl Acad Sci U S A 97:8352–8357
5. Bennett EP, Mandel U, Clausen H et al (2012) Control of mucin-type O-glycosylation: a classification of the polypeptide GalNAc-transferase gene family. Glycobiology 22:736–756
6. Biber J, Hernando N, Forster I (2013) Phosphate transporters and their function. Annu Rev Physiol 75: 535–550
7. Bon N, Couasnay G, Bourgine A et al (2018a) Phosphate (Pi)-regulated heterodimerization of the high-affinity sodium-dependent Pi transporters PiT1/Slc20a1 and PiT2/Slc20a2 underlies extracellular Pi sensing independently of Pi uptake. J Biol Chem 293:2102–2114
8. Bon N, Frangi G, Sourice S et al (2018b) Phosphate-dependent FGF23 secretion is modulated by PiT2/Slc20a2. Mol Metab 11:197–204
9. Brown EM, Macleod RJ (2001) Extracellular calcium sensing and extracellular calcium signaling. Physiol Rev 81:239–297
10. Brown EM, Gamba G, Riccardi D et al (1993) Cloning and characterization of an extracellular ca(2+)-sensing receptor from bovine parathyroid. Nature 366:575–580
11. Centeno PP, Herberger A, Mun HC et al (2019) Phosphate acts directly on the calcium-sensing receptor to stimulate parathyroid hormone secretion. Nat Commun 10:4693
12. Comps-Agrar L, Dunshee DR, Eaton DL et al (2015) Unliganded fibroblast growth factor receptor 1 forms density-independent dimers. J Biol Chem 290:24166–24177
13. Drüeke TB (2004) Modulation and action of the calcium-sensing receptor. Nephrol Dial Transplant 19 (Suppl 5):V20–V26
14. Feng JQ, Ward LM, Liu S et al (2006) Loss of DMP1 causes rickets and osteomalacia and identifies a role for osteocytes in mineral metabolism. Nat Genet 38: 1310–1315
15. Ferrari SL, Bonjour JP, Rizzoli R (2005) Fibroblast growth factor-23 relationship to dietary phosphate and renal phosphate handling in healthy young men. J Clin Endocrinol Metab 90:1519–1524
16. Florenzano P, Cipriani C, Roszko KL et al (2020) Approach to patients with hypophosphataemia. Lancet Diabetes Endocrinol 8:163–174
17. Frishberg Y, Ito N, Rinat C et al (2007) Hyperostosis-hyperphosphatemia syndrome: a congenital disorder of O-glycosylation associated with augmented processing of fibroblast growth factor 23. J Bone Miner Res 22:235–242
18. Fukumoto S, Yamashita T (2007) FGF23 is a hormone-regulating phosphate metabolism--unique biological characteristics of FGF23. Bone 40:1190–1195
19. Furdui CM, Lew ED, Schlessinger J et al (2006) Autophosphorylation of FGFR1 kinase is mediated by a sequential and precisely ordered reaction. Mol Cell 21:711–717
20. Goetz R, Mohammadi M (2013) Exploring mechanisms of FGF signalling through the lens of structural biology. Nat Rev Mol Cell Biol 14:166–180
21. Gotoh N (2008) Regulation of growth factor signaling by FRS2 family docking/scaffold adaptor proteins. Cancer Sci 99:1319–1325
22. Guagnano V, Furet P, Spanka C et al (2011) Discovery of 3-(2,6-dichloro-3,5-dimethoxy-phenyl)-1-{6-[4-(4-ethyl-piperazin-1-yl)-phenylamin o]-pyrimidin-4-yl}-1-methyl-urea (NVP-BGJ398), a potent and selective inhibitor of the fibroblast growth factor receptor family of receptor tyrosine kinase. J Med Chem 54:7066–7083
23. Jono S, Mckee MD, Murry CE et al (2000) Phosphate regulation of vascular smooth muscle cell calcification. Circ Res 87:E10–E17
24. Khoshniat S, Bourgine A, Julien M et al (2011) The emergence of phosphate as a specific signaling molecule in bone and other cell types in mammals. Cell Mol Life Sci 68:205–218
25. Kimata M, Michigami T, Tachikawa K et al (2010) Signaling of extracellular inorganic phosphate

up-regulates cyclin D1 expression in proliferating chondrocytes via the Na+/Pi cotransporter Pit-1 and Raf/MEK/ERK pathway. Bone 47:938–947
26. Kinoshita Y, Fukumoto S (2018) X-linked hypophosphatemia and FGF23-related Hypophosphatemic diseases: prospect for new treatment. Endocr Rev 39:274–291
27. Kuro-O M, Matsumura Y, Aizawa H et al (1997) Mutation of the mouse klotho gene leads to a syndrome resembling ageing. Nature 390:45–51
28. Kurosu H, Ogawa Y, Miyoshi M et al (2006) Regulation of fibroblast growth factor-23 signaling by klotho. J Biol Chem 281:6120–6123
29. Lew ED, Furdui CM, Anderson KS et al (2009) The precise sequence of FGF receptor autophosphorylation is kinetically driven and is disrupted by oncogenic mutations. Sci Signal 2:ra6
30. Lin CC, Melo FA, Ghosh R et al (2012) Inhibition of basal FGF receptor signaling by dimeric Grb2. Cell 149:1514–1524
31. Liu S, Zhou J, Tang W et al (2006) Pathogenic role of Fgf23 in Hyp mice. Am J Physiol Endocrinol Metab 291:E38–E49
32. Mckay MM, Morrison DK (2007) Integrating signals from RTKs to ERK/MAPK. Oncogene 26:3113–3121
33. Michigami T, Kawai M, Yamazaki M et al (2018) Phosphate as a signaling molecule and its sensing mechanism. Physiol Rev 98:2317–2348
34. Nielsen PK, Feldt-Rasmussen U, Olgaard K (1996) A direct effect in vitro of phosphate on PTH release from bovine parathyroid tissue slices but not from dispersed parathyroid cells. Nephrol Dial Transplant 11:1762–1768
35. Nishino J, Yamazaki M, Kawai M et al (2017) Extracellular phosphate induces the expression of dentin matrix protein 1 through the FGF receptor in osteoblasts. J Cell Biochem 118:1151–1163
36. Ornitz DM, Itoh N (2015) The fibroblast growth factor signaling pathway. Wiley Interdiscip Rev Dev Biol 4: 215–266
37. Perwad F, Azam N, Zhang MY et al (2005) Dietary and serum phosphorus regulate fibroblast growth factor 23 expression and 1,25-dihydroxyvitamin D metabolism in mice. Endocrinology 146:5358–5364
38. Saito H, Maeda A, Ohtomo S et al (2005) Circulating FGF-23 is regulated by 1alpha,25-dihydroxyvitamin D3 and phosphorus in vivo. J Biol Chem 280:2543–2549
39. Sarabipour S, Hristova K (2016) Mechanism of FGF receptor dimerization and activation. Nat Commun 7: 10262
40. Shimada T, Mizutani S, Muto T et al (2001) Cloning and characterization of FGF23 as a causative factor of tumor-induced osteomalacia. Proc Natl Acad Sci U S A 98:6500–6505
41. Shimada T, Muto T, Urakawa I et al (2002) Mutant FGF-23 responsible for autosomal dominant hypophosphatemic rickets is resistant to proteolytic cleavage and causes hypophosphatemia in vivo. Endocrinology 143:3179–3182
42. Shimada T, Hasegawa H, Yamazaki Y et al (2004) FGF-23 is a potent regulator of vitamin D metabolism and phosphate homeostasis. J Bone Miner Res 19: 429–435
43. Slatopolsky E, Finch J, Denda M et al (1996) Phosphorus restriction prevents parathyroid gland growth. High phosphorus directly stimulates PTH secretion in vitro. J Clin Invest 97:2534–2540
44. Tagliabracci VS, Engel JL, Wiley SE et al (2014) Dynamic regulation of FGF23 by Fam20C phosphorylation, GalNAc-T3 glycosylation, and furin proteolysis. Proc Natl Acad Sci U S A 111:5520–5525
45. Takashi Y, Fukumoto S (2018) FGF23 beyond Phosphotropic hormone. Trends Endocrinol Metab 29:755–767
46. Takashi Y, Kosako H, Sawatsubashi S et al (2019) Activation of unliganded FGF receptor by extracellular phosphate potentiates proteolytic protection of FGF23 by its O-glycosylation. Proc Natl Acad Sci U S A 116:11418–11427
47. Thomas L, Bettoni C, Knöpfel T et al (2017) Acute adaption to Oral or intravenous phosphate requires parathyroid hormone. J Am Soc Nephrol 28:903–914
48. Topaz O, Shurman DL, Bergman R et al (2004) Mutations in GALNT3, encoding a protein involved in O-linked glycosylation, cause familial tumoral calcinosis. Nat Genet 36:579–581
49. Trokovic R, Trokovic N, Hernesniemi S et al (2003) FGFR1 is independently required in both developing mid- and hindbrain for sustained response to isthmic signals. EMBO J 22:1811–1823
50. Urakawa I, Yamazaki Y, Shimada T et al (2006) Klotho converts canonical FGF receptor into a specific receptor for FGF23. Nature 444:770–774
51. Virkki LV, Biber J, Murer H et al (2007) Phosphate transporters: a tale of two solute carrier families. Am J Physiol Renal Physiol 293:F643–F654
52. White KE, Cabral JM, Davis SI et al (2005) Mutations that cause osteoglophonic dysplasia define novel roles for FGFR1 in bone elongation. Am J Hum Genet 76: 361–367
53. Yamazaki Y, Tamada T, Kasai N et al (2008) Anti-FGF23 neutralizing antibodies show the physiological role and structural features of FGF23. J Bone Miner Res 23:1509–1518
54. Yamazaki M, Ozono K, Okada T et al (2010) Both FGF23 and extracellular phosphate activate Raf/MEK/ERK pathway via FGF receptors in HEK293 cells. J Cell Biochem 111:1210–1221
55. Zhang M, Xuan S, Bouxsein ML et al (2002) Osteoblast-specific knockout of the insulin-like growth factor (IGF) receptor gene reveals an essential role of IGF signaling in bone matrix mineralization. J Biol Chem 277:44005–44012

Vitamin D and Phosphate Interactions in Health and Disease

5

Nuraly S. Akimbekov, Ilya Digel, Dinara K. Sherelkhan, and Mohammed S. Razzaque

Abstract

Vitamin D plays an essential role in calcium and inorganic phosphate (Pi) homeostasis, maintaining their optimal levels to assure adequate bone mineralization. Vitamin D, as calcitriol ($1,25(OH)_2D$), not only increases intestinal calcium and phosphate absorption but also facilitates their renal reabsorption, leading to elevated serum calcium and phosphate levels. The interaction of $1,25(OH)_2D$ with its receptor (VDR) increases the efficiency of intestinal absorption of calcium to 30–40% and phosphate to nearly 80%. Serum phosphate levels can also influence $1,25(OH)_2D$ and fibroblast growth factor 23 (FGF23) levels, i.e., higher phosphate concentrations suppress vitamin D activation and stimulate parathyroid hormone (PTH) release, while a high FGF23 serum level leads to reduced vitamin D synthesis. In the vitamin D-deficient state, the intestinal calcium absorption decreases and the secretion of PTH increases, which in turn causes the stimulation of $1,25(OH)_2D$ production, resulting in excessive urinary phosphate loss. Maintenance of phosphate homeostasis is essential as hyperphosphatemia is a risk factor of cardiovascular calcification, chronic kidney diseases (CKD), and premature aging, while hypophosphatemia is usually associated with rickets and osteomalacia. This chapter elaborates on the possible interactions between vitamin D and phosphate in health and disease.

Keywords

Vitamin D · PTH · FGF23 · Klotho · Phosphate · Kidney · Intestine · Bone

5.1 Introduction

Vitamin D research has more than 100 years of history since McCollum and Davis's discovered the "growth-promoting fat-soluble vitamin" that was found in cod liver oil [20]. The effect of this growth-promoting factor in the treatment of rickets was so effective that cod liver oil was regarded as a panacea and gave a powerful impetus to further research on vitamin D throughout the world [71]. In the last 20 years, it has been shown that vitamin D's biological activities extend far beyond its involvement in calcium

N. S. Akimbekov (✉) · D. K. Sherelkhan
Department of Biotechnology, Al-Farabi Kazakh National University, Almaty, Kazakhstan
e-mail: Akimbekov.Nuraly@kaznu.kz; Sherelkhan.Dinara@med-kaznu.com

I. Digel
Institute for Bioengineering FH Aachen University of Applied Sciences, Jülich, Germany
e-mail: digel@fh-aachen.de

M. S. Razzaque
Department of Pathology, Lake Erie College of Osteopathic Medicine, Erie, PA, USA
e-mail: mrazzaque@lecom.edu; msr.nagasaki@gmail.com

M. S. Razzaque (ed.), *Phosphate Metabolism*, Advances in Experimental Medicine and Biology 1362,
https://doi.org/10.1007/978-3-030-91623-7_5

metabolism. Along with proven efficacy in pathological conditions and diseases such as rickets, bone loss, and osteomalacia, some novel effects of vitamin D on very diverse physiological processes have been well established [8, 39]. Vitamin D deficiency remains a critical health issue worldwide, and it has been estimated that around one billion people suffer from various vitamin D-related disorders [35].

The biological effects of $1,25(OH)_2D$ can be divided into two types: skeletal (primarily related to calcemic and phosphatemic activities) and non-skeletal, typically not associated with mineral metabolism [15]. The homeostasis of serum phosphate mediated by vitamin D is of paramount importance for adequate bone mineralization, muscle contraction, nerve conduction, and many other vital functions [26]. This brief chapter reviews our understanding of vitamin D-mediated regulation of phosphate homeostasis in health and diseases.

5.2 Physiological Regulation of Phosphate Homeostasis

Phosphorus is the sixth most abundant chemical element in the body [34]. In nature it mainly exists as phosphates, the form most suitable for living organisms [14]. In mammals, the phosphate group is primarily concentrated (~85%) in bones and teeth as hydroxyapatite. The remaining ~15% are distributed in the other tissues as intracellular ortho- and pyrophosphate groups, either free ("inorganic") or as a part of nucleotides, coenzymes, and high-energy phosphate compounds. (referred to as "organophosphates"). Inorganic phosphates exist in two forms: monovalent dihydrogen phosphate ($H_2PO_4^-$) and divalent hydrogen phosphate (HPO_4^{2-}). In the cytosol dihydrogen phosphate is contributing bulk amounts (62% of all cytosolic phosphates).

The extracellular fluid contains only <1% of the whole pool of body's inorganic phosphates [27, 33]. Interestingly, compared to the cytosol, the proportion $H_2PO_4^-$/ HPO_4^{2-} is inverted, so that the major component is now hydrogen phosphate (61% of all phosphates). In general, a 70-kilogram adult with 25% body fat content would have total body phosphorus of approximately 630 g (~21 mol) [34].

Due to its unique chemical structure, various phosphate groups (especially as nucleoside triphosphates) are key players in cellular energy metabolism, in genetic information storage, in signaling pathways, and as phospholipid components of the cell membranes [37]. Inorganic phosphates, together with bicarbonate and protein buffer systems, constitute the basis of the acid-base homeostasis of the body [42].

A healthy adult consumes 1000 mg on average of dietary phosphate per day (Fig. 5.1). Of this amount 700 mg. is absorbed in the small intestine through passive and active pathways [97]. The unabsorbed phosphate is excreted in the feces. Approximately 150 mg. phosphate is secreted into the gut in the saliva, intestinal and pancreatic secretions, while some of it is reabsorbed [47]. Although dietary phosphate intake differs from day to day, principally, phosphate homeostasis is adjusted by intestinal absorption, renal reabsorption, and skeletal resorption. The average serum phosphate concentration in healthy adults is 2.5–4.9 mg/dl [67].

The kidneys filter about 9000 mg. of phosphate daily, 80–90% of which is reabsorbed mainly in the proximal tubule [68]. At least three distinct cotransporters are involved for phosphate transcellular reabsorption in the proximal tubule, namely NaPi-IIa (SLC34A1), NaPi-IIc (SLC34A3), and PiT-2 (SLC20A2) [7] (Fig. 5.2). Phosphate reabsorption is coupled with sodium-dependent (Na^+) transport. Type NaPi II cotransporters are capable of transporting both $H_2PO_4^-$ and HPO_4^{2-} across brush border membrane (BBM) of the proximal tubules [90]. In contrast, in the small intestine, phosphate is absorbed by both transcellular (active) and paracellular (passive) processes, with the active transport being mainly mediated by NaPi-IIb [55].

Given the generally acknowledged role of phosphate in almost every molecular and cellular function, altered phosphate balance can lead to untoward effects. The serum phosphate homeostasis is firmly regulated by endocrine

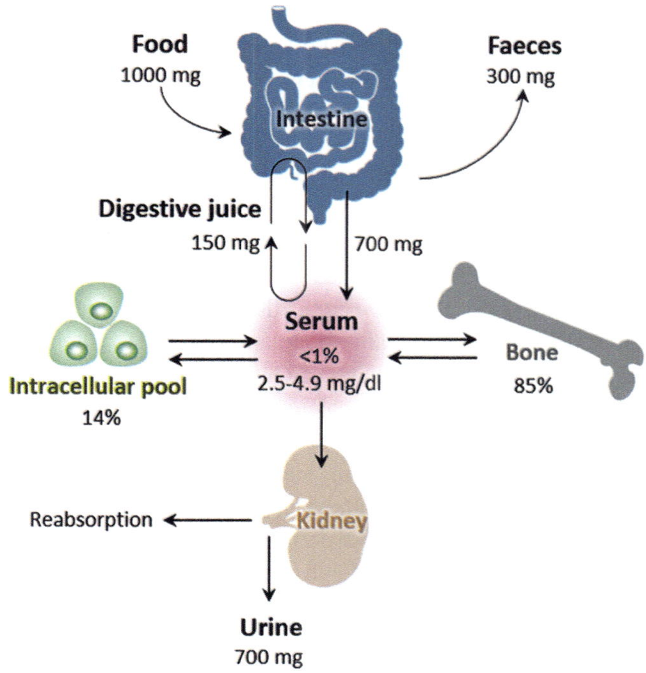

Fig. 5.1 Phosphate flows and balances in the human body [66, 83]

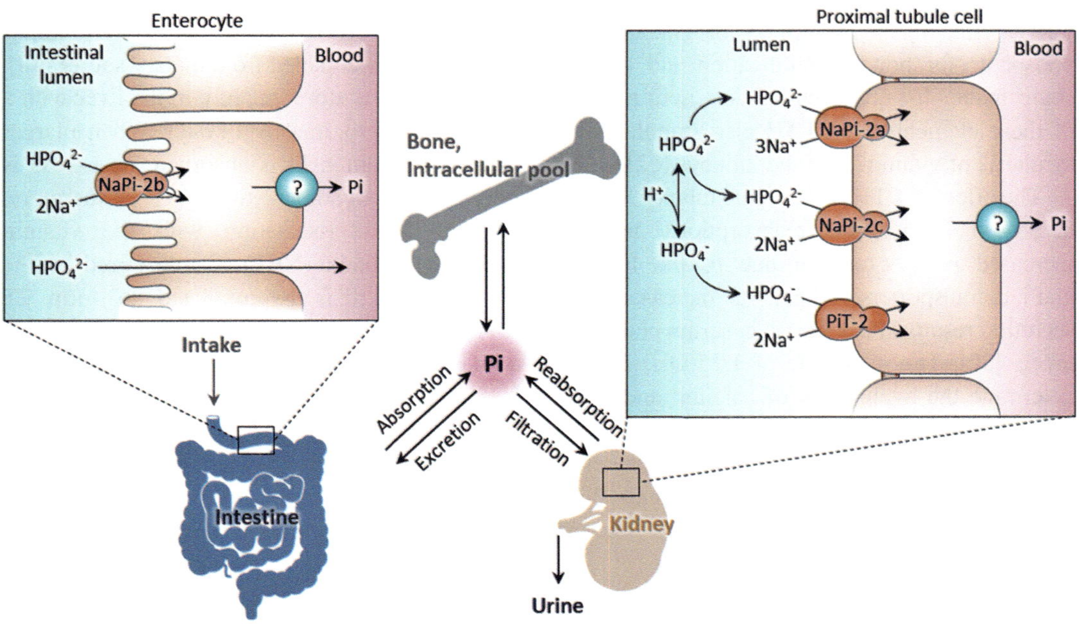

Fig. 5.2 Main transcellular phosphate traffic mechanisms

communication among parathyroid hormone (PTH), calcitriol (1,25(OH)$_2$D), and fibroblast growth factor 23 (FGF-23) [5, 11].

5.2.1 Parathyroid Hormone (PTH)

PTH, a polypeptide containing 84 amino acids with MW 9500 Da, is secreted by chief cells of parathyroid glands [92]. Extracellular calcium concentration is the main modulator of PTH secretion [60]. PTH stimulates calcium resorption from bone tissue, increases calcium reabsorption in the renal tubules, facilitates hydroxylation of 25(OH)D to 1,25(OH)$_2$D in the kidneys, and induces renal excretion of phosphate [50, 69].

In bone tissue, PTH at a permissive level of 1,25(OH)$_2$D promotes calcium resorption by activating osteoclasts [93]. In the intestine, PTH increases the reabsorption of calcium and phosphate by enhancing 1,25(OH)$_2$D synthesis [69]. High serum PTH levels and hypophosphatemia lead to activation of vitamin D-activating enzyme 1α-hydroxylase [57]. 1,25 (OH)$_2$D facilitates absorption of calcium and phosphate for bone mineralization and homeostatic metabolism, preventing low serum levels of these elements [43]. PTH also stimulates the synthesis of vitamin D in the kidneys [52].

The effect of PTH on the renal tubules leads to decreased phosphate reabsorption and its increased renal excretion due to the lowered NaPi cotransporters. In general increased PTH secretion results in a decrease in serum phosphate levels [30]. The main role of 1,25(OH)$_2$D is to determine the availability of calcium and phosphate to form new bone and prevent the development of hypocalcemia and hypophosphatemia [3, 30]. This hormone increases intestinal phosphate absorption elevating its serum concentration.

Secretion PTH by the parathyroid glands is mainly triggered by low extracellular calcium by acting on Ca-sensing receptors (CaSR) [85]. Stimulation of CaSR (they belong to the class of G-protein-coupled receptors) activates multiple heterotrimeric G-proteins, in turn passing the signal to mitogen-activated protein kinase (MAPK) pathways. This cascade of reactions ultimately leads to the suppression of PTH secretion by a negative feedback loop. It has been shown that 1,25(OH)$_2$D upregulates the transcription of the gene encoding the CaSR in the parathyroid gland [13]. Additionally, a low level of calcium indirectly induces parathyroid hyperplasia [23]. However, there is also evidence of the opposite effect of stimulation of parathyroid cell proliferation in response to a high calcium concentration [81].

Interestingly, high serum phosphate levels (hyperphosphatemia) also increase PTH secretion independently of shifts in extracellular calcium [41, 86]. The further secretion of PTH is directly suppressed by 1,25(OH)$_2$D, acting on VDR of parathyroid glands [79].

5.2.2 Vitamin D (Calcitriol)

From a biological point of view, vitamin D is a steroid hormone, as it is synthesized in the body and has a highly specific receptor (VDR). Most vitamin D (90–95%) is formed in the skin under the influence of UVB light, and only a minor fraction of it is obtained from dietary sources [8].

Vitamin D is stored mainly in the liver with a half-life of approximately 14 days. When a larger amount of vitamin D is absorbed, its excess is stored mainly in adipose tissue [1]. Furthermore, vitamin D in association with the vitamin D-binding protein (VBP) is transferred to the liver, where it is hydroxylated to form 25 (OH)D, which subsequently undergoes 1-α-hydroxylation in the renal tubules, turning into 1,25(OH)$_2$D. This biologically active form of vitamin D is under control by serum PTH, phosphate, and FGF23 concentrations. The synthesis of 1,25(OH)$_2$D is stimulated by low serum phosphate levels and high PTH concentrations [78].

Vitamin D promotes the intestinal absorption of calcium and phosphate, significantly increases their renal reabsorption, and also inhibits the PTH secretion [40] (Fig. 5.3). Thus the major effects of 1,25(OH)$_2$D are to augment the intestinal absorption of both calcium and phosphate for proper bone mineral matrix formation [40]. In the intestine and kidneys, 1,25(OH)$_2$D increases the

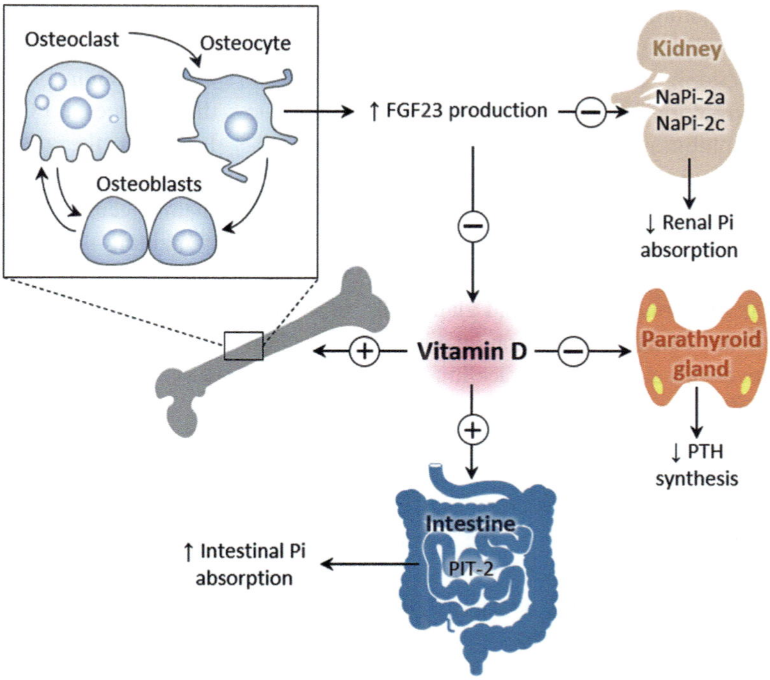

Fig. 5.3 Possible regulation of phosphate homeostasis by vitamin D

formation of calcium-binding proteins (calbidins), which promote transmembrane calcium transport to control homeostasis [2]. In bone, 1,25$(OH)_2$D potentiates the effects of PTH, stimulates bone resorption by osteoclasts, and promotes maturation of monocytes into osteoclasts [70, 84]. In parathyroid glands, 1,25 $(OH)_2$D binds to the VDR, resulting in the suppression of PTH production [96]. The optimal level of serum phosphate is maintained by the interaction of hormones; lowering serum phosphate level by PTH and FGF23, while, increasing serum phosphate level by elevating its absorption in the intestine (1,25$(OH)_2$D) and its resorption from bones (PTH, 1.25$(OH)_2$D) [37]. PTH directly activates osteoclasts and causes phosphate resorption, and indirectly enhances intestinal phosphate absorption by stimulating 1,25 $(OH)_2$D production [44].

Activation of the VDR is a potent and rapid modulator of FGF23 expression, thus forming a "classical" endocrine negative feedback loop between FGF23 and vitamin D [17]. In addition, 1,25$(OH)_2$D is a potent suppressor of PTH gene expression [9].

5.2.3 Fibroblast Growth Factor 23 (FGF23)

FGF23, secreted in bone (osteocytes, osteoblasts, and odontoblasts), is an around 32 kDa glycoprotein, which can be converted in its inactive form through cleavage by a proconvertase-type enzyme into two smaller fragments: 18 kDa (amino fragment) and 12 kDa (carboxy fragment) [32].

FGF23, like PTH, reduces renal phosphate reabsorption, which leads to a drop-in plasma phosphate levels [18]. This hormone also suppresses the secretion of PTH and inhibits the 1α-hydroxylase activity of the kidneys, thus reducing the synthesis of 1,25$(OH)_2$D [46, 51]. FGF23 acts by stimulating its receptors, for the normal function of which a cofactor is needed, i.e. the Klotho protein, synthesized, mostly in the kidneys [87]. The transmembrane Klotho protein is essential for FGF23 to exert its phosphaturic effects in the kidney [72–74, 89].

A decrease in serum phosphate under the FGF23 is achieved by inhibiting phosphate reabsorption in the renal tubules, as well as by stimulating PTH secretion and suppressing 1,25 $(OH)_2D$ synthesis [12, 51, 56, 72, 91]. In contrast, calcitonin, is another hormone produced by the thyroid gland, slightly lowers serum calcium due to inhibition of renal and intestinal calcium reabsorption, reducing calcium and phosphate resorption from bones [36]. Plasma calcium is regulated by a complex system involving PTH and $1,25(OH)_2D$ on the intestine, bones, and kidneys. As mentioned, parathyroid gland cells respond to serum calcium concentration via CaSR. A high level of calcium in extracellular fluid stimulates CaSR receptors and activates cellular mechanisms, which ultimately leads to inhibition of PTH release [6].

Imbalance of calcium and phosphate is manifested as a shift in the calcium, phosphate levels in serum and the levels of serum hormones [PTH and 25(OH)D], as well as the development of bone pathology and cardiovascular calcification with soft anomalies [76, 88]. The exact etiology and pathogenesis of serum phosphate derangements (hyperphosphatemia and hypophosphatemia) will need further studies.

5.3 Hyperphosphatemia

Renal failure is the most common cause of hyperphosphatemia [80]. The decline in estimated glomerular filtration rate disrupts phosphate homeostasis: when it falls below 30 mL/min/1.73 m^2, the reabsorption of phosphate is maximally suppressed and fractional excretion markedly reduced. As a result, the serum level of phosphate increases [16, 21]. A primary increase in tubular reabsorption of phosphate is less common and can be observed in hypoparathyroidism, acromegaly, and tumoral calcification [38].

Excessive phosphate can be released from the intracellular compartment, which is observed in acute tumor lysis syndrome, rhabdomyolysis, hemolysis, hyperthermia, profound catabolic stress, and acute leukemia. Tumor lysis syndrome is commonly observed in malignant hematological patients, particularly non-Hodgkin's lymphoma and acute leukemia, following chemotherapy [4]. Risk factors for developing the syndrome include impaired renal function, increased levels of lactate dehydrogenase, and hyperuricemia [95]. The latter is caused by the disturbances in FGF23-mediated phosphate regulation in the proximal tubule of the kidney [10]. Increased intestinal phosphate absorption is mainly caused either by the use of phosphate-containing oral laxative, or by vitamin D overdoses [59].

5.4 Hypophosphatemia

Hypophosphatemia may be a consequence of the decreased intestinal absorption, internal redistribution, and increased urinary loss of phosphate [31]. The acute shift of phosphate from the extracellular to the intracellular compartment is most often caused by respiratory alkalosis and refeeding syndrome in hospitalized patients [19, 54]. Respiratory alkalosis causes an increase in intracellular pH, which stimulates phosphofructokinase, leading to severe hypophosphatemia with plasma phosphate of >0.32 mmol/L [82]. The intracellular shift of phosphate is also observed in the treatment of diabetic ketoacidosis and hungry bone syndrome, which occurs after parathyroidectomy performed for patients with long-standing hyperparathyroidism [31]. At the same time, in the postoperative period, serum calcium and phosphate concentrations significantly decrease.

Low phosphate intake rarely causes hypophosphatemia, probably because the phosphate content in the diet almost always exceeds the phosphate loss through the gastrointestinal tract, and the kidneys can reabsorb nearly all of the filtered phosphate [24]. Excessive urinary loss of phosphate is observed in both primary and secondary hyperparathyroidism caused by impaired vitamin D metabolism, Fanconi syndrome, diuretics, and tumor-induced osteomalacia (TIO) [31, 48]. TIO is a rare paraneoplastic syndrome characterized by hypophosphatemia, phosphaturia, decreased $1,25(OH)_2D$ level, normal 25(OH)D levels, and osteomalacia [29]. Overproduction of FGF23 caused by TIO reduces tubular phosphate reabsorption and $1,25(OH)_2D$ production [58].

5.5 Genetic Disorders Associated with Hypophosphatemia

Several inherited abnormalities are characterized by phosphate-wasting syndromes, commonly mediated by FGF23. These diseases, resulted by impaired FGF23 metabolism, include autosomal dominant hypophosphatemic rickets (ADHR), X-linked hypophosphatemic rickets (XLHR), and autosomal recessive hypophosphatemic rickets (ARHR) [94].

ADHR (OMIM 193100) is produced by *FGF23* gain-of-function mutation, which causes the resistance of the mutant FGF23 to proteolytic degradation [22]. ADHR manifests as a defect in renal phosphate transport, associated with decreased 1,25(OH)2D levels, while the PTH levels remain normal. ADHR is characterized by hypophosphatemia, renal phosphate loss, short stature, and bone disorders [25].

ARHR (OMIM 241520) is caused by mutations in the *DMP1* gene (located on chromosome locus 4q21). Patients with ARHR suffer from decreased renal phosphate reabsorption and typically display hyperphosphaturia, hypophosphatemia, reduced 1,25(OH)$_2$D concentration, with PTH values remaining normal [28, 49].

XLHR (OMIM 307800) appears as a result of mutations inactivating *PHEX* (phosphate-regulating gene with homologies to endopeptidases located on the X-chromosome). The *PHEX* gene encodes a zinc-dependent metalloproteinase, and is strongly expressed in osteoblasts, osteocytes, and odontoblasts [53]. The XLHR symptoms include growth retardation, hypophosphatemia, osteomalacia, and defective renal phosphate reabsorption. The diseased state is resistant to phosphate and vitamin D therapy [63].

5.6 Conclusions

Serum phosphate levels are tightly regulated by hormonal and metabolic factors mainly related to the triad "vitamin D-PTH-FGF23" as well as dietary phosphate. Experimental studies have convincingly shown that disorders and disturbances in phosphate regulation can lead to serious systemic complications [45, 61, 62, 64, 65, 75, 77]. Particular attention should be placed on the central activity of vitamin D in phosphate metabolism, as 1,25(OH)2D both, directly and indirectly, impact serum phosphate levels. However, despite the well-studied pivotal roles of vitamin D in phosphate homeostasis, many aspects remain unclear. For instance, what are the underlying mechanisms by which vitamin D acts on renal phosphate reabsorption, and how exactly do calcium and vitamin D modulate FGF23 production? A better understanding of these processes and interactions would help to develop more efficient strategies for the treatment of phosphate-related disorders.

Acknowledgement The authors would like to thank Dr. Margo Wolfe for carefully reading the manuscript and providing useful suggestions.

References

1. Abbas MA (2017) Physiological functions of Vitamin D in adipose tissue. J Steroid Biochem Mol Biol 165: 369–381
2. Armbrecht HJ, Boltz MA, Christakos S, Bruns ME (1998) Capacity of 1,25-dihydroxyvitamin D to stimulate expression of calbindin D changes with age in the rat. Arch Biochem Biophys 352:159–164
3. Avcil S, Uysal P, Yilmaz M, Erge D, Demirkaya SK, Eren E (2017) Vitamin D deficiency and a blunted parathyroid hormone response in children with attention-deficit/hyperactivity disorder. Clin Lab 63: 435–443
4. Belay Y, Yirdaw K, Enawgaw B (2017) Tumor lysis syndrome in patients with hematological malignancies. J Oncol 2017:9684909
5. Bergwitz C, Jüppner H (2010) Regulation of phosphate homeostasis by PTH, vitamin D, and FGF23. Annu Rev Med 61:91–104
6. Berridge MJ, Bootman MD, Roderick HL (2003) Calcium signalling: dynamics, homeostasis and remodelling. Nat Rev Mol Cell Biol 4:517–529
7. Biber J, Hernando N, Forster I, Murer H (2009) Regulation of phosphate transport in proximal tubules. Pflugers Arch 458:39–52
8. Bikle DD (2014) Vitamin D metabolism, mechanism of action, and clinical applications. Chem Biol 21: 319–329

9. Bohnert BN, Daniel C, Amann K, Voelkl J, Alesutan I, Lang F, Heyne N, Häring H-U, Artunc F (2015) Impact of phosphorus restriction and vitamin D-substitution on secondary hyperparathyroidism in a proteinuric mouse model. Kidney Blood Press Res 40: 153–165
10. Boyce AM, Lee AE, Roszko KL, Gafni RI (2020) Hyperphosphatemic Tumoral calcinosis: pathogenesis, clinical presentation, and challenges in management. Front Endocrinol (Lausanne) 11:293
11. Brown RB, Haq A, Stanford CF, Razzaque MS (2015) Vitamin D, phosphate, and vasculotoxicity. Can J Physiol Pharmacol 93:1077–1082
12. Buchanan S, Combet E, Stenvinkel P, Shiels PG (2020) Klotho, aging, and the failing kidney. Front Endocrinol (Lausanne) 11:560
13. Canaff L, Hendy GN (2002) Human calcium-sensing receptor gene. Vitamin D response elements in promoters P1 and P2 confer transcriptional responsiveness to 1,25-dihydroxyvitamin D. J Biol Chem 277:30337–30350
14. Chapuy MC, Schott AM, Garnero P, Hans D, Delmas PD, Meunier PJ (1996) Healthy elderly French women living at home have secondary hyperparathyroidism and high bone turnover in winter. EPIDOS Study Group. J Clin Endocrinol Metab 81:1129–1133
15. Charoenngam N, Shirvani A, Holick MF (2019) Vitamin D for skeletal and non-skeletal health: what we should know. J Clin Orthop Trauma 10:1082–1093
16. Chen W, Bushinsky D (2017) Chronic kidney disease–mineral and bone disorder. In: Nissenson, A.R., Fine, R.N.B.T. Handbook of Dialysis Therapy, 5th edn. Elsevier, pp 685–697.e1
17. Clinkenbeard EL, White KE (2016) Systemic control of bone homeostasis by FGF23 signaling. Curr Mol Biol Rep 2:62–71
18. Coyac BR, Hoac B, Chafey P, Falgayrac G, Slimani L, Rowe PS, Penel G, Linglart A, McKee MD, Chaussain C, Bardet C (2018) Defective mineralization in X-linked hypophosphatemia dental pulp cell cultures. J Dent Res 97:184–191
19. De Marchi S, Cecchin E, Basile A, Bertotti A, Nardini R, Bartoli E (1993) Renal tubular dysfunction in chronic alcohol abuse–effects of abstinence. N Engl J Med 329:1927–1934
20. Deluca HF (2014) History of the discovery of vitamin D and its active metabolites. Bonekey Rep 3:479
21. Diez C, Mohr P, Koch D, Silber R-E, Schmid C, Hofmann H-S (2009) Age- and gender-specific values of estimated glomerular filtration rate among 6232 patients undergoing cardiac surgery. Interact Cardiovasc Thorac Surg 9:593–597
22. Drezner MK, Whyte MP (2018) Chapter 40: Heritable renal phosphate wasting disorders. In: Thakker RV, Whyte MP, Eisman JA, Igarashi TBT (eds) Genetics of bone biology and skeletal disease, 2nd edn. Academic, pp 761–782
23. Drüeke TB (2001) Parathyroid gland hyperplasia in uremia. Kidney Int 59:1182–1183
24. Dvm SDF, Moreland KJ (1989) Hypophosphatemia. J Vet Intern Med 3:149–159
25. Econs MJ (2005) Chapter 70: Disorders of phosphate metabolism: autosomal dominant hypophosphatemic rickets, tumor induced osteomalacia, fibrous dysplasia, and the pathophysiological relevance of FGF23. In: Feldman DBT (ed) Vitamin D, 2nd edn. Academic, Burlington, pp 1189–1195
26. Erem S, Razzaque MS (2018) Dietary phosphate toxicity: an emerging global health concern. Histochem Cell Biol 150:711–719
27. Farrow EG, White KE (2010) Recent advances in renal phosphate handling. Nat Rev Nephrol 6:207–217
28. Feng JQ, Ward LM, Liu S, Lu Y, Xie Y, Yuan B, Yu X, Rauch F, Davis SI, Zhang S, Rios H, Drezner MK, Quarles LD, Bonewald LF, White KE (2006) Loss of DMP1 causes rickets and osteomalacia and identifies a role for osteocytes in mineral metabolism. Nat Genet 38:1310–1315
29. Florenzano P, Gafni RI, Collins MT (2017) Tumor-induced osteomalacia. Bone Rep 7:90–97
30. Fukumoto S (2014) Phosphate metabolism and vitamin D. Bonekey Rep 3:497
31. Gaasbeek A, Meinders AE (2005) Hypophosphatemia: an update on its etiology and treatment. Am J Med 118:1094–1101
32. Goetz R, Nakada Y, Hu MC, Kurosu H, Wang L, Nakatani T, Shi M, Eliseenkova AV, Razzaque MS, Moe OW, Kuro-o M, Mohammadi M (2010) Isolated C-terminal tail of FGF23 alleviates hypophosphatemia by inhibiting FGF23-FGFR-Klotho complex formation. Proc Natl Acad Sci U S A 107:407–412
33. Goretti Penido M, Alon US (2012) Phosphate homeostasis and its role in bone health. Pediatr Nephrol 27: 2039–2048
34. Heaney RP (2011) Chapter 34: Vitamin D: role in the calcium and phosphorus economies. In: Feldman D, Pike JW, Adams JSBT (eds) Vitamin D, 3rd edn. Academic, San Diego, pp 607–624
35. Holick MF (2007) Vitamin D deficiency. N Engl J Med 357:266–281
36. Ikegame M, Ejiri S, Ozawa H (2004) Calcitonin-induced change in serum calcium levels and its relationship to osteoclast morphology and number of calcitonin receptors. Bone 35:27–33
37. Jacquillet G, Unwin RJ (2019) Physiological regulation of phosphate by vitamin D, parathyroid hormone (PTH) and phosphate (Pi). Pflugers Arch 471:83–98
38. Kamenický P, Blanchard A, Gauci C, Salenave S, Letierce A, Lombès M, Brailly-Tabard S, Azizi M, Prié D, Souberbielle J-C, Chanson P (2012) Pathophysiology of renal calcium handling in acromegaly: what lies behind hypercalciuria? J Clin Endocrinol Metab 97:2124–2133
39. Khammissa RAG, Fourie J, Motswaledi MH, Ballyram R, Lemmer J, Feller L (2018) The biological activities of vitamin D and its receptor in relation to calcium and bone homeostasis, cancer, immune and cardiovascular systems, skin biology, and Oral health. Biomed Res Int 2018:9276380
40. Khazai N, Judd SE, Tangpricha V (2008) Calcium and vitamin D: skeletal and extraskeletal health. Curr Rheumatol Rep 10:110–117

41. Kilav R, Silver J, Naveh-Many T (1995) Parathyroid hormone gene expression in hypophosphatemic rats. J Clin Invest 96:327–333
42. Klumpp S, Krieglstein J (2002) Phosphorylation and dephosphorylation of histidine residues in proteins. Eur J Biochem 269:1067–1071
43. Korvala J, Hartikka H, Pihlajamäki H, Solovieva S, Ruohola J-P, Sahi T, Barral S, Ott J, Ala-Kokko L, Männikkö M (2010) Genetic predisposition for femoral neck stress fractures in military conscripts. BMC Genet 11:95
44. Koumakis E, Cormier C, Roux C, Briot K (2021) The causes of hypo- and hyperphosphatemia in humans. Calcif Tissue Int 108:41–73
45. Lanske B, Razzaque MS (2007) Premature aging in klotho mutant mice: cause or consequence? Ageing Res Rev 6:73–79
46. Lederer E (2014) Regulation of serum phosphate. J Physiol 592:3985–3995
47. Leung J, Crook M (2019) Disorders of phosphate metabolism. J Clin Pathol 72:741–747
48. Levi M (2001) Post-transplant hypophosphatemia. Kidney Int 59:2377–2387
49. Levy-Litan V, Hershkovitz E, Avizov L, Leventhal N, Bercovich D, Chalifa-Caspi V, Manor E, Buriakovsky S, Hadad Y, Goding J, Parvari R (2010) Autosomal-recessive hypophosphatemic rickets is associated with an inactivation mutation in the ENPP1 gene. Am J Hum Genet 86:273–278
50. Lips P (2006) Vitamin D physiology. Prog Biophys Mol Biol 92:4–8
51. Liu S, Tang W, Zhou J, Stubbs JR, Luo Q, Pi M, Quarles LD (2006) Fibroblast growth factor 23 is a counter-regulatory phosphaturic hormone for vitamin D. J Am Soc Nephrol 17:1305–1315
52. Lofrese JJ, Basit H, Lappin SL (2021) Physiology, parathyroid. Treasure Island (FL)
53. Magen D, Zelikovic I (2012) Chapter 27: Hereditary tubular disorders of mineral handling. In: Glorieux FH, Pettifor JM, Jüppner HBT (eds) Pediatric bone, 2nd edn. Academic, San Diego, pp 727–770
54. Marik PE, Bedigian MK (1996) Refeeding hypophosphatemia in critically ill patients in an intensive care unit. A prospective study. Arch Surg 131:1043–1047
55. Marks J (2019) The role of SLC34A2 in intestinal phosphate absorption and phosphate homeostasis. Pflugers Arch 471:165–173
56. Marthi A, Donovan K, Haynes R, Wheeler DC, Baigent C, Rooney CM, Landray MJ, Moe SM, Yang J, Holland L, di Giuseppe R, Bouma-de Krijger A, Mihaylova B, Herrington WG (2018) Fibroblast growth Factor-23 and risks of cardiovascular and noncardiovascular diseases: a meta-analysis. J Am Soc Nephrol 29:2015–2027
57. Miller WL (2017) Genetic disorders of vitamin D biosynthesis and degradation. J Steroid Biochem Mol Biol 165:101–108
58. Minisola S, Peacock M, Fukumoto S, Cipriani C, Pepe J, Tella SH, Collins MT (2017) Tumour-induced osteomalacia. Nat Rev Dis Prim 3:17044
59. Moe SM, Daoud JR (2014) Chapter 11: Disorders of mineral metabolism: calcium, phosphorus, and magnesium. In: Gilbert SJ, Weiner DE (eds) National Kidney Foundation Primer on Kidney Diseases, 6th edn, Philadelphia, pp 100–112
60. Muresan Z, MacGregor RR (1994) The release of parathyroid hormone and the exocytosis of a proteoglycan are modulated by extracellular Ca2+ in a similar manner. Mol Biol Cell 5:725–737
61. Nakatani T, Ohnishi M, Razzaque MS (2009a) Inactivation of klotho function induces hyperphosphatemia even in presence of high serum fibroblast growth factor 23 levels in a genetically engineered hypophosphatemic (Hyp) mouse model. FASEB J 23:3702–3711
62. Nakatani T, Sarraj B, Ohnishi M, Densmore MJ, Taguchi T, Goetz R, Mohammadi M, Lanske B, Razzaque MS (2009b) In vivo genetic evidence for klotho-dependent, fibroblast growth factor 23 (Fgf23) -mediated regulation of systemic phosphate homeostasis. FASEB J 23:433–441
63. Neves RL, Chiarantin GMD, Nascimento FD, Pesquero JB, Nader HB, Tersariol ILS, McKee MD, Carmona AK, Barros NMT (2016) Expression and inactivation of osteopontin-degrading PHEX enzyme in squamous cell carcinoma. Int J Biochem Cell Biol 77:155–164
64. Ohnishi M, Razzaque MS (2010) Dietary and genetic evidence for phosphate toxicity accelerating mammalian aging. FASEB J 24:3562–3571
65. Ohnishi M, Nakatani T, Lanske B, Razzaque MS (2009) In vivo genetic evidence for suppressing vascular and soft-tissue calcification through the reduction of serum phosphate levels, even in the presence of high serum calcium and 1,25-dihydroxyvitamin d levels. Circ Cardiovasc Genet 2:583–590
66. Osuka S, Razzaque MS (2012) Can features of phosphate toxicity appear in normophosphatemia? J Bone Miner Metab 30:10–18
67. Peacock M (2015) Chapter 31: Primary hyperparathyroidism and the kidney. In: Bilezikian JPBT (ed) The parathyroids, 3rd edn. Academic, San Diego, pp 455–467
68. Prasad N, Bhadauria D (2013) Renal phosphate handling: physiology. Indian J Endocrinol Metab 17: 620–627
69. Pu F, Chen N, Xue S (2016) Calcium intake, calcium homeostasis and health. Food Sci Hum Wellness 5:8–16
70. Quinn JM, Fujikawa Y, McGee JO, Athanasou NA (1997) Rodent osteoblast-like cells support osteoclastic differentiation of human cord blood monocytes in the presence of M-CSF and 1,25 dihydroxyvitamin D3. Int J Biochem Cell Biol 29:173–179
71. Rajakumar K (2003) Vitamin D, cod-liver oil, sunlight, and rickets: a historical perspective. Pediatrics 112:e132–e135
72. Razzaque MS (2009a) FGF23-mediated regulation of systemic phosphate homeostasis: is klotho an essential player? Am J Physiol Renal Physiol 296:F470–F476

73. Razzaque MS (2009b) Does FGF23 toxicity influence the outcome of chronic kidney disease? Nephrol Dial Transplant 24:4–7
74. Razzaque MS (2014) Bone-kidney axis in systemic phosphate turnover. Arch Biochem Biophys 561: 154–158
75. Razzaque MS, Lanske B (2006) Hypervitaminosis D and premature aging: lessons learned from Fgf23 and Klotho mutant mice. Trends Mol Med 12:298–305
76. Razzaque MS, St-Arnaud R, Taguchi T, Lanske B (2005) FGF-23, vitamin D and calcification: the unholy triad. Nephrol Dial Transplant 20:2032–2035
77. Razzaque MS, Sitara D, Taguchi T, St-Arnaud R, Lanske B (2006) Premature aging-like phenotype in fibroblast growth factor 23 null mice is a vitamin D-mediated process. FASEB J 20:720–722
78. Rigo J, Pieltain C, Viellevoye R, Bagnoli F (2018) Calcium and phosphorus homeostasis: pathophysiology. In: Buonocore G, Bracci R, Weindling M (eds) Neonatology: a practical approach to neonatal diseases. Springer International Publishing, Cham, pp 639–668
79. Ritter CS, Armbrecht HJ, Slatopolsky E, Brown AJ (2006) 25-Hydroxyvitamin D3 suppresses PTH synthesis and secretion by bovine parathyroid cells. Kidney Int 70:654–659
80. Ritz E, Gross M-L (2005) Hyperphosphatemia in renal failure. Blood Purif 23:6–9
81. Roussanne MC, Lieberherr M, Souberbielle JC, Sarfati E, Drüeke T, Bourdeau A (2001) Human parathyroid cell proliferation in response to calcium, NPS R-467, calcitriol and phosphate. Eur J Clin Investig 31: 610–616
82. Rudolph EH, Gonin JM (2012) Chapter 79: Disorders of phosphorus metabolism. In: Lerma EV, Nissenson ARBT (eds) Nephrology secrets, 3rd edn. Mosby, Saint Louis, pp 551–559
83. Schiavi SC, Kumar R (2004) The phosphatonin pathway: new insights in phosphate homeostasis. Kidney Int 65:1–14
84. Schmitt A, Ehnert S, Schyschka L, Buschner P, Kühnl A, Döbele S, Siebenlist S, Lucke M, Stöckle U, Nussler AK (2012) Monocytes do not transdifferentiate into proper osteoblasts. Sci World J 2012:384936
85. Shoback DM, Bilezikian JP, Turner SA, McCary LC, Guo MD, Peacock M (2003) The calcimimetic cinacalcet normalizes serum calcium in subjects with primary hyperparathyroidism. J Clin Endocrinol Metab 88:5644–5649
86. Slatopolsky E, Brown A, Dusso A (2001) Role of phosphorus in the pathogenesis of secondary hyperparathyroidism. Am J Kidney Dis 37:S54–S57
87. Stefanopoulos D, Nasiri-Ansari N, Dontas I, Vryonidou A, Galanos A, Psaridi L, Fatouros IG, Mastorakos G, Papavassiliou AG, Kassi E, Tournis S (2020) Fibroblast growth factor 23 (FGF23) and klotho protein in Beta-thalassemia. Horm Metab Res – Horm und Stoffwechselforsch 52:194–201
88. Sun M, Wu X, Yu Y, Wang L, Xie D, Zhang Z, Chen L, Lu A, Zhang G, Li F (2020) Disorders of calcium and phosphorus metabolism and the proteomics/metabolomics-based research. Front Cell Dev Biol 8:576110
89. Urakawa I, Yamazaki Y, Shimada T, Iijima K, Hasegawa H, Okawa K, Fujita T, Fukumoto S, Yamashita T (2006) Klotho converts canonical FGF receptor into a specific receptor for FGF23. Nature 444:770–774
90. Villa-Bellosta R, Ravera S, Sorribas V, Stange G, Levi M, Murer H, Biber J, Forster IC (2009) The Na+-Pi cotransporter PiT-2 (SLC20A2) is expressed in the apical membrane of rat renal proximal tubules and regulated by dietary Pi. Am J Physiol Renal Physiol 296:F691–F699
91. Vogt I, Haffner D, Leifheit-Nestler M (2019) FGF23 and phosphate-cardiovascular toxins in CKD. Toxins (Basel) 11:647
92. Voinescu A, Martin KJ (2013) Chapter 19: Calcium, phosphate, PTH, Vitamin D and FGF-23 in chronic kidney disease. In: Kopple JD, Massry SG, Kalantar-Zadeh KBT (eds) Nutritional management of renal disease. Academic, pp 263–283
93. Wald A, Narasimhan S, Nieves-Arriba L, Waggoner S (2009) Prolonged hypercalcemia following resection of dysgerminoma: a case report. Obstet Gynecol Int 2009:956935
94. White KE, Evans WE, O'Riordan JLH, Speer MC, Econs MJ, Lorenz-Depiereux B, Grabowski M, Meitinger T, Strom TM (2000) Autosomal dominant hypophosphataemic rickets is associated with mutations in FGF23. Nat Genet 26:345–348
95. Will A, Tholouli E (2011) The clinical management of tumour lysis syndrome in haematological malignancies. Br J Haematol 154:3–13
96. Xiao Z, Dallas M, Qiu N, Nicolella D, Cao L, Johnson M, Bonewald L, Quarles LD (2011) Conditional deletion of Pkd1 in osteocytes disrupts skeletal mechanosensing in mice. FASEB J 25:2418–2432
97. Yamaguchi T, Sugimoto T, Chihara K (2002) Intestinal absorption of phosphate. In: Morii H, Nishizawa Y, Massry SG (eds) Calcium in internal medicine. Springer, London, pp 123–135

Fibroblast Growth Factor 23 as Regulator of Vitamin D Metabolism

6

Shinya Nakatani, Ayumi Nakatani, Katsuhito Mori, Masanori Emoto, Masaaki Inaba, and Mohammed S. Razzaque

Abstract

Fibroblast growth factor 23 (FGF23) is a hormone produced by osteocytes in bone that acts on the kidneys to regulate phosphate and vitamin D metabolism.FGF23 levels were shown to be increased in the early stage of chronic kidney disease (CKD), with a slight decline in estimated glomerular filtration rate (eGFR) even when the range was restricted to above 60 mL/min/1.73 m^2, indicating that subtle phosphate load is a stimulator of FGF23 in serum. FGF23 is also known to inhibit vitamin D activation from 25-hydroxyvitamin D (25-OH-D) to 1,25-dihydroxyvitamin D [1,25 $(OH)_2D$], while it stimulates its degradation from 25-OH-D to 24,25-dihydroxyvitamin D [24,25$(OH)_2D$]. Previously, we demonstrated a significant and negative association of serum FGF23 with serum 1,25$(OH)_2D$ and 1,25 $(OH)_2D$/25-OH-D ratio, a putative parameter for CYP27B1, and confirmed the physiological effects of FGF23 on phosphate and vitamin D metabolism in non-CKD subjects. Elevated FGF23 by itself is reported to be associated with various adverse outcomes, including left ventricular hypertrophy, endothelial dysfunction, and activation of the renin-angiotensin-aldosterone system, leading to increased mortality even in non-CKD individuals. On the other hand, our previous study showed that the impaired incremental response of serum FGF23 in response to oral phosphate load in diabetic patients can help to significantly increase serum phosphate (Yoda et al., J Clin Endocrinol Metab 97:E2036–43, 2012) and thus may contribute to progression of vascular calcification in those patients (personal observation). It is suggested that increased serum FGF23 might be an important indicator of adverse outcomes in non-CKD as well as CKD patients.

Authors Shinya Nakatani and Masaaki Inaba have equally contributed to this chapter.

S. Nakatani (✉) · A. Nakatani · M. Emoto
Department of Metabolism, Endocrinology, and Molecular Medicine, Osaka City University Graduate School of Medicine, Osaka, Japan
e-mail: m2026719@med.osaka-cu.ac.jp; ayumi.diamant.xxx1118@gmail.com; memoto@med.osaka-cu.ac.jp

K. Mori
Department of Nephrology, Osaka City University Graduate School of Medicine, Osaka, Japan
e-mail: ktmori@med.osaka-cu.ac.jp

M. Inaba
Department of Metabolism, Endocrinology, and Molecular Medicine, Osaka City University Graduate School of Medicine, Osaka, Japan

Department of Nephrology, Osaka City University Graduate School of Medicine, Osaka, Japan
e-mail: inaba-m@med.osaka-cu.ac.jp

M. S. Razzaque
Department of Pathology, Lake Erie College of Osteopathic Medicine, Erie, PA, USA
e-mail: mrazzaque@lecom.edu; msr.nagasaki@gmail.com

M. S. Razzaque (ed.), *Phosphate Metabolism*, Advances in Experimental Medicine and Biology 1362,
https://doi.org/10.1007/978-3-030-91623-7_6

Keywords

FGF23 · Vitamin D · Phosphate · CKD-MBD · Klotho

6.1 Physiology of FGF23

Fibroblast growth factor 23 (FGF23) is a hormone produced by osteocytes in bone that acts on the kidneys to regulate phosphate and vitamin D metabolism through activation of the FGF receptor (FGFR)/α-klotho co-receptor complex [31]. This hormone induces phosphaturia by decreasing phosphate reabsorption in the proximal tubule through down-regulation of luminal sodium-phosphate co-transporters [48, 53]. Furthermore, it inhibits the activation step of vitamin D from 25-hydroxyvitamin D (25-OH-D) to 1,25-dihydroxyvitamin D [1,25$(OH)_2$D] [6, 52], while it stimulates the degradation step of 25-OH-D to 24,25-dihydroxyvitamin D [24,25$(OH)_2$D] [53, 54, 62], thus inhibiting intestinal phosphate absorption (Fig. 6.1). Consequently, FGF23 has been proposed as a major regulator of phosphate and vitamin D metabolism.

6.2 Direct and Indirect Effects of FGF23 on Various Factors

Chronic kidney disease (CKD) is known to be complicated with CKD-mineral and bone disorder (CKD-MBD), which is mainly comprised of

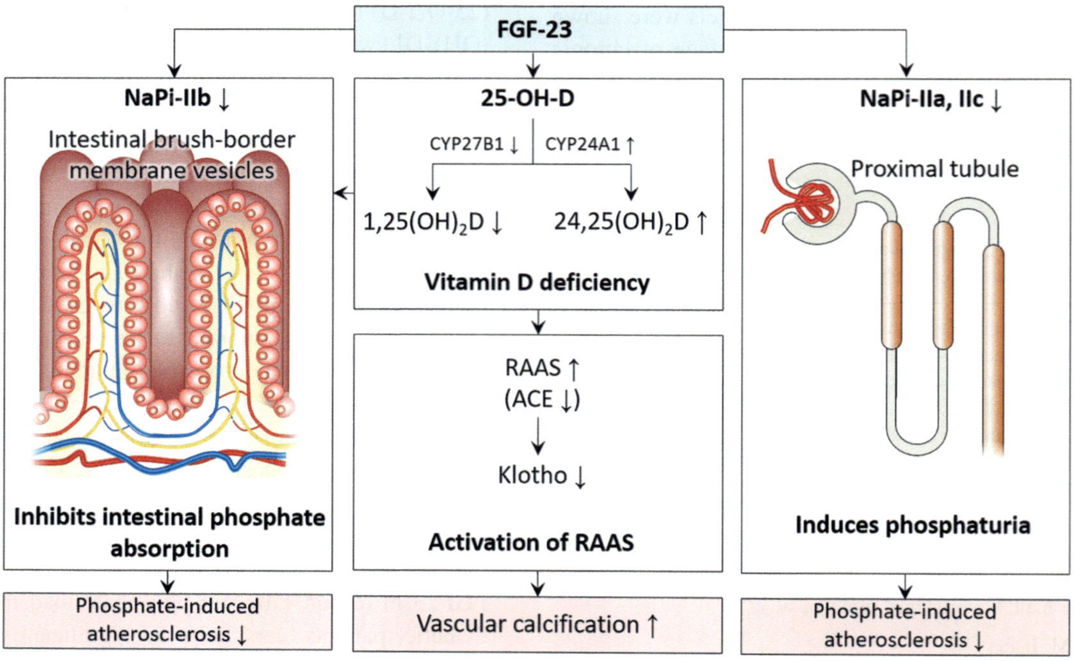

Fig. 6.1 Putative mechanisms of the effect of elevated FGF23 in non-CKD subjects
Even in non-CKD subjects, FGF23 inhibits CYP27B1, leading to decreased serum 1,25$(OH)_2$D. Elevated FGF23 and decreased serum 1,25$(OH)_2$D induces activation of RAAS, while RAAS decreases klotho expression, causing vascular calcification
On the other hand, FGF23 decreases phosphate reabsorption in the kidneys and intestines via down-regulation of the cotransporters NaPi-IIa and NaPi-IIc in the proximal tubule, and NaPi-IIb in the intestinal brush-border membrane vesicles. This protects against development of phosphate-induced atherosclerosis and particularly vascular calcification
ACE angiotensin-converting enzyme, *FGF23*, fibroblast growth factor 23, *NaPi* sodium dependent inorganic phosphate, *RAAS* renin-angiotensin-aldosterone system, *25-OH-D* 25-hydroxyvitamin D, *1,25(OH)₂D* 1,25-dihydroxyvitamin D, *24,25(OH)₂D* 24,25-dihydroxyvitamin D

renal bone disease and vascular calcification, leading to increased cardiovascular morbidity and mortality [29, 34, 39]. Among the various abnormalities related to CKD-MBD, phosphate load has been hypothesized as an initial causative factor [10, 27]. Since FGF23 regulates phosphate metabolism, it is considered to be a predictor of cardiovascular outcome in CKD patients [27]. Findings supportive of that view include the association of elevated FGF23 level with vascular calcification [28, 57], left ventricular hypertrophy (LVH) [11, 16], and increased arterial stiffness and endothelial dysfunction [21, 37], leading to increased mortality in patients with CKD and end-stage kidney disease (ESKD) [12, 15, 24, 26]. Of importance, even in non-CKD subjects, elevated FGF23 levels have been shown to be associated with LVH [51], endothelial function [37], and increased mortality [2, 45, 56] (Fig. 6.2).

In addition, animal experiment results suggest that FGF23 directly stimulates the renin-angiotensin-aldosterone system (RAAS) by suppressing angiotensin-converting enzyme (ACE) expression in the kidneys, independent of other abnormalities related to mineral and bone disorders [8] (Fig. 6.1). Activation of RAAS has been linked with numerous adverse consequences, such as hypertension, endothelial dysfunction, progression of atherosclerosis, and diabetic nephropathy [46] (Fig. 6.2). Indeed, FGF23-mediated activation of local RASS in the heart was reported to promote cardiac hypertrophy and fibrosis [3], while activation of RAAS has also been found to reduce the expression of klotho in the kidneys [9]. Thus, the effect of FGF23 on RAAS may be a potential cause of various adverse outcomes.

Controversy remains regarding whether FGF23 is a contributor to vascular calcification. This hormone helps to maintain phosphate levels within a normal range in response to phosphate overload, and we previously reported that oral phosphate loading significantly increases serum

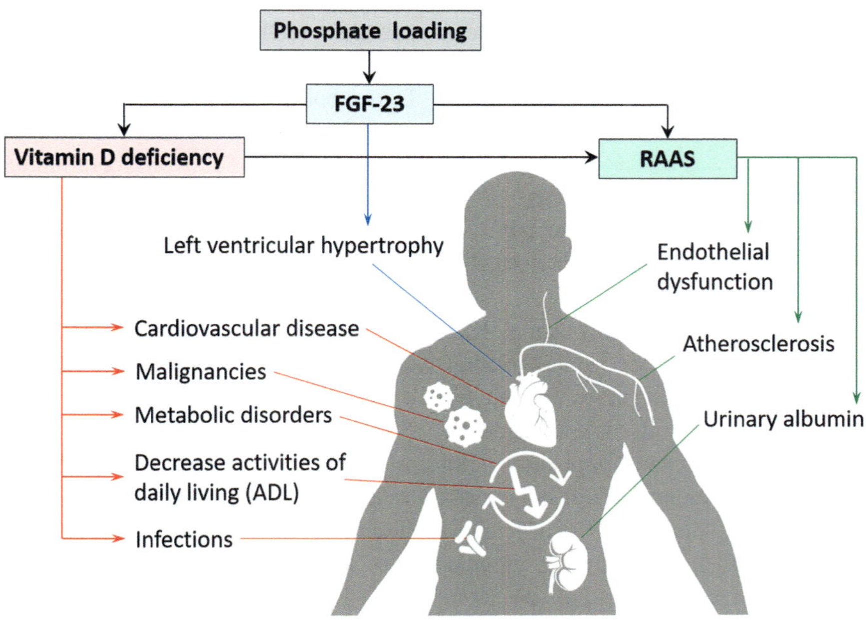

Fig. 6.2 Putative mechanisms of effects of elevated FGF23 in non-CKD subjects
Elevated FGF23 alone, decreased $1,25(OH)_2D$, and activated RAAS are each associated with numerous adverse outcomes, leading to high mortality
ACE angiotensin-converting enzyme, *ADL* activities of daily living, *CVD* cardiovascular disease, *FGF23* fibroblast growth factor 23, *P* phosphate, *RAAS* renin-angiotensin-aldosterone system

FGF23 and parathyroid hormone (PTH) to attenuate an increased serum phosphate level [60]. In patients with type 2 diabetes mellitus (DM), impaired bone formation resulting from osteoblast/osteocyte deficit is the main feature of DM bone abnormalities. Our prior study found impaired incremental responses of serum FGF23 and PTH in DM patients, which was in contrast to their increases in non-DM subjects. As a result, a significant rise in serum phosphate is observed in type 2 DM but not non-DM patients, clearly indicating the protective effect of FGF23 against the increase of serum phosphate after oral phosphate load in non-DM individuals, which protects against development of phosphate-induced atherosclerosis and particularly vascular calcification [7] (Fig. 6.1). Furthermore, a cohort study that included 1501 patients revealed that baseline plasma FGF23 level was not associated with the severity of calcium (Ca) content in the coronary artery [49]. Serum FGF23 level by itself may be a surrogate marker for vascular calcification in CKD patients [23], because it reflects phosphate load. It is therefore considered that increased serum FGF23 might serve as an important indicator of adverse outcomes in CKD as well as non-CKD patients.

6.3 Significance of Inhibition of Vitamin D Activation by FGF23 in Non-CKD Individuals

Vitamin D, specifically biologically active 1,25 $(OH)_2D$, plays a key role in bone and mineral metabolism. Vitamin D deficiency, defined by serum 25-OH-D < 20 ng/mL, is estimated to occur in 36–57% of the general population [18] and 50–86% of CKD patients [33]. It is increasingly recognized that vitamin D insufficiency (25-OH-D:20-29 ng/mL), in addition to vitamin D deficiency, is a risk factor for various diseases, such as diabetes, infections, cardiovascular disease, and cancer [19, 20]. Furthermore, harmful effects of vitamin D deficiency have recently been recognized even in the general population, based on an association with increased risk of all-cause mortality and reduced activities of daily living [36, 47, 55, 64] (Fig. 6.2). Together with the notion that vitamin D might be a primitive steroid-like hormone, because the nuclear vitamin D receptor belongs to the steroid receptor superfamily, these observations support the importance of vitamin D.

6.4 Potential Role of FGF23 in Regulation of Vitamin D Metabolism in Non-CKD Individuals

Serum FGF23 level has been reported to be increased even at an early stage of CKD in patients with an eGFR $\geq$60 mL/min/1.73 m^2 [24]. While both FGF23 and PTH have potent phosphaturic activities, they have opposite effects in terms of vitamin D metabolism, as FGF23 inhibits [6, 52] and PTH stimulates [5, 14, 17] vitamin D activation from 25-OH-D to 1,25 $(OH)_2D$. To examine the physiological role of FGF23 in regulation of vitamin D metabolism and thus phosphate metabolism in non-CKD patients (eGFR$\geq$60 mL/min/1.73 m^2), we measured serum FGF23 along with three vitamin D metabolites, 25-OH-D, 1,25$(OH)_2D$, and 24,25 $(OH)_2D$, in subjects with eGFR $\geq$60 mL/min/ 1.73 m^2 [40]. To avoid confounding factors, subjects with DM as well as those taking corticosteroids, vitamin D, phosphate-binder, supplementary Ca, estrogen, or thyroid hormone were excluded from the study. The results showed that serum FGF23 was significantly and inversely correlated with serum 1,25$(OH)_2D$ and 1,25 $(OH)_2D$/25-OH-D ratio, putative parameters for CYP27B1. Additionally, in multiple regression analysis that included FGF23, PTH, and eGFR as independent variables, FGF23 emerged as a significant factor showing an independent negative association with 1,25$(OH)_2D$ and 1,25 $(OH)_2D$/25-OH-D ratio. These results suggested that FGF23 inhibits CYP27B1, leading to decreased serum 1,25$(OH)_2D$ even in non-CKD individuals. In addition, FGF23 was found to be significantly and independently associated in a positive manner with 24,25$(OH)_2D$ /1,25 $(OH)_2D$ ratio. Thus, based on the findings showing that FGF23 plays an important role in vitamin D catabolism in non-CKD individuals, it was speculated that phosphate overload exists even in non-CKD patients, and that an increase in serum FGF23 acts directly to attenuate increased serum phosphate, and indirectly to stimulate phosphaturia and inhibit activation of vitamin D, resulting in suppression of intestinal phosphate absorption, based on the effect of 1,25 $(OH)_2D$ to stimulate Na-Pi co-transport in intestinal brush-border membrane vesicles [38, 50, 59].

6.5 Potential Role of FGF23 in Regulation of Phosphate Metabolism in Non-CKD Individuals

A recent cohort study found that serum FGF23, but not phosphate, was negatively correlated with eGFR in non-CKD postmenopausal female subjects with eGFR $\geq$60 mL/min/1.73 m^2 [43]. Serum phosphate levels begin to increase at a later stage of CKD [24], thus it is probable that an increase in FGF23 acts to protect against development of hyperphosphatemia, as noted in our previous study [60]. Serum FGF23 levels have been shown to be not correlated with serum phosphate in subjects with normal kidney function [35, 40]. Furthermore, we found a significant and independent association of serum FGF23 with urinary albumin-to-creatinine excretion in non-CKD postmenopausal female subjects with eGFR $\geq$60 mL/min/1.73 m^2 [43]. In CKD mice, transcriptome analysis revealed that FGF23 regulates several genes associated with kidney injuries, including Neutrophil Gelatinase-Associated Lipocalin (NGAL), and carbonic anhydrase 14 (Car 14) [8]. Together, these observations suggest that FGF23 exerts a direct effect to cause kidney damage. Alternatively, phosphate overload, as reflected by increased serum FGF23, might induce kidney injury, as noted in previous reports [1, 41]. Therefore, it is considered that FGF23 may play an important role in regulation of phosphate metabolism, based on its effect on vitamin D metabolism as well as the established role of PTH in individuals with normal kidney function.

6.6 Vitamin D Metabolism and FGF23-Klotho Axis in Non-CKD Individuals

Klotho, a 130-kDa transmembrane β-glucuronidase that catalyzes the hydrolysis of steroid β-glucuronides [32], is required for FGF23 to activate FGFRs and their downstream molecules [58]. Since the klotho/FGFR complex binds to FGF23 with higher affinity than either alone, FGF23 exerts its biological effects via activation of FGFR in a klotho-dependent manner [58]. Interestingly, it has been demonstrated that klotho deficiency causes vascular calcification in CKD patients [22] and that vitamin D deficiency is associated with RAAS activation [44]. In addition, angiotensin II has been found to negatively regulate renal expression of klotho [61, 63] (Fig. 6.1). Thus, activation of RAAS and klotho deficiency by the effects of FGF23 on vitamin D metabolism may lead to vascular calcification in CKD patients as well as non-CKD individuals.

6.7 Future Prospects

Therapeutic strategies to control serum FGF23 level and improve adverse outcomes in non-CKD patients remain to be clarified. Clinical trials will be needed to examine the effects of therapeutic intervention, such as vitamin D, phosphate binders, or dietary phosphate restriction, on serumFGF23 levels, and also to determine how such changes impact adverse outcomes in CKD as well as non-CKD cases. Restriction of dietary phosphate, particularly phosphate additives, may be important, because a large amount of rapidly absorbable inorganic phosphate can be found in various food additives and preservatives [42]. A small-scale short-term pilot study of CKD stage 3–4 patients showed a significant decrease in FGF23 in those who consumed a diet that included the combination of lanthanum carbonate and 900 mg phosphate, though that level was not significantly decreased in those with a lanthanum carbonate and *ad libitum* diet [25]. Another report summarized the results of several interventional clinical trials that used phosphate binders and/or a phosphate restriction diet for examination of FGF23 levels, though most of those studies had a relatively short duration [4]. Long-term large-scale clinical trials focused on phosphate restriction in healthy and non-CKD subjects are needed. Administration of phosphate binders, such as lanthanum carbonate and sevelamer hydrochloride, has been reported to lower serum FGF23 levels in patients with CKD stage 3 and ESKD, respectively [13, 30]. Thus, it may be possible to decrease the FGF23 level in serum and improve adverse outcomes with phosphate binders in patients without CKD.

FGF23 has direct and/or indirect effects outside of the field of MBD, which suggests additional possibilities for treatment of adverse outcomes associated with an elevated level. Since FGF23 stimulates RAAS, administration of ACE inhibitors, angiotensin-receptor blockers, and aldosterone receptor antagonists may be justified.

6.8 Conclusion

Even in individuals with normal kidney function, FGF23 plays an important role in regard to vitamin D metabolism. An elevated FGF23 level and vitamin D deficiency are associated with various adverse events. Therefore, it is possible that an increase in vitamin D and decrease in FGF23 could be instrumental in engendering adverse outcomes, even in non-CKD individuals.

Acknowledgments We want to express our sincere gratitude to Dr. Nuraly Akimbekov, Ph.D. (Al-Farabi Kazakh National University, Kazakhstan), to help draw the illustrations.

Disclosure None of the authors has a conflict of interest to declare.

References

1. Arai-Nunota N, Mizobuchi M, Ogata H, Yamazaki-Nakazawa A, Kumata C, Kondo F et al (2014) Intravenous phosphate loading increases fibroblast growth factor 23 in uremic rats. PLoS One 9(3):e91096
2. Arnlov J, Carlsson AC, Sundstrom J, Ingelsson E, Larsson A, Lind L et al (2013) Higher fibroblast growth factor-23 increases the risk of all-cause and cardiovascular mortality in the community. Kidney Int 83(1):160–166
3. Bockmann I, Lischka J, Richter B, Deppe J, Rahn A, Fischer DC et al (2019) FGF23-mediated activation of local RAAS promotes cardiac hypertrophy and fibrosis. Int J Mol Sci 20(18):4634
4. Bouma-De Krijger A, Vervloet MG (2020) Fibroblast growth factor 23: are we ready to use it in clinical practice? J Nephrol 33(3):509–527
5. Brenza HL, Kimmel-Jehan C, Jehan F, Shinki T, Wakino S, Anazawa H et al (1998) Parathyroid hormone activation of the 25-hydroxyvitamin D3-1alpha-hydroxylase gene promoter. Proc Natl Acad Sci U S A 95(4):1387–1391
6. Chanakul A, Zhang MY, Louw A, Armbrecht HJ, Miller WL, Portale AA et al (2013) FGF-23 regulates CYP27B1 transcription in the kidney and in extra-renal tissues. PLoS One 8(9):e72816
7. Chonchol M, Dale R, Schrier RW, Estacio R (2009) Serum phosphorus and cardiovascular mortality in type 2 diabetes. Am J Med 122(4):380–386
8. Dai B, David V, Martin A, Huang J, Li H, Jiao Y et al (2012) A comparative transcriptome analysis identifying FGF23 regulated genes in the kidney of a mouse CKD model. PLoS One 7(9):e44161
9. De Borst MH, Vervloet MG, Ter Wee PM, Navis G (2011) Cross talk between the renin-angiotensin-aldosterone system and vitamin D-FGF-23-klotho in chronic kidney disease. J Am Soc Nephrol 22(9): 1603–1609
10. Emrich IE, Brandenburg V, Sellier AB, Schauerte J, Wiedenroth J, Untersteller K et al (2019) Strength of fibroblast growth factor 23 as a cardiovascular risk predictor in chronic kidney disease weaken by ProBNP adjustment. Am J Nephrol 49(3):203–211
11. Faul C, Amaral AP, Oskouei B, Hu MC, Sloan A, Isakova T et al (2011) FGF23 induces left ventricular hypertrophy. J Clin Invest 121(11):4393–4408
12. Fliser D, Kollerits B, Neyer U, Ankerst DP, Lhotta K, Lingenhel A et al (2007) Fibroblast growth factor 23 (FGF23) predicts progression of chronic kidney disease: the Mild to Moderate Kidney Disease (MMKD) Study. J Am Soc Nephrol 18(9):2600–2608
13. Gansevoort RT, Matsushita K, Van Der Velde M, Astor BC, Woodward M, Levey AS et al (2011)

Lower estimated GFR and higher albuminuria are associated with adverse kidney outcomes. A collaborative meta-analysis of general and high-risk population cohorts. Kidney Int 80(1):93–104
14. Garabedian M, Holick MF, Deluca HF, Boyle IT (1972) Control of 25-hydroxycholecalciferol metabolism by parathyroid glands. Proc Natl Acad Sci U S A 69(7):1673–1676
15. Gutierrez OM, Mannstadt M, Isakova T, Rauh-Hain JA, Tamez H, Shah A et al (2008) Fibroblast growth factor 23 and mortality among patients undergoing hemodialysis. N Engl J Med 359(6):584–592
16. Gutierrez OM, Januzzi JL, Isakova T, Laliberte K, Smith K, Collerone G et al (2009) Fibroblast growth factor 23 and left ventricular hypertrophy in chronic kidney disease. Circulation 119(19):2545–2552
17. Hoenderop JG, Chon H, Gkika D, Bluyssen HA, Holstege FC, St-Arnaud R et al (2004) Regulation of gene expression by dietary Ca2+ in kidneys of 25-hydroxyvitamin D3-1 alpha-hydroxylase knockout mice. Kidney Int 65(2):531–539
18. Holick MF (2006) High prevalence of vitamin D inadequacy and implications for health. Mayo Clin Proc 81(3):353–373
19. Holick MF (2007) Vitamin D deficiency. N Engl J Med 357(3):266–281
20. Holick MF, Chen TC (2008) Vitamin D deficiency: a worldwide problem with health consequences. Am J Clin Nutr 87(4):1080S–1086S
21. Hsu JJ, Katz R, Ix JH, De Boer IH, Kestenbaum B, Shlipak MG (2014) Association of fibroblast growth factor-23 with arterial stiffness in the multi-ethnic study of atherosclerosis. Nephrol Dial Transplant 29(11):2099–2105
22. Hu MC, Shi M, Zhang J, Quinones H, Griffith C, Kuro OM et al (2011) Klotho deficiency causes vascular calcification in chronic kidney disease. J Am Soc Nephrol 22(1):124–136
23. Inaba M, Okuno S, Imanishi Y, Yamada S, Shioi A, Yamakawa T et al (2006) Role of fibroblast growth factor-23 in peripheral vascular calcification in non-diabetic and diabetic hemodialysis patients. Osteoporos Int 17(10):1506–1513
24. Isakova T, Wahl P, Vargas GS, Gutierrez OM, Scialla J, Xie H et al (2011) Fibroblast growth factor 23 is elevated before parathyroid hormone and phosphate in chronic kidney disease. Kidney Int 79(12): 1370–1378
25. Isakova T, Barchi-Chung A, Enfield G, Smith K, Vargas G, Houston J et al (2013) Effects of dietary phosphate restriction and phosphate binders on FGF23 levels in CKD. Clin J Am Soc Nephrol 8(6): 1009–1018
26. Jean G, Terrat JC, Vanel T, Hurot JM, Lorriaux C, Mayor B et al (2009) High levels of serum fibroblast growth factor (FGF)-23 are associated with increased mortality in long haemodialysis patients. Nephrol Dial Transplant 24(9):2792–2796
27. Kendrick J, Cheung AK, Kaufman JS, Greene T, Roberts WL, Smits G et al (2011) FGF-23 associates with death, cardiovascular events, and initiation of chronic dialysis. J Am Soc Nephrol 22(10):1913–1922
28. Khan AM, Chirinos JA, Litt H, Yang W, Rosas SE (2012) FGF-23 and the progression of coronary arterial calcification in patients new to dialysis. Clin J Am Soc Nephrol 7(12):2017–2022
29. Kidney Disease: Improving Global Outcomes, CKD-MBD Update WG (2017) KDIGO 2017 clinical practice guideline update for the diagnosis, evaluation, prevention, and treatment of chronic kidney disease-mineral and bone disorder (CKD-MBD). Kidney Int Suppl (2011) 7(1):1–59
30. Koiwa F, Kazama JJ, Tokumoto A, Onoda N, Kato H, Okada T et al (2005) Sevelamer hydrochloride and calcium bicarbonate reduce serum fibroblast growth factor 23 levels in dialysis patients. Ther Apher Dial 9(4):336–339
31. Kovesd CP, Quarles LD (2013) Fibroblast growth factor-23: what we know, what we don't know, and what we need to know. Nephrol Dial Transplant 28(9): 2228–2236
32. Kuro-O M, Matsumura Y, Aizawa H, Kawaguchi H, Suga T, Utsugi T et al (1997) Mutation of the mouse klotho gene leads to a syndrome resembling ageing. Nature 390(6655):45–51
33. Levin A, Bakris GL, Molitch M, Smulders M, Tian J, Williams LA et al (2007) Prevalence of abnormal serum vitamin D, PTH, calcium, and phosphorus in patients with chronic kidney disease: results of the study to evaluate early kidney disease. Kidney Int 71(1):31–38
34. London GM, Guerin AP, Marchais SJ, Metivier F, Pannier B, Adda H (2003) Arterial media calcification in end-stage renal disease: impact on all-cause and cardiovascular mortality. Nephrol Dial Transplant 18(9):1731–1740
35. Marsell R, Grundberg E, Krajisnik T, Mallmin H, Karlsson M, Mellstrom D et al (2008) Fibroblast growth factor-23 is associated with parathyroid hormone and renal function in a population-based cohort of elderly men. Eur J Endocrinol 158(1):125–129
36. Melamed ML, Michos ED, Post W, Astor B (2008) 25-hydroxyvitamin D levels and the risk of mortality in the general population. Arch Intern Med 168(15): 1629–1637
37. Mirza MA, Larsson A, Lind L, Larsson TE (2009) Circulating fibroblast growth factor-23 is associated with vascular dysfunction in the community. Atherosclerosis 205(2):385–390
38. Miyamoto K, Haito-Sugino S, Kuwahara S, Ohi A, Nomura K, Ito M et al (2011) Sodium-dependent phosphate cotransporters: lessons from gene knockout and mutation studies. J Pharm Sci 100(9):3719–3730
39. Moe S, Drueke T, Cunningham J, Goodman W, Martin K, Olgaard K et al (2006) Definition, evaluation, and classification of renal osteodystrophy: a position statement from kidney disease: improving global outcomes (KDIGO). Kidney Int 69(11):1945–1953
40. Nakatani S, Nakatani A, Tsugawa N, Yamada S, Mori K, Imanishi Y et al (2015) Fibroblast growth factor-23 and

Vitamin D metabolism in subjects with eGFR >/=60 ml/min/1.73 m2. Nephron 1308(2):119–126
41. Nishida Y, Taketani Y, Yamanaka-Okumura H, Imamura F, Taniguchi A, Sato T et al (2006) Acute effect of oral phosphate loading on serum fibroblast growth factor 23 levels in healthy men. Kidney Int 70(12):2141–2147
42. Noori N, Sims JJ, Kopple JD, Shah A, Colman S, Shinaberger CS et al (2010) Organic and inorganic dietary phosphorus and its management in chronic kidney disease. Iran J Kidney Dis 4(2):89–100
43. Ozaki E, Yamada S, Kuriyama N, Matsui D, Watanabe I, Koyama T et al (2018) Association of BAP with urinary albumin excretion in postmenopausal, but not premenopausal, non-CKD Japanese women. Sci Rep 8(1):82
44. Park CW, Oh YS, Shin YS, Kim CM, Kim YS, Kim SY et al (1999) Intravenous calcitriol regresses myocardial hypertrophy in hemodialysis patients with secondary hyperparathyroidism. Am J Kidney Dis 33(1):73–81
45. Parker BD, Schurgers LJ, Brandenburg VM, Christenson RH, Vermeer C, Ketteler M et al (2010) The associations of fibroblast growth factor 23 and uncarboxylated matrix Gla protein with mortality in coronary artery disease: the Heart and Soul Study. Ann Intern Med 152(10):640–648
46. Perazella MA, Setaro JF (2003) Renin-angiotensin-aldosterone system: fundamental aspects and clinical implications in renal and cardiovascular disorders. J Nucl Cardiol 10(2):184–196
47. Pilz S, Tomaschitz A, Marz W, Drechsler C, Ritz E, Zittermann A et al (2011) Vitamin D, cardiovascular disease and mortality. Clin Endocrinol 75(5):575–584
48. Saito H, Kusano K, Kinosaki M, Ito H, Hirata M, Segawa H et al (2003) Human fibroblast growth factor-23 mutants suppress Na+^-dependent phosphate co-transport activity and 1alpha,25-dihydroxyvitamin D3 production. J Biol Chem 278(4):2206–2211
49. Scialla JJ, Lau WL, Reilly MP, Isakova T, Yang HY, Crouthamel MH et al (2013) Fibroblast growth factor 23 is not associated with and does not induce arterial calcification. Kidney Int 83(6):1159–1168
50. Segawa H, Kaneko I, Yamanaka S, Ito M, Kuwahata M, Inoue Y et al (2004) Intestinal Na-P(i) cotransporter adaptation to dietary P(i) content in vitamin D receptor null mice. Am J Physiol Renal Physiol 287(1):F39–F47
51. Seiler S, Cremers B, Rebling NM, Hornof F, Jeken J, Kersting S et al (2011) The phosphatonin fibroblast growth factor 23 links calcium-phosphate metabolism with left-ventricular dysfunction and atrial fibrillation. Eur Heart J 32(21):2688–2696
52. Shigematsu T, Kazama JJ, Yamashita T, Fukumoto S, Hosoya T, Gejyo F et al (2004) Possible involvement of circulating fibroblast growth factor 23 in the development of secondary hyperparathyroidism associated with renal insufficiency. Am J Kidney Dis 44(2): 250–256
53. Shimada T, Hasegawa H, Yamazaki Y, Muto T, Hino R, Takeuchi Y et al (2004) FGF-23 is a potent regulator of vitamin D metabolism and phosphate homeostasis. J Bone Miner Res 19(3):429–435
54. Shimada T, Yamazaki Y, Takahashi M, Hasegawa H, Urakawa I, Oshima T et al (2005) Vitamin D receptor-independent FGF23 actions in regulating phosphate and vitamin D metabolism. Am J Physiol Renal Physiol 289(5):F1088–F1095
55. Skalska A, Galas A, Grodzicki T et al (2012) 25-hydroxyvitamin D and physical and cognitive performance in older people with chronic conditions. Pol Arch Med Wewn 122(4):162–169
56. Souma N, Isakova T, Lipiszko D, Sacco RL, Elkind MS, Derosa JT et al (2016) Fibroblast growth factor 23 and cause-specific mortality in the general population: the northern Manhattan study. J Clin Endocrinol Metab 101(10):3779–3786
57. Turan MN, Kircelli F, Yaprak M, Sisman AR, Gungor O, Bayraktaroglu S et al (2016) FGF-23 levels are associated with vascular calcification, but not with atherosclerosis, in hemodialysis patients. Int Urol Nephrol 48(4):609–617
58. Urakawa I, Yamazaki Y, Shimada T, Iijima K, Hasegawa H, Okawa K et al (2006) Klotho converts canonical FGF receptor into a specific receptor for FGF23. Nature 444(7120):770–774
59. Xu H, Bai L, Collins JF, Ghishan FK (2002) Age-dependent regulation of rat intestinal type IIb sodium-phosphate cotransporter by 1,25-(OH)(2) vitamin D(3). Am J Physiol Cell Physiol 282(3):C487–C493
60. Yoda K, Imanishi Y, Yoda M, Mishima T, Ichii M, Yamada S et al (2012) Impaired response of FGF-23 to oral phosphate in patients with type 2 diabetes: a possible mechanism of atherosclerosis. J Clin Endocrinol Metab 97(11):E2036–E2043
61. Yoon HE, Ghee JY, Piao S, Song JH, Han DH, Kim S et al (2011) Angiotensin II blockade upregulates the expression of Klotho, the anti-ageing gene, in an experimental model of chronic cyclosporine nephropathy. Nephrol Dial Transplant 26(3):800–813
62. Yuan B, Xing Y, Horst RL, Drezner MK (2004) Evidence for abnormal translational regulation of renal 25-hydroxyvitamin D-1alpha-hydroxylase activity in the hyp-mouse. Endocrinology 145(8):3804–3812
63. Zhou Q, Lin S, Tang R, Veeraragoo P, Peng W, Wu R (2010) Role of Fosinopril and valsartan on klotho gene expression induced by angiotensin II in rat renal tubular epithelial cells. Kidney Blood Press Res 33(3): 186–192
64. Zittermann A, Iodice S, Pilz S, Grant WB, Bagnardi V, Gandini S (2012) Vitamin D deficiency and mortality risk in the general population: a meta-analysis of prospective cohort studies. Am J Clin Nutr 95(1):91–100

Phosphate and Cellular Senescence

7

Ming Chang Hu and Orson W. Moe

Abstract

Cellular senescence is one type of permeant arrest of cell growth and one of increasingly recognized contributor to aging and age-associated disease. High phosphate and low Klotho individually and synergistically lead to age-related degeneration in multiple organs. Substantial evidence supports the causality of high phosphate in cellular senescence, and potential contribution to human aging, cancer, cardiovascular, kidney, neurodegenerative, and musculoskeletal diseases. Phosphate can induce cellular senescence both by direct phosphotoxicity, and indirectly through downregulation of Klotho and upregulation of plasminogen activator inhibitor-1. Restriction of dietary phosphate intake and blockage of intestinal absorption of phosphate help suppress cellular senescence. Supplementation of Klotho protein, cellular senescence inhibitor, and removal of senescent cells with senolytic agents are potential novel strategies to attenuate phosphate-induced cellular senescence, retard aging, and ameliorate age-associated, and phosphate-induced disorders.

Keywords

Aging · Age-associated disease · Cellular senescence · Fibrosis · Klotho · Phosphate · Phosphorus · Phosphotoxicity · Plasminogen activator inhibitor-1 · p16 · p21

M. C. Hu (✉)
Charles and Jane Pak Center for Mineral Metabolism and Clinical Research, University of Texas Southwestern Medical Center, Dallas, TX, USA

Departments of Internal Medicine, University of Texas Southwestern Medical Center, Dallas, TX, USA
e-mail: ming-chang.hu@utsouthwestern.edu

O. W. Moe
Charles and Jane Pak Center for Mineral Metabolism and Clinical Research, University of Texas Southwestern Medical Center, Dallas, TX, USA

Departments of Internal Medicine, University of Texas Southwestern Medical Center, Dallas, TX, USA

Departments of Physiology, University of Texas Southwestern Medical Center, Dallas, TX, USA
e-mail: Orson.moe@utsouthwestern.edu

7.1 Introduction

Phosphorus, the element of phosphate biologic moiety, is the sixth most abundant element after hydrogen, oxygen, carbon, nitrogen, and calcium in the human body. While the term "phosphorus" is used in conventional clinical laboratory reporting, there is no elemental phosphorus in mammalian biology, the term phosphate (inorganic or organic) will be used despite the accepted clinical parlance. In the human body, 85% of phosphate is in bone and teeth as hydroxyapatite, 14% is located intracellularly as various organic phosphate compounds and some inorganic phosphate; and only 1% extracellularly [72, 123, 168]. Of the 1% located in extracellular

M. S. Razzaque (ed.), *Phosphate Metabolism*, Advances in Experimental Medicine and Biology 1362,
https://doi.org/10.1007/978-3-030-91623-7_7

space, 20% is protein bound [72, 123, 168] and the inorganic phosphate is distributed as pyrophosphate (minor) and orthophosphate (major) in various valences determined by the plasma pH.

Serum phosphate serves as an exchange pool among various phosphate-containing and phosphate-regulating organs [72, 123, 168]. Phosphate homeostasis is principally maintained by gut absorption and urine phosphate excretion which is controlled by a complex but tightly and efficiently regulated network consisting of several calciophosphotropic hormones including parathyroid hormone, 1,25-$(OH)_2$ vitamin D, fibroblast growth factor (FGF)-23, and Klotho [65, 72].

The terminology in this field can be confusing and necessitates some front-end clarification. *Hyperphosphatemia* refers to elevated serum or plasma phosphate level (measured as the mass of phosphorus constituent per volume) (Fig. 7.1). It is a "state" and may or may not be associated with disease. *Phosphate loading* refers to an amount of phosphate intake into a system, with overload being a flux that is higher than what is necessary for health. This may or may not be associated with hyperphosphatemia or ill effects. *Positive balance* occurs when intake exceed excretion at the organism level. *Phosphotoxicity* refers to a cellular, organ, or whole organism condition where excess phosphate (extracellular or intracellular) begets an undesirable phenotype- one that predisposes to or actually constitute disease.

A positive imbalance can result from insufficient renal phosphate excretion and/or excessive phosphate intake, and is emerging as a novel detrimental contributor to aging and age-associated disease [25, 94–96, 123, 131, 132]. As stated above, the ill effects of excessive phosphate is collectively termed "phosphotoxicity" with no designation of the cause or specific resultant phenotypes. Aging is a progressive and inevitable process with multi-organ deterioration. Aging is triggered and exacerbated by numerous factors including genetic, and epigenetic factors. Among those identified factors such as diabetes, hypertension and others, the role of phosphate in cellular senescence is incompletely elucidated [147].

Cellular senescence is a type of permanent arrest of cell growth, which was initially identified as a defense mechanism to inhibit tumorigenesis and metastasis [17, 26, 37, 109, 124, 154]. Similar to autophagy and apoptosis, an appropriate cellular senescence activity is therefore required to maintain tissue function and regeneration after tissue damage [180]. But chronic and severe cellular senescence also contributes to aging and age-associated diseases.

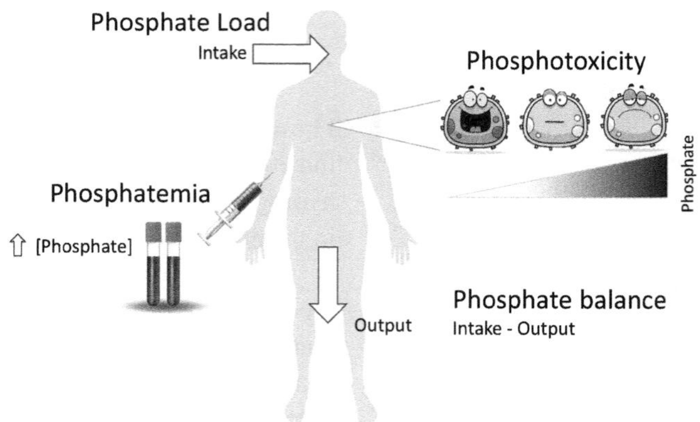

Fig. 7.1 Concepts about phosphate metabolism and balance
Hyperphosphatemia refers to elevated levels of serum or plasma phosphate, which may or may not be associated with disease. External phosphate balance is the amount of phosphate intake minus urinary phosphate excretion. The increased flux of phosphate through the organism may or may not be associated with hyperphosphatemia or ill effects. Phosphotoxicity refers to a state that excess phosphate (extracellular or intracellular) causes an undesirable phenotype at a cellular, organ, or whole organism level

Aging is triggered, driven, and promoted by multiple genetic and acquired detrimental factors.

The cellular and molecular mechanisms whereby phosphate accelerates aging, and excerbates age-associated diseases are complex and multifactorial [68, 122, 123, 135]. High phosphate can reduce autophagy [147] and Klotho [68, 176], induce cell apoptosis [36, 101], and activate cellular senescence [106]. Klotho was discovered in 1997 by Kuro-o and colleagues. Klotho was originally identified as an anti-aging protein and later on was found to directly inhibit cellular senescence, reduce serum phosphate, and prevent phosphate-induced cellular senescence in a tripartite relationship [106]. Therefore, abnormal cellular senescence is attributable in part to the deleterious actions of Klotho deficiency from phosphotoxicity [44, 139] and not necessarily phosphate *per se*.

In this monograph, we will first update current understanding of cellular senescence in aging and age-associated disease, review the effect of phosphate on the activation of cellular senescence and to aging and human disease, and finally discuss the potential of targeting cellular senescence in prevention and treatment of phosphotoxicity.

7.2 Cellular Senescence

7.2.1 Cellular Senescence and Its Signaling Pathways

Cellular senescence was first described by Hayflick and Moorhead when they established an immortal cell line [61]. They found that the cells assume a flattened and enlarged morphology, ceased to proliferate without any responses to growth factors, but were still alive in cultured dishes. Those cells are called senescent cells because they have irreversible growth arrest, which is different from quiescent cells, because quiescent cells are still able to re-enter the cell cycle.

Senescent cells have remarkable morphological and metabolic changes. Those changes include reorganizing chromatin, reprogramming gene expression, and endlessly producing many pro-inflammatory and pro-fibrotic growth factors and cytokines, a phenotype called the senescence-associated secretory phenotype (SASP) [29, 45, 162]. Cellular senescence conferred both protective and deleterious effects depending on the physiological and pathophysiologic scenarios. Although cellular senescence likely functions as a defense mechanism to inhibit malignant transformation of damaged cells, persistent and chronic senescence may promote aging and age-associated pathologies including tissue degeneration, cell dysfunction, and chronic inflammation in the tissues [17, 26, 29, 37, 38, 45, 66, 82, 92, 108, 109, 124, 143, 151, 154, 162].

Cellular senescence is activated and maintained by at least 2 cell signaling pathway: p53/p21 and p16^{Ink4a}/retinoblastoma-1 (RB-1). Both are tumor suppressive proteins. These two pathways are activated by DNA damage, reactive metabolites, oncogenic mutations, high mitogen signals, proteotoxic stress, and other yet-to-be-identified factors [17, 18, 26, 66, 82, 143, 154]. Furthermore, cellular senescence is also induced by the perturbation of metabolism including abnormal phosphate metabolism, which may accelerate age-related phenotypes [17, 26, 66, 106, 143, 154].

7.2.2 Senescence-Associated Secretory Phenotype

Senescent cells have a profound phenotype of endless production and secretion of proinflammatory and profibrotic growth factors referred as SASP [29, 45]. The number of senescent cells is widely variable from <1% to >15% depending on physiological and pathophysiologic context, species, and tissue origins, and cell origins [18]. However, a few senescent cells are able to initiate a detrimental positive circle through spreading senescence signal to neighbor cells [45] and to propagate pathologic actions in whole tissue and organ through SASP and even exert systemic effects [125].

Chronic inflammation induced by cellular senescence is the result of unrestrained SASP in the tissue and organ [79]. Obesity-induced senescent cells can maintain chronic and low-grade inflammation in the pancreas [125]. Therefore, the persistence of senescent cells can promote inflammation and tissue disruption present in chronic disease [124]. Cellular senescence occurs in many types of cells including endocrine cells, endothelial cells, epithelial cells, inflammatory cells, and even stem cells in almost every tissue and organ [18, 26, 66, 154, 157].

7.2.3 Cellular Senescence Effect on Aging

Aging is a continuous and progressive sequence of changes in any organisms that leads to biological dysfunctions and morphological tissue destruction, and causes degenerative pathology [17, 26, 38]. Cellular senescence is present throughout lifetime, occurring in embryogenesis and activated to protect cells against a variety of insults throughout life [17]. With time, the number of senescent cells increases and when coupled with depletion of stem cells and progenitor cells in tissue, causes a decline in tissue regeneration [170].

7.2.4 Cellular Senescence in Human Disease

Cellular senescence has been implicated in many age-associated degenerative phenotypes (Fig. 7.2). In most cases, senescent cells drive chronic degeneration mainly through the secretion of proinflammatory and profibrotic cytokines, growth factors, and proteinases via SASP, inducing inflammation, disrupting tissue structures, and leading to fibrosis. High cellular senescence also causes stem cell depletion [28, 55, 81] and impairs tissue regeneration [13].

7.2.4.1 Tumorigenesis

Cellular senescence was initially proposed to serve as a tumor suppressive mechanism since its discovery in the 1960s [45], but it is now demonstrated that highly active cellular senescence actually drives tumorigenesis.

Cells carrying activated oncogenes such as p53 [58, 90], p21 [1, 91, 174] and p16 [140, 185, 186] were shown to either die through apoptosis or enter stable cell cycle arrest that defines cellular senescence; either way, they may be prohibited from exerting further harm. More importantly, the implication of those oncogenes in activation of cellular senescence and in tumor development have been confirmed both in animal models and in human samples [45, 169]. In addition to tumor initiation, SASP can induce extracellular matrix remodeling through release of numerous inflammatory and growth factors [45] to promote metastasis and induce resistance to therapy [98]. Therefore, cellular senescence is a novel therapeutic target for cancer treatment.

7.2.4.2 Chronic Non-neoplastic Diseases

The morbidity and mortality of many chronic degenerative diseases increases with age, including Alzheimer's disease, Parkinson's diseases, Down Syndrome [9], macular degeneration [104], pulmonary disease [156], cardiovascular disease [27], kidney disease [154], liver and digestive disease [6, 12, 46, 50, 87, 88, 164, 193], saropenia [119, 155, 191], osteoporosis and osteoarthritis [75], metabolic disorders such as diabetes [87, 88, 125], autoimmune disease and rheumatic disease [75], and other age-related pathologies, even infectious disease [7, 56, 73, 184] (Fig. 7.2).

7.3 Phosphotoxicity in Aging and Age-Associated Disease

7.3.1 Phosphate Effect on Aging and Klotho

In patients after myocardial infarction, high baseline serum phosphate was associated with high all-cause death after 60 months of follow-up [166]. Since this paper, there are many studies describing the link between serum phosphate and cardiovascular disease in healthy human beings in

Cellular senescence	Phosphotoxicity	Organ
	√	Parathyroid
√	√	Heart
√	√	Liver
√	√	Kidney
		Tumor
		Bone
		Vessel

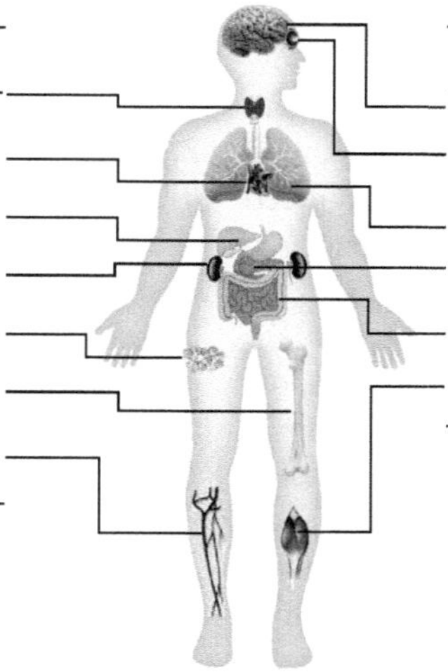

Organ	Cellular senescence	Phosphotoxicity
CNS	√	√
Eye	√	
Lung	√	√
Pancreas	√	√
Intestines	√	
Muscle	√	√

Fig. 7.2 Cellular senescence and phosphate toxicity in human disease
A summary of human acute and chronic diseases which are confirmed or proposed to be associated with high cellular senescence and phosphotoxicity

the Framingham Offspring Study [35], the Coronary Artery Risk Development Study [48], as well as intermediate phenotypes of vascular and endothelial dysfunction [74, 150], vascular and valvular calcification [3], and greater left ventricular mass [189]. The serum levels of phosphate negatively correlate with anticipated lifespan in animals and human beings [93, 96]. The observational studies in human populations are very compelling but do not provide proof of causality of high phosphate and aging.

Direct evidence to support the detrimental effect of high phosphate on aging comes from animal experiments. The correction of high serum phosphate with low phosphate diet or genetic deletion of sodium-dependent phosphate co-transporter-2a (NaP-2a) in kidney tubules leading to urinary phosphate leak effectively prolongs lifespan and rescues almost all phenotypes in Klotho deficient mice [122, 133]. More importantly, feeding high phosphate food abolishes that benefit conferred by NaPi-2a deletion and brings the premature aging phenotypes back. The longer lifespan associated with high autophagy in mice is at least in part mediated by higher urinary phosphate excretion due to reduced sodium-dependent phosphate cotransporters type II in the kidney [147]. The long-term challenge with high dietary phosphate dramatically abolished the beneficial effects of high autophagy activity [147], further supporting the role of high phosphate in promoting aging.

Klotho is also known as an inhibitor of cellular senescence [16, 20, 102, 105, 106, 126, 129]. High phosphate significantly decreases kidney and circulating Klotho [67–69, 71, 106, 117, 147]. However, this finding has not been reproduced in humans consuming a high phosphate diet [114, 141]. It is conceivable that high phosphate-induced short lifespan is multi-factorial and results at least in part from the reduction of Klotho production. Klotho deficiency may be one of molecular mechanisms behind high phosphate effect on cellular senescence. Moreover, higher levels of baseline serum phosphate are also associated senescence and aging.

7.3.2 Phosphotoxicity and Human Disease

Although phosphate is essential for body structure and function, excessive accumulation of phosphate in the body due to impaired phosphate homeostasis can cause functional and morphologic changes in almost every organ/tissue and system (Fig. 7.2). All of the chronic diseases associated with high phosphate discussed below are more prominent in the senior population. Interestingly, these diseases are important components in aging which are also triggered and driven by over active cellular senescence.

7.3.2.1 Cancer

Phosphotoxicity has been proposed to promote tumorigenesis (Fig. 7.2). Epidemiologic findings show association between high incidence of some types of cancers and abnormal phosphate metabolism [15, 181], and between high levels of serum phosphate and low survival rate of several types of cancer [167, 190]. Tumor cells store more intracellular phosphate through higher activity of sodium-dependent phosphate transport han in normal cells [41]. High phosphate can act as a mitogenic factor to induce tumor cell proliferation and activate tumor cell growth through induction of tumorigenic signaling [15]. As cellular senescence is one contributor to tumorigenesis, the direct effect of high phosphate on induction of cancer development and metastasis needs to be explored. Furthermore, Klotho is a tumor suppressor through multiple cellular signaling pathways [2, 24, 39, 136, 160, 182], and phosphate-indued reduction in Klotho can also contribute to tumorigenesis.

7.3.2.2 Cardiovascular Disease

Clinical observational studies showed that high serum phosphate is identified as an independent contributor to cardiovascular morbidity and mortality in both chronic kidney disease (CKD) patients and in non-kidney disease subjects [40, 47, 80, 86, 95, 115, 135, 142, 152, 166] (Fig. 7.2) There is also a strong correlation between high serum phosphate and high morbidity and mortality of cardiovascular events such as cardiomyopathy, vascular calcification, arterial stiffness, and hypertension, in dialysis and non-dialysis CKD and experimental animals as well [5, 32, 68, 71, 80, 83, 107, 111, 115, 175]. Furthermore, higher serum phosphate at baseline is related to an increased risk of *de novel* onset of heart failure, myocardial infarction, and other cardiovascular events [166]. Therefore, high phosphate is an indicator and detrimental contributor of cardiovascular disease.

7.3.2.3 Kidney Disease

It has been known for long time that high phosphate contributes to acute kidney damage after large acute phosphate loading in short-term, and leads to kidney fibrosis in normal animals, and promotes progression of CKD animals after long-term phosphate loading [68, 69, 86, 106, 115, 137, 144]. High phosphate is closely associated with complications including cardiovascular disease and metabolic bone disease in CKD. Phosphate binders effectively decrease cellular senescence in vascular smooth muscle cells induced by high phosphate and vascular calcification in uremic rodent model [188]. Phosphotoxicity in the kidney is attributable to high phosphate-induced cell apoptosis, decreased autophagy, and activated cellular senescence [106].

7.3.2.4 Metabolism

High dietary phosphate loading causes abnormal mineral metabolism [22, 67, 94, 114, 141, 152] and disturbed fatty acid metabolism [128] in humans and experimental animals. Disturbed metabolism can be one of underlying mechanisms for phosphotoxicity in some tissues and organs.

7.3.2.5 Other Diseases

Hyperphosphatemia is also thought to lead to damage in other organs and tissues (Fig. 7.2) including the lung [23, 76, 78], bone [22, 60, 152], skeletal muscle [15, 128], diabetes [14], liver injury [97], and impaired brain development [77]. Patients with inflammatory disease and tumor in central nervous system have significantly higher levels of phosphate in cerebrospinal fluid; and ones with an intracranial hemorrhage also have elevation of phosphate in cerebrospinal fluid [63].

It is estimated that nearly 1% of total body phosphate is stored within cells as well as in extracellular fluid. Currently, measuring serum phosphate level only estimates the overall status of phosphate in the body, but the levels of serum phosphate may not be a good indicator to always reflect total amount of phosphate in the body.

Sometimes, the early or mild phosphate toxicity might not be associated with detectably high levels of serum phosphate [123]. In fact, several clinical observational studies showed that a very small increase in serum phosphate within normal range of clinical laboratories (3.4–4.5 mg/dl or1.12–1.45 mmol/L), is significantly associated with higher cardiovascular events in non-kidney disease subjects [35, 48, 166]. Therefore, the measurement of tissue or intracellular phosphate may provide more precise and timely status of phosphate metabolism.

7.4 Stimulation of Cellular Senescence by Phosphate

Phosphate-induced cellular senescence is well conserved across species from plants to vertebrates. This effect is found in plant leaf [31, 153]; prokaryotes [33], and cells and tissues in mammals [106, 188]. But the cellular and molecular mechanisms whereby phosphate activates cellular senescence are largely incompletely illustrated. The experiments performed in authors' laboratories showed both direct and indirect effects of high phosphate on activation of cellular senescence in the kidney [106, 188].

7.4.1 Direct Stimulation of Cellular Senescence

After one week of high dietary phosphate treatment (2.0% phosphate w:w), the mice had modest increase in plasma phosphate and nearly normal plasma Klotho. However, the mice had higher p16/p21 expression in the kidneys. Those results imply that high phosphate induces senescence in the kidney which is independent of plasma Klotho (Fig. 7.3). High phosphate (2.0 mM) media induce cellular senescence in cultured kidney cell line without endogenous Klotho expression, supporting that in the absence of Klotho, high phosphate can directly stimulate cellular senescence through p16/p21 signaling pathway [106]. However, whether the phosphate effect on cellular senescence is mediated through an elevation of intracellular phosphate via sodium-dependent phosphate co-transport [100, 112, 173, 177, 187] and/or through other phosphate transport-independent signaling pathway [11] needs to be explored.

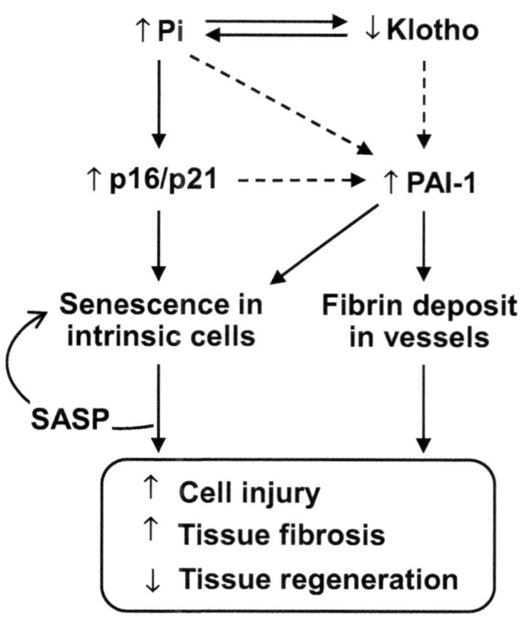

Fig. 7.3 Potential cellular mechanisms of phosphate-induced cellular senescence
High phosphate upregulates p16 and p21 and induces cellular senescence (early phase). High phosphate also increases PAI-1 expression and reduces Klotho protein production (late phase). It is unclear whether high phosphate upregulates PAI-1 in a dependent or independent manner of higher p16/p21 and/or lower Klotho (dash line). Moreover, high PAI-1 activity initiates and accelerates fibrin deposits in vessels, activates cellular senescence, and induces tissue injury and fibrosis. High phosphate-induced Klotho deficiency increases serum phosphate which further reduces Klotho production. Cellular senescence increases senescent cells through activation of a vicious cycle of secreted pro-inflammatory and pro-fibrotic growth factors, and proteases (SASP). Chronic senescence activation in intrinsic cells and vessels, and fibrin deposits in vessels initiate and/or promote fibrosis, hence destructing tissue structure. Senescence in stem cells and progenitor cells impair tissue regeneration

7.4.2 Indirect Stimulation of Cellular Senescence

7.4.2.1 Downregulation of Klotho

It is documented in rodents that high dietary phosphate loading reduces kidney and circulating Klotho [51, 67, 68, 117, 122, 130, 147, 148, 179, 192] although the cellular and molecular mechanism of how phosphate inhibits Klotho is still elusive.

Klotho protein was been shown to suppress cellular senescence (Fig. 7.3) in the kidney of several rodent models including CKD, glomerulonephritis [59], genetic Klotho deficiency [43], and in several cultured cell lines. Cellular senescence induced Klotho deficiency is associated with high Wnt signaling activity [103]. Wnt is a potent inducer of cellular senescence [103]. It has been shown that intracellular, but not extracellular Klotho protein interplays with retinoic-acid-inducible gene-I to suppress SASP and consequently inhibit inflammation [102]. But, other studies confirmed that extracellular domain of Klotho protein is still able to inhibit cellular senescence [24, 92, 106], which provides therapeutic potential to inhibit phosphate-induced cellular senescence with soluble Klotho protein.

7.4.2.2 Upregulation of Plasminogen Activator Inhibitor Type-1

Plasminogen activator inhibitor type-1 (PAI-1) belongs to the superfamily of serine-protease inhibitors, and inhibits both tissue-type and urinary-type plasminogen activators whose primary function is to activate plasminogen. So PAI-1 participates in control of fibrinolysis [21, 64]. The *in vivo* experiments showed that high PAI-1 expression suppresses fibrinolysis, and results in the massive fibrin deposition in the tissue [21, 106]. PAI-1 also regulates endothelial cell replication and angiogenesis [34, 62]. Emerging evidences showed that PAI-1 also makes major contribution to other pathologic processes independently from modulation of fibrinolysis process. PAI-1 is now considered as an inducer of cellular senescence through unclarified mechanism [43, 44, 159, 172].

High dietary phosphate upregulates PAI-1 expression in the kidney of mice *in vivo* and high phosphate media increases PAI-1 in cultured kidney cells *in vitro* [106]. In addition, Klotho deficiency also induces PAI-1 over-expression in the kidney [44, 159]. Therefore, high phosphate works synergistically with Klotho and PAI-1 to amplify cellular senescence.

7.4.2.3 Activation of SASP

High phosphate induces the secretion of pro-inflammatory growth factors such as TNFα and TGFβ, and promotes oxidative stress in experimental animals *in vivo* and in cultured cells *in vitro* [68, 106, 110, 116, 165, 194]. These factors released from damaged tissues or cells would reduce Klotho expression in the kidney in a paracrine or endocrine manner. The experiments conducted in the authors' laboratories showed that senescence markers are first elevated, followed by higher levels of oxidation, SASP, and lower Klotho protein expression. It is conceivable that high phosphate induces cellular senescence, activates oxidative stress, stimulates SASP, and decreases Klotho protein expression [106]. This hypothesis is about to be confirmed.

Taken together, high phosphate activates cellular senescence through upregulating p16/p21 and/or downregulating Klotho production. Klotho deficiency may secondarily induce PAI-1 signaling, which induces fibrin deposition in the vessel, causes tissue fibrosis, and enhances cellular senescence [43, 89] (Fig. 7.3).

7.5 Cellular Senescence: A Novel Downstream Target for the Treatment of Phosphotoxicity

Conventional prevention or treatment of phosphotoxicity consists of tilting the balance between phosphate intake and excretion towards the negative side. Equipped with the knowledge that cellular senescence is one downstream effector of phosphotoxicity, one opens up a novel array of therapeutic options.

7.5.1 Control of Dietary Phosphate Intake and Reduction of Serum Phosphate

Controlling phosphate absorption is usually prescribed for patients with CKD by restriction of phosphate intake and administration of phosphate binders (Fig. 7.4).

Reducing dietary phosphate intake is a simple and easy strategy to keep normal phosphate homeostasis. While reduction in total phosphate ingestion can be challenging, the form of digested phosphate in food can be modified because bioavailability of different kinds of ingestible phosphate differ tremendously [10, 145]. Inorganic phosphate, which is rare in nature, is much more bioavailable than organic phosphate present in living matter. The most common storage form of phosphorus in plant is phytate, which is very poorly absorbed as humans lack phytases to release the phosphate from the inositol ring [134, 171, 178]. A plant-based diet is able to provide enough dietary protein requirement without increasing risk of high phosphate intake for CKD subjects [53] and lower risk of severity of CKD-mineral bone disorder compared to animal-based protein [113]. Most importantly, one needs to account for the added inorganic phosphate into food mainly from food processing. Almost all processed foods contain phosphate additives [10, 42]. Inorganic phosphate in the preservatives is very readily absorbed across the digestive tract to enter into blood circulation.

Phosphate binders reduce bioavailability of dietary phosphate and minimize phosphate absorption from the intestine [10, 19, 30, 49, 142]. The inhibition of sodium-dependent phosphate co-transporter type 2b, which controls transcellular phosphate absorption and secondarily reduces paracellular transport from gut [85] can complement low phosphate diet and phosphate binders [30, 57] to enhance the control of phosphate balance. It has been shown that phosphate binder effectively blocks cellular senescence activation by high phosphate in vascular smooth muscle cells and vascular calcification in CKD rats [188].

7.5.2 Augmentation of Phosphate Excretion from the Kidney

Klotho is shown to induce phosphaturia by inhibiting sodium-dependent phosphate co-transporters in the kidney tubules through FGF23 dependent pathway via FGF23-FGFR-

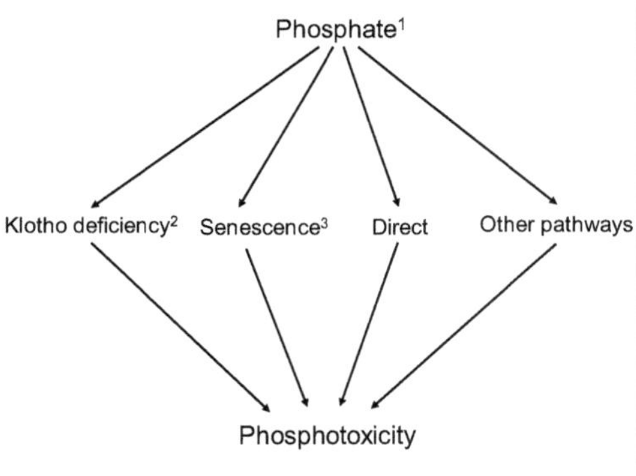

1. Reduce phosphate load	Decrease intestinal absorption • Decrease total dietary phosphate • Decrease inorganic phosphate in diet • Phosphate binders • Inhibition of transcellular phosphate transport • Inhibition of paracellular phosphate transport
2. Correct Klotho deficiency	Increase endogenous production • Treat kidney disease • Vitamin D • Off-label use of approved drugs - PPARγ agonists - Antioxidants - RAAS blockade
	Exogenous Klotho • Recombinant protein • Various forms of gene therapy
3. Target cellular senescence	Suppression of senescence induction Removal of senescent cells Interruption of SASP

Fig. 7.4 Potential strategies to treat phosphotoxicity Attenuation of phosphotoxicity can be achieved through reducing phosphate load to the body (Strategy 1), correction of Klotho deficiency (Strategy 2), and direct target of cellular senescence (Strategy 3). Since phosphotoxicity is induced and mediated through multiple cellular and molecular mechanisms, direct inhibition of phosphate-associated intracellular signaling pathways, and modulation of other downstream biologic effectors such as autophagy, mitochondria, and oxidative stress would also participate in alleviation of phosphotoxicity

Klotho complex formation [54] and FGF23-independent pathway via Klotho's enzymatic function [70]. However, clinical utility remains to be confirmed.

7.5.3 Restoration of Plasma Klotho

Elevation of plasma Klotho (Fig. 7.4) to counteract Klotho deficiency that occurs in kidney disease and after high dietary phosphate loading represents a feasible strategy to reduce blood phosphate through induction of phosphaturia, protect kidney disease from chronic progression, and also directly suppress cellular senescence [69, 70, 106, 147].

7.5.3.1 Delivery of Klotho cDNA

One strategy which has been repeatedly approved to be successful to replete Klotho in the research animals is the use of virus, plasmid or minicircle-based vectors to deliver Klotho cDNA. Delivery of Klotho cDNA in rodent can improve kidney function, ameliorate endothelial function, and attenuate uremic cardiomyopathy in acute kidney injury, CKD, and hypertensive models [138, 149, 183, 195].

7.5.3.2 Administration of Recombinant Klotho Protein

Exogenous recombinant Klotho protein has been shown to have at least two effects: direct restoration of serum Klotho levels, and stimulation of endogenous Klotho production in the kidney [69, 70, 146, 148]. Administration of Klotho protein is the only practical and feasible method to elevate serum Klotho to precise and safe levels to date. While Klotho replacement therapy is successful in correction of Klotho deficiency and improvement of renal, cardiovascular, and other diseases in many experimental animals, its clinical application still faces certain hurdles. Obviously ample animal studies have provided solid and convincing evident to support the concept that Klotho protein replacement is far more effective than virus-based Klotho [120].

7.5.3.3 Disinhibition of Endogenous Klotho Production

Another potential strategy to increase serum Klotho is to re-activate or stimulate endogenous Klotho production in the kidney by reversing mechanisms which inhibit Klotho production and/or release of extracellular domain of membrane Klotho protein in diseased kidney or after exposure to dietary phosphate overloading. In experimental CKD rodents, vitamin D receptor agonist augmented serum Klotho levels and reduced the levels of serum phosphate, but the origin of increased soluble Klotho in serum and urine was not identified, indicating that an increase in shedding membrane Klotho may be the source [99]. Experimental study showed that 1,25-$(OH)_2$-vitamin D_3 increased Klotho expression in the kidney [118, 158]. Other medications that can potentially stimulate endogenous Klotho protein production in the kidney include androgen, angiotensin II receptor antagonists, PPAR-γ agonists, and statins [120, 127]. Off-label use of existing medications to pharmacologically activate or de-suppress endogenous Klotho protein production is a viable and safe option. However, to date, there is no clinical trials to test the efficacy of pharmacological interventions in correction of Klotho deficiency. It is also unknown whether these FDA-approved medication can increase endogenous Klotho production or stimulate release of extracellular domain of membrane Klotho in high phosphate-fed animals whose kidney has low Klotho expression is not known.

Another strategy to increase endogenous Klotho expression is epigenetic approach. It has been shown that DNA demethylating agents increase Klotho expression in non-Klotho expressing cells [8]. While theoretically feasible, the effect of modulation of methylation and acetylation of the Klotho gene promoter on upregulation of Klotho in the kidney needs further confirmation.

7.5.4 Direct Suppression of Senescence Signaling Pathway and Removal of Senescent Cells to Ameliorate Phosphotoxicity

Phosphotoxicity exerts its ill effects along multiple pathways, many of which are still not defined. However, among which is the promotion of cellular senescence. Because cellular senescence is characterized by replicative arrest, suppressed apoptosis, and typical secretory phenotype, blockage of cellular senescence activation, induction of apoptosis in senescent cells and interruption of SASP are three targets for intervention of cellular senescence to treat phosphate-induced diseases (Fig. 7.4).

7.5.4.1 Removal of Senescence Stimuli

The inhibition of senescent cells formation needs to remove any senescence stimuli. The control of blood phosphate, treatment of chronic inflammation and use of Klotho protein were shown to inhibit cellular senescence as we discussed above.

7.5.4.2 Removal of Senescent Cells

Because senescent cells are typically considered cells with irreversible fates that cannot re-enter cell cycle, there is increasing interest in targeting and cleaning senescent cells from diseased organs and damaged tissues. The most efforts that scientist have made are largely to repurpose FDA-approved medications or to explore pharmacological agents that can induce senescent cell death. These compounds are together called senolytics [124].

Ideally, selectively stimulating apoptosis with senolytic drugs which block pro-survival pathways including Bcl2, p53/21, is able to remove senescent cells [124]. Quercetin can block Bcl2 activity and probably also p53 signaling to induce apoptosis and remove senescent cells. Some compounds (ABT-737, ABT-262 and A1331852) also induce apoptosis through inhibiting Bcl2 although most of them are only relatively specific.

Other options to clear senescent cells include: (1) Potentiate an immune response with monoclonal antibodies such as anti-PD1, anti-DPP4, and anti-vimentin. These immune-based interventions are most used in cancer therapy [124]; (2) Use silica nanoparticles coated with galacto-oligosaccharides [4]. Cellular uptake and digestion of coated particles are in senescent cells with expression of senescence-associated β-galactosidase. Therefore, the content will be released from the nanoparticle within cells and apoptosis induced in senescent cells [4]. Senolytic application based on senescence-associated β-galactosidase has not been tested *in vivo* [124]. Whether the above agents can ameliorate phosphotoxicity still remains to be determined.

7.5.4.3 Interruption of SASP

If one cannot remove senescent cells from diseased tissue or organ, prevention of release of proinflammatory cytokines, pro-fibrotic growth factors, or matrix-remodeling proteases and/or blockage of their actions can serve as an alternative strategy to interrupt SASP, to reduce senescence amplification, and to limit detrimental and non-cell-autonomous effects of senescent cells [124]. Rapamycin, a well-known inhibitor of the mammalian target of rapamycin signaling pathway [52, 147, 163] and metformin also functioning as mTOR inhibitor [84, 121] effectively suppress inflammation, reduce NF-κB signaling activity, prolong lifespan, and improve health in aged animals [124]. In fact, rapamycin also reduces serum phosphate and increases Klothoexpression [147]. Those two events should consequently suppress cellular senescence [161].

Overall, cellular senescence activation is one of mechanisms mediatin phosphotoxicity. Thus, targeting cellular senescence is one of strategies to reduce phosphotoxicity and treat phosphate-associated diseases, which complements the current therapy consisting of restriction of dietary phosphate intake and administration of phosphate binders to enhance the efficacy (Fig. 7.4).

7.6 Conclusion

With aging, chronic non-communicable diseases are strongly associated with shortened lifespan and diminished life quality in senior population. The detrimental effect of chronically excessive phosphate intake (e.g. processed food) and/or deficiency in phosphate excretion (e.g. CKD) on aging and age-associated diseases draw more and more attention of nephrologists, nutritionists, and basic researchers. Phosphotoxicity has been shown to induce oxidative stress, DNA damage, chronic inflammation, disturbed mineral hormones homeostasis, abnormal autophagy, and cellular senescence.

Cellular senescence is evolutionarily conserved across plant to animal kingdom as a defense mechanism to respond to cellular insults and to prevent the cells from necrosis. However persistent and over active cellular senescence can impair tissue regeneration and amplify tissue damage through SASP. The discovery of phosphate effect on cellular senescence leads to develop novel strategies to prevent or treat phosphotoxicity and phosphate-associated chronic diseases. In addition to the control of phosphate, the successful development of senolytics and inhibition of the SASP of senescent cells called senostatics would render possible to target senescent cells or senescence-related downstream signal pathways for the treatment of many age and phosphate-related diseases that are approximately involved in every organ/tissue/system in humans (Fig. 7.2).

Because phosphate-induced cellular senescence is a chronic physiologic and pathophysiologic process driven by multiple signaling pathways, better understanding of cellular senescence in phosphotoxicity would open a novel horizon to treat phosphate-associated diseases by directly targeting downstream cellular senescence which will be adjoined with effective control of serum phosphate to optimize the efficacy in attenuating phosphotoxicity, extending life span and improving health span (Fig. 7.4). Moreover, every intervention aimed at improving the quantity and quality of human life through targeting phosphate-induced cellular senescence should be thoroughly explored and validated clinically.

Acknowledgement The authors are in part supported by the National Institutes of Health (R01-DK091392 and R01-DK092461 to M.C.H. and O.W.M.), the UT Southwestern George O'Brien Kidney Research Center (P30-DK-07938 to O.W.M.), and Endowed Professors' Collaborative Research Support from the Charles Y.C Pak Foundation (to M.C.H. and O.W.M.).

References

1. Abbas T, Dutta A (2009) p21 in cancer: intricate networks and multiple activities. Nat Rev Cancer 9: 400–414
2. Abolghasemi M, Yousefi T, Maniati M et al (2019) The interplay of Klotho with signaling pathway and microRNAs in cancers. J Cell Biochem 120:14306–14317
3. Adeney KL, Siscovick DS, Ix JH et al (2009) Association of serum phosphate with vascular and valvular calcification in moderate CKD. J Am Soc Nephrol 20:381–387
4. Agostini A, Mondragon L, Bernardos A et al (2012) Targeted cargo delivery in senescent cells using capped mesoporous silica nanoparticles. Angew Chem Int Ed Engl 51:10556–10560
5. Alesutan I, Feger M, Tuffaha R et al (2016) Augmentation of phosphate-induced osteo−/chondrogenic transformation of vascular smooth muscle cells by homoarginine. Cardiovasc Res 110:408–418
6. Aravinthan A, Scarpini C, Tachtatzis P et al (2013) Hepatocyte senescence predicts progression in non-alcohol-related fatty liver disease. J Hepatol 58:549–556
7. Asghar M, Hasselquist D, Hansson B et al (2015) Chronic infection. Hidden costs of infection: chronic malaria accelerates telomere degradation and senescence in wild birds. Science 347:436–438
8. Azuma M, Koyama D, Kikuchi J et al (2012) Promoter methylation confers kidney-specific expression of the Klotho gene. FASEB J 26:4264–4274
9. Baker DJ, Petersen RC (2018) Cellular senescence in brain aging and neurodegenerative diseases: evidence and perspectives. J Clin Invest 128:1208–1216
10. Barreto FC, Barreto DV, Massy ZA et al (2019) Strategies for phosphate control in patients with CKD. Kidney Int Rep 4:1043–1056
11. Beck L, Leroy C, Salaun C et al (2009) Identification of a novel function of PiT1 critical for cell proliferation and independent of its phosphate transport activity. J Biol Chem 284:31363–31374
12. Bird TG, Muller M, Boulter L et al (2018) TGFbeta inhibition restores a regenerative response in acute liver injury by suppressing paracrine senescence. Sci Transl Med 10

13. Braun H, Schmidt BM, Raiss M et al (2012) Cellular senescence limits regenerative capacity and allograft survival. J Am Soc Nephrol 23:1467–1473
14. Brown RB (2020) Diabetes, diabetic complications, and phosphate toxicity: a scoping review. Curr Diabetes Rev 16:674–689
15. Brown RB, Razzaque MS (2018) Phosphate toxicity and tumorigenesis. Biochim Biophys Acta Rev Cancer 1869:303–309
16. Buendia P, Carracedo J, Soriano S et al (2015) Klotho prevents NFkappaB translocation and protects endothelial cell from senescence induced by Uremia. J Gerontol A Biol Sci Med Sci 70:1198–1209
17. Calcinotto A, Kohli J, Zagato E et al (2019) Cellular senescence: aging, cancer, and injury. Physiol Rev 99:1047–1078
18. Campisi J, D'adda Di Fagagna F (2007) Cellular senescence: when bad things happen to good cells. Nat Rev Mol Cell Biol 8:729–740
19. Candeal E, Caldas YA, Guillen N et al (2017) Intestinal phosphate absorption is mediated by multiple transport systems in rats. Am J Physiol Gastrointest Liver Physiol 312:G355–G366
20. Carracedo J, Buendia P, Merino A et al (2012) Klotho modulates the stress response in human senescent endothelial cells. Mech Ageing Dev 133: 647–654
21. Cesari M, Pahor M, Incalzi RA (2010) Plasminogen activator inhibitor-1 (PAI-1): a key factor linking fibrinolysis and age-related subclinical and clinical conditions. Cardiovasc Ther 28:e72–e91
22. Chande S, Bergwitz C (2018) Role of phosphate sensing in bone and mineral metabolism. Nat Rev Endocrinol 14:637–655
23. Chang SH, Yu KN, Lee YS et al (2006) Elevated inorganic phosphate stimulates Akt-ERK1/2-Mnk1 signaling in human lung cells. Am J Respir Cell Mol Biol 35:528–539
24. Chen B, Liang Y, Chen L et al (2018) Overexpression of Klotho inhibits HELF fibroblasts SASP-related protumoral effects on non-small cell lung cancer cells. J Cancer 9:1248–1258
25. Chen TH, Kuro OM, Chen CH et al (2013) The secreted Klotho protein restores phosphate retention and suppresses accelerated aging in Klotho mutant mice. Eur J Pharmacol 698:67–73
26. Childs BG, Durik M, Baker DJ et al (2015) Cellular senescence in aging and age-related disease: from mechanisms to therapy. Nat Med 21: 1424–1435
27. Childs BG, Li H, Van Deursen JM (2018) Senescent cells: a therapeutic target for cardiovascular disease. J Clin Invest 128:1217–1228
28. Cianflone E, Torella M, Biamonte F et al (2020) Targeting cardiac stem cell senescence to treat cardiac aging and disease. Cell 9:1558
29. Coppe JP, Desprez PY, Krtolica A et al (2010) The senescence-associated secretory phenotype: the dark side of tumor suppression. Annu Rev Pathol 5: 99–118
30. Cozzolino M, Ketteler M, Wagner CA (2020) An expert update on novel therapeutic targets for hyperphosphatemia in chronic kidney disease: preclinical and clinical innovations. Expert Opin Ther Targets 24:477–488
31. Crafts-Brandner SJ (1992) Phosphorus nutrition influence on leaf senescence in soybean. Plant Physiol 98:1128–1132
32. Crouthamel MH, Lau WL, Leaf EM et al (2013) Sodium-dependent phosphate cotransporters and phosphate-induced calcification of vascular smooth muscle cells: redundant roles for PiT-1 and PiT-2. Arterioscler Thromb Vasc Biol 33:2625–2632
33. Damoglou AP, Dawes EA (1967) Effect of phosphate limitation on the senescence of glucose-and acetate-grown Escherichia coli. Biochem J 104:48P–49P
34. Daniel AE, Timmerman I, Kovacevic I et al (2015) Plasminogen activator Inhibitor-1 controls vascular integrity by regulating VE-cadherin trafficking. PLoS One 10:e0145684
35. Dhingra R, Sullivan LM, Fox CS et al (2007) Relations of serum phosphorus and calcium levels to the incidence of cardiovascular disease in the community. Arch Intern Med 167:879–885
36. Di Marco GS, Hausberg M, Hillebrand U et al (2008) Increased inorganic phosphate induces human endothelial cell apoptosis in vitro. Am J Physiol Renal Physiol 294:F1381–F1387
37. Docherty MH, O'sullivan ED, Bonventre JV et al (2019) Cellular senescence in the kidney. J Am Soc Nephrol 30:726–736
38. Dodig S, Cepelak I, Pavic I (2019) Hallmarks of senescence and aging. Biochem Med (Zagreb) 29: 030501
39. Doi S, Zou Y, Togao O et al (2011) Klotho inhibits transforming growth factor-beta1 (TGF-beta1) signaling and suppresses renal fibrosis and cancer metastasis in mice. J Biol Chem 286:8655-8665
40. Eddington H, Hoefield R, Sinha S et al (2010) Serum phosphate and mortality in patients with chronic kidney disease. Clin J Am Soc Nephrol 5:2251–2257
41. Elser JJ, Kyle MM, Smith MS et al (2007) Biological stoichiometry in human cancer. PLoS One 2:e1028
42. Erem S, Razzaque MS (2018) Dietary phosphate toxicity: an emerging global health concern. Histochem Cell Biol 150:711–719
43. Eren M, Boe AE, Klyachko EA et al (2014a) Role of plasminogen activator inhibitor-1 in senescence and aging. Semin Thromb Hemost 40:645–651
44. Eren M, Boe AE, Murphy SB et al (2014b) PAI-1-regulated extracellular proteolysis governs senescence and survival in Klotho mice. Proc Natl Acad Sci U S A 111:7090–7095
45. Faget DV, Ren Q, Stewart SA (2019) Unmasking senescence: context-dependent effects of SASP in cancer. Nat Rev Cancer 19:439–453
46. Faggioli F, Palagano E, Di Tommaso L et al (2018) B lymphocytes limit senescence-driven fibrosis resolution and favor hepatocarcinogenesis in mouse liver injury. Hepatology 67:1970–1985

47. Foley RN (2009) Phosphate levels and cardiovascular disease in the general population. Clin J Am Soc Nephrol 4:1136–1139
48. Foley RN, Collins AJ, Herzog CA et al (2009) Serum phosphorus levels associate with coronary atherosclerosis in young adults. J Am Soc Nephrol 20:397–404
49. Forster I, Hernando N, Sorribas V et al (2011) Phosphate transporters in renal, gastrointestinal, and other tissues. Adv Chronic Kidney Dis 18:63–76
50. Frey N, Venturelli S, Zender L et al (2018) Cellular senescence in gastrointestinal diseases: from pathogenesis to therapeutics. Nat Rev Gastroenterol Hepatol 15:81–95
51. Fukuda-Tatano S, Yamamoto H, Nakahashi O et al (2019) Regulation of alpha-Klotho expression by dietary phosphate during growth periods. Calcif Tissue Int 104:667–678
52. Garbers C, Kuck F, Aparicio-Siegmund S et al (2013) Cellular senescence or EGFR signaling induces Interleukin 6 (IL-6) receptor expression controlled by mammalian target of rapamycin (mTOR). Cell Cycle 12:3421–3432
53. Garcia-Torres R, Young L, Murray DP et al (2020) Dietary protein source and phosphate levels in patients on hemodialysis. J Ren Nutr 30:423–429
54. Goetz R, Nakada Y, Hu MC et al (2010) Isolated C-terminal tail of FGF23 alleviates hypophosphatemia by inhibiting FGF23-FGFR-Klotho complex formation. Proc Natl Acad Sci U S A 107:407–412
55. Gonzalez-Meljem JM, Haston S, Carreno G et al (2017) Stem cell senescence drives age-attenuated induction of pituitary tumours in mouse models of paediatric craniopharyngioma. Nat Commun 8:1819
56. Gowen JW (1962) Genetic patterns in senescence and infection. J Am Geriatr Soc 10:107–124
57. Gryshkova V, Lituiev D, Savinska L et al (2011) Generation of monoclonal antibodies against tumor-associated antigen MX35/sodium-dependent phosphate transporter NaPi2b. Hybridoma (Larchmt) 30: 37–42
58. Gunia S, Kakies C, Erbersdobler A et al (2012) Expression of p53, p21 and cyclin D1 in penile cancer: p53 predicts poor prognosis. J Clin Pathol 65: 232–236
59. Haruna Y, Kashihara N, Satoh M et al (2007) Amelioration of progressive renal injury by genetic manipulation of Klotho gene. Proc Natl Acad Sci U S A 104:2331–2336
60. Haussler MR, Whitfield GK, Kaneko I et al (2012) The role of vitamin D in the FGF23, Klotho, and phosphate bone-kidney endocrine axis. Rev Endocr Metab Disord 13:57–69
61. Hayflick L, Moorhead PS (1961) The serial cultivation of human diploid cell strains. Exp Cell Res 25: 585–621
62. Heberlein KR, Straub AC, Best AK et al (2010) Plasminogen activator inhibitor-1 regulates myoendothelial junction formation. Circ Res 106: 1092–1102
63. Heipertz R, Eickhoff K, Karstens KH (1979) Magnesium and inorganic phosphate content in CSF related to blood-brain barrier function in neurological disease. J Neurol Sci 40:87–95
64. Hekman CM, Loskutoff DJ (1987) Fibrinolytic pathways and the endothelium. Semin Thromb Hemost 13:514–527
65. Hernando N, Gagnon K, Lederer E (2021) Phosphate transport in epithelial and nonepithelial tissue. Physiol Rev 101:1–35
66. Herranz N, Gil J (2018) Mechanisms and functions of cellular senescence. J Clin Invest 128:1238–1246
67. Hu MC, Scanni R, Ye J et al (2019) Dietary vitamin D interacts with high phosphate-induced cardiac remodeling in rats with normal renal function. Nephrol Dial Transplant
68. Hu MC, Shi M, Cho HJ et al (2015) Klotho and phosphate are modulators of pathologic uremic cardiac remodeling. J Am Soc Nephrol 26:1290–1302
69. Hu MC, Shi M, Gillings N et al (2017) Recombinant alpha-Klotho may be prophylactic and therapeutic for acute to chronic kidney disease progression and uremic cardiomyopathy. Kidney Int 91:1104–1114
70. Hu MC, Shi M, Zhang J et al (2010) Klotho: a novel phosphaturic substance acting as an autocrine enzyme in the renal proximal tubule. FASEB J 24: 3438–3450
71. Hu MC, Shi M, Zhang J et al (2011) Klotho deficiency causes vascular calcification in chronic kidney disease. J Am Soc Nephrol 22:124–136
72. Hu MC, Shiizaki K, Kuro-O M et al (2013) Fibroblast growth factor 23 and Klotho: physiology and pathophysiology of an endocrine network of mineral metabolism. Annu Rev Physiol 75:503–533
73. Ibler AEM, Elghazaly M, Naylor KL et al (2019) Typhoid toxin exhausts the RPA response to DNA replication stress driving senescence and Salmonella infection. Nat Commun 10:4040
74. Ix JH, De Boer IH, Peralta CA et al (2009) Serum phosphorus concentrations and arterial stiffness among individuals with normal kidney function to moderate kidney disease in MESA. Clin J Am Soc Nephrol 4:609–615
75. Jeon OH, David N, Campisi J et al (2018) Senescent cells and osteoarthritis: a painful connection. J Clin Invest 128:1229–1237
76. Jin H, Chang SH, Xu CX et al (2007) High dietary inorganic phosphate affects lung through altering protein translation, cell cycle, and angiogenesis in developing mice. Toxicol Sci 100:215–223
77. Jin H, Hwang SK, Yu K et al (2006) A high inorganic phosphate diet perturbs brain growth, alters Akt-ERK signaling, and results in changes in cap-dependent translation. Toxicol Sci 90:221–229
78. Jin H, Xu CX, Lim HT et al (2009) High dietary inorganic phosphate increases lung tumorigenesis

and alters Akt signaling. Am J Respir Crit Care Med 179:59–68
79. Jin H, Zhang Y, Ding Q et al (2019) Epithelial innate immunity mediates tubular cell senescence after kidney injury. JCI Insight 4:e125490
80. Kestenbaum B, Sampson JN, Rudser KD et al (2005) Serum phosphate levels and mortality risk among people with chronic kidney disease. J Am Soc Nephrol 16:520–528
81. Khademi-Shirvan M, Ghorbaninejad M, Hosseini S et al (2020) The importance of stem cell senescence in regenerative medicine. Adv Exp Med Biol 1288: 87–102
82. Khosla S, Farr JN, Kirkland JL (2018) Inhibiting cellular senescence: a new therapeutic paradigm for age-related osteoporosis. J Clin Endocrinol Metab 103:1282–1290
83. Kim HK, Mizuno M, Vongpatanasin W (2019) Phosphate, the forgotten mineral in hypertension. Curr Opin Nephrol Hypertens 28:345–351
84. Kim HS, Ren G, Kim T et al (2020) Metformin reduces saturated fatty acid-induced lipid accumulation and inflammatory response by restoration of autophagic flux in endothelial cells. Sci Rep 10: 13523
85. King AJ, Siegel M, He Y et al (2018) Inhibition of sodium/hydrogen exchanger 3 in the gastrointestinal tract by tenapanor reduces paracellular phosphate permeability. Sci Transl Med 10:eaam6474
86. Komaba H, Fukagawa M (2016) Phosphate-a poison for humans? Kidney Int 90:753–763
87. Kondo Y, Hasegawa G, Okada H et al (2013) Lepr (db/db) mice with senescence marker protein-30 knockout (Lepr(db/db)Smp30(Y/−)) exhibit increases in small dense-LDL and severe fatty liver despite being fed a standard diet. PLoS One 8:e65698
88. Kondo Y, Ishigami A (2016) Involvement of senescence marker protein-30 in glucose metabolism disorder and non-alcoholic fatty liver disease. Geriatr Gerontol Int 16(Suppl 1):4–16
89. Kortlever RM, Bernards R (2006) Senescence, wound healing and cancer: the PAI-1 connection. Cell Cycle 5:2697–2703
90. Kortlever RM, Higgins PJ, Bernards R (2006) Plasminogen activator inhibitor-1 is a critical downstream target of p53 in the induction of replicative senescence. Nat Cell Biol 8:877–884
91. Kumar R, Gururaj AE, Barnes CJ (2006) p21-activated kinases in cancer. Nat Rev Cancer 6: 459–471
92. Kuro-O M (2008) Klotho as a regulator of oxidative stress and senescence. Biol Chem 389:233–241
93. Kuro-O M (2010) A potential link between phosphate and aging–lessons from Klotho-deficient mice. Mech Ageing Dev 131:270–275
94. Kuro-O M (2013) Klotho, phosphate and FGF-23 in ageing and disturbed mineral metabolism. Nat Rev Nephrol 9:650–660
95. Kuro OM (2013) A phosphate-centric paradigm for pathophysiology and therapy of chronic kidney disease. Kidney Int Suppl 3:420–426
96. Kuro OM (2020) Phosphate as a pathogen of arteriosclerosis and aging. J Atheroscler Thromb 28:203–213
97. Lakshmi CP, Goel A, Basu D (2014) Cholestatic presentation of yellow phosphorus poisoning. J Pharmacol Pharmacother 5:67–69
98. Lasry A, Ben-Neriah Y (2015) Senescence-associated inflammatory responses: aging and cancer perspectives. Trends Immunol 36:217–228
99. Lau WL, Leaf EM, Hu MC et al (2012) Vitamin D receptor agonists increase Klotho and osteopontin while decreasing aortic calcification in mice with chronic kidney disease fed a high phosphate diet. Kidney Int 82:1261–1270
100. Levi M, Gratton E, Forster IC et al (2019) Mechanisms of phosphate transport. Nat Rev Nephrol 15:482–500
101. Li Z, Wiernek S, Patterson C et al (2018) MicroRNA-21 mediates high phosphate-induced endothelial cell apoptosis. Am J Physiol Cell Physiol 315:C830–C838
102. Liu F, Wu S, Ren H et al (2011) Klotho suppresses RIG-I-mediated senescence-associated inflammation. Nat Cell Biol 13:254–262
103. Luo C, Zhou S, Zhou Z et al (2018) Wnt9a promotes renal fibrosis by accelerating cellular senescence in tubular epithelial cells. J Am Soc Nephrol 29:1238–1256
104. Luu J, Palczewski K (2018) Human aging and disease: lessons from age-related macular degeneration. Proc Natl Acad Sci U S A 115:2866–2872
105. Maekawa Y, Ohishi M, Ikushima M et al (2011) Klotho protein diminishes endothelial apoptosis and senescence via a mitogen-activated kinase pathway. Geriatr Gerontol Int 11:510–516
106. Maique J, Flores B, Shi M et al (2020) High phosphate induces and Klotho attenuates kidney epithelial senescence and fibrosis. Front Pharmacol 11:1273
107. Martin KJ, Gonzalez EA (2011) Prevention and control of phosphate retention/hyperphosphatemia in CKD-MBD: what is normal, when to start, and how to treat? Clin J Am Soc Nephrol 6:440–446
108. Mccarthy CG, Wenceslau CF, Webb RC et al (2019) Novel contributors and mechanisms of cellular senescence in hypertension-associated premature vascular aging. Am J Hypertens 32:709–719
109. Mchugh D, Gil J (2018) Senescence and aging: causes, consequences, and therapeutic avenues. J Cell Biol 217:65–77
110. Mitobe M, Yoshida T, Sugiura H et al (2005) Oxidative stress decreases klotho expression in a mouse kidney cell line. Nephron Exp Nephrol 101:e67–e74
111. Mizuno M, Mitchell JH, Crawford S et al (2016) High dietary phosphate intake induces hypertension and augments exercise pressor reflex function in rats.

Am J Physiol Regul Integr Comp Physiol 311:R39–R48
112. Moe OW (2009) PiT-2 coming out of the pits. Am J Physiol Renal Physiol 296:F689–F690
113. Moe SM, Chen NX, Seifert MF et al (2009) A rat model of chronic kidney disease-mineral bone disorder. Kidney Int 75:176–184
114. Mohammad J, Scanni R, Bestmann L et al (2018) A controlled increase in dietary phosphate elevates BP in healthy human subjects. J Am Soc Nephrol 29:2089–2098
115. Moon H, Chin HJ, Na KY et al (2019) Hyperphosphatemia and risks of acute kidney injury, end-stage renal disease, and mortality in hospitalized patients. BMC Nephrol 20:362
116. Moreno JA, Izquierdo MC, Sanchez-Nino MD et al (2011) The inflammatory cytokines TWEAK and TNFalpha reduce renal klotho expression through NFkappaB. J Am Soc Nephrol 22:1315–1325
117. Morishita K, Shirai A, Kubota M et al (2001) The progression of aging in Klotho mutant mice can be modified by dietary phosphorus and zinc. J Nutr 131:3182–3188
118. Munoz-Castaneda JR, Herencia C, Pendon-Ruiz De Mier MV et al (2017) Differential regulation of renal Klotho and FGFR1 in normal and uremic rats. FASEB J 31:3858–3867
119. Murton AJ, Greenhaff PL (2009) Muscle atrophy in immobilization and senescence in humans. Curr Opin Neurol 22:500–505
120. Neyra JA, Hu MC, Moe OW (2020) Klotho in clinical nephrology: diagnostic and therapeutic implications. Clin J Am Soc Nephrol 16:162–176
121. Noren Hooten N, Martin-Montalvo A, Dluzen DF et al (2016) Metformin-mediated increase in DICER1 regulates microRNA expression and cellular senescence. Aging Cell 15:572–581
122. Ohnishi M, Razzaque MS (2010) Dietary and genetic evidence for phosphate toxicity accelerating mammalian aging. FASEB J 24:3562–3571
123. Osuka S, Razzaque MS (2012) Can features of phosphate toxicity appear in normophosphatemia? J Bone Miner Metab 30:10–18
124. Ovadya Y, Krizhanovsky V (2018) Strategies targeting cellular senescence. J Clin Invest 128:1247–1254
125. Palmer AK, Tchkonia T, Lebrasseur NK et al (2015) Cellular senescence in type 2 diabetes: a therapeutic opportunity. Diabetes 64:2289–2298
126. Pan J, Zhong J, Gan LH et al (2011) Klotho, an anti-senescence related gene, is frequently inactivated through promoter hypermethylation in colorectal cancer. Tumour Biol 32:729–735
127. Pastor J, Moe OW (2019) Treating systemic klotho deficiency. Am J Nephrol 49:410–412
128. Peri-Okonny P, Baskin KK, Iwamoto G et al (2019) High-phosphate diet induces exercise intolerance and impairs fatty acid metabolism in mice. Circulation 139:1422–1434
129. Perveez M, Ajaz M, Afroze D (2015) Promoter hypermethylation of Klotho; an anti-senescence related gene in colorectal cancer patients of Kashmir valley. Mol Biol Res Commun 4:217–224
130. Perwad F, Azam N, Zhang MY et al (2005) Dietary and serum phosphorus regulate fibroblast growth factor 23 expression and 1,25-dihydroxyvitamin D metabolism in mice. Endocrinology 146:5358–5364
131. Razzaque MS (2011) Phosphate toxicity: new insights into an old problem. Clin Sci (Lond) 120:91–97
132. Razzaque MS (2013) Phosphate toxicity and vascular mineralization. Contrib Nephrol 180:74–85
133. Razzaque MS, Sitara D, Taguchi T et al (2006) Premature aging-like phenotype in fibroblast growth factor 23 null mice is a vitamin D-mediated process. FASEB J 20:720–722
134. Reddy NR, Sathe SK, Salunkhe DK (1982) Phytates in legumes and cereals. Adv Food Res 28:1–92
135. Ritter CS, Slatopolsky E (2016) Phosphate toxicity in CKD: the killer among us. Clin J Am Soc Nephrol 11:1088–1100
136. Rubinek T, Wolf I (2016) The role of alpha-Klotho as a universal tumor suppressor. Vitam Horm 101:197–214
137. Rubio-Aliaga I (2020) Phosphate and kidney healthy aging. Kidney Blood Press Res 45:1–10
138. Saito Y, Nakamura T, Ohyama Y et al (2000) In vivo Klotho gene delivery protects against endothelial dysfunction in multiple risk factor syndrome. Biochem Biophys Res Commun 276:767–772
139. Sato S, Kawamata Y, Takahashi A et al (2015) Ablation of the p16(INK4a) tumour suppressor reverses ageing phenotypes of klotho mice. Nat Commun 6:7035
140. Satyanarayana A, Rudolph KL (2004) p16 and ARF: activation of teenage proteins in old age. J Clin Invest 114:1237–1240
141. Scanni R, Vonrotz M, Jehle S et al (2014) The human response to acute enteral and parenteral phosphate loads. J Am Soc Nephrol 25:2730–2739
142. Scialla JJ, Kendrick J, Uribarri J et al (2020) State-of-the-art Management of hyperphosphatemia in patients with CKD: an NKF-KDOQI controversies perspective. Am J Kidney Dis 77:132–141
143. Sharpless NE, Sherr CJ (2015) Forging a signature of in vivo senescence. Nat Rev Cancer 15:397–408
144. Shen ZJ, Hu J, Shiizaki K et al (2016) Phosphate-induced renal fibrosis requires the prolyl isomerase Pin1. PLoS One 11:e0150093
145. Sherman RA, Mehta O (2009) Dietary phosphorus restriction in dialysis patients: potential impact of processed meat, poultry, and fish products as protein sources. Am J Kidney Dis 54:18–23
146. Shi M, Flores B, Gillings N et al (2016) alphaKlotho mitigates progression of AKI to CKD through activation of autophagy. J Am Soc Nephrol 27:2331–2345
147. Shi M, Maique J, Shaffer J et al (2020) The tripartite interaction of phosphate, autophagy, and alphaKlotho in health maintenance. FASEB J 34:3129–3150

148. Shi M, Mcmillan KL, Wu J et al (2018) Cisplatin nephrotoxicity as a model of chronic kidney disease. Lab Investig 98:1105–1121
149. Shin YJ, Luo K, Quan Y et al (2019) Therapeutic challenge of minicircle vector encoding Klotho in animal model. Am J Nephrol 49:413–424
150. Shuto E, Taketani Y, Tanaka R et al (2009) Dietary phosphorus acutely impairs endothelial function. J Am Soc Nephrol 20:1504–1512
151. Sosa Pena MDP, Lopez-Soler R, Melendez JA (2018) Senescence in chronic allograft nephropathy. Am J Physiol Renal Physiol 315:F880–F889
152. Stevens LA, Djurdjev O, Cardew S et al (2004) Calcium, phosphate, and parathyroid hormone levels in combination and as a function of dialysis duration predict mortality: evidence for the complexity of the association between mineral metabolism and outcomes. J Am Soc Nephrol 15:770–779
153. Stigter KA, Plaxton WC (2015) Molecular mechanisms of phosphorus metabolism and transport during Leaf senescence. Plants (Basel) 4:773–798
154. Sturmlechner I, Durik M, Sieben CJ et al (2017) Cellular senescence in renal ageing and disease. Nat Rev Nephrol 13:77–89
155. Sugihara H, Teramoto N, Nakamura K et al (2020) Cellular senescence-mediated exacerbation of Duchenne muscular dystrophy. Sci Rep 10:16385
156. Sugimoto K, Yokokawa T, Misaka T et al (2019) Senescence marker protein 30 deficiency exacerbates pulmonary hypertension in hypoxia-exposed mice. Int Heart J 60:1430–1434
157. Sun Y, Coppe JP, Lam EW (2018) Cellular senescence: the sought or the unwanted? Trends Mol Med 24:871–885
158. Takenaka T, Inoue T, Ohno Y et al (2014) Calcitriol supplementation improves endothelium-dependent vasodilation in rat hypertensive renal injury. Kidney Blood Press Res 39:17–27
159. Takeshita K, Yamamoto K, Ito M et al (2002) Increased expression of plasminogen activator inhibitor-1 with fibrin deposition in a murine model of aging, "Klotho" mouse. Semin Thromb Hemost 28:545–554
160. Tang X, Wang Y, Fan Z et al (2016) Klotho: a tumor suppressor and modulator of the Wnt/beta-catenin pathway in human hepatocellular carcinoma. Lab Investig 96:197–205
161. Tataranni T, Biondi G, Cariello M et al (2011) Rapamycin-induced hypophosphatemia and insulin resistance are associated with mTORC2 activation and Klotho expression. Am J Transplant 11:1656–1664
162. Tchkonia T, Zhu Y, Van Deursen J et al (2013) Cellular senescence and the senescent secretory phenotype: therapeutic opportunities. J Clin Invest 123: 966–972
163. Thapa RK, Nguyen HT, Jeong JH et al (2017) Progressive slowdown/prevention of cellular senescence by CD9-targeted delivery of rapamycin using lactose-wrapped calcium carbonate nanoparticles. Sci Rep 7: 43299
164. Thomas H (2018) Senescence prevents regeneration after acute liver injury. Nat Rev Gastroenterol Hepatol 15:582
165. Thurston RD, Larmonier CB, Majewski PM et al (2010) Tumor necrosis factor and interferon-gamma down-regulate Klotho in mice with colitis. Gastroenterology 138:1384–1394, 1394 e1381–1382
166. Tonelli M, Sacks F, Pfeffer M et al (2005) Relation between serum phosphate level and cardiovascular event rate in people with coronary disease. Circulation 112:2627–2633
167. Umeda M, Okuda S, Izumi H et al (2006) Prognostic significance of the serum phosphorus level and its relationship with other prognostic factors in multiple myeloma. Ann Hematol 85:469–473
168. Uribarri J (2007) Phosphorus homeostasis in normal health and in chronic kidney disease patients with special emphasis on dietary phosphorus intake. Semin Dial 20:295–301
169. Valenzuela CA, Quintanilla R, Moore-Carrasco R et al (2017) The potential role of senescence as a modulator of platelets and tumorigenesis. Front Oncol 7:188
170. Van Deursen JM (2014) The role of senescent cells in ageing. Nature 509:439–446
171. Van Doorn DA, Everts H, Wouterse H et al (2004) The apparent digestibility of phytate phosphorus and the influence of supplemental phytase in horses. J Anim Sci 82:1756–1763
172. Vaughan DE, Rai R, Khan SS et al (2017) Plasminogen activator Inhibitor-1 is a marker and a mediator of senescence. Arterioscler Thromb Vasc Biol 37:1446–1452
173. Villa-Bellosta R, Levi M, Sorribas V (2009) Vascular smooth muscle cell calcification and SLC20 inorganic phosphate transporters: effects of PDGF, TNF-alpha, and Pi. Pflugers Arch 458:1151–1161
174. Viola MV, Fromowitz F, Oravez S et al (1986) Expression of ras oncogene p21 in prostate cancer. N Engl J Med 314:133–137
175. Voelkl J, Lang F, Eckardt KU et al (2019) Signaling pathways involved in vascular smooth muscle cell calcification during hyperphosphatemia. Cell Mol Life Sci 76:2077–2091
176. Voelkl J, Tuffaha R, Luong TTD et al (2018) Zinc inhibits phosphate-induced vascular calcification through TNFAIP3-mediated suppression of NF-kappaB. J Am Soc Nephrol 29:1636–1648
177. Wallingford MC, Giachelli CM (2014) Loss of PiT-1 results in abnormal endocytosis in the yolk sac visceral endoderm. Mech Dev 133:189–202
178. Wang XX, Hoffland E, Feng G et al (2017) Phosphate uptake from Phytate due to hyphae-mediated phytase activity by arbuscular mycorrhizal maize. Front Plant Sci 8:684
179. Watanabe R, Fujita N, Sato Y et al (2017) Enpp1 is an anti-aging factor that regulates Klotho under phosphate overload conditions. Sci Rep 7:7786
180. White E, Lowe SW (2009) Eating to exit: autophagy-enabled senescence revealed. Genes Dev 23:784–787

181. Wilson KM, Shui IM, Mucci LA et al (2015) Calcium and phosphorus intake and prostate cancer risk: a 24-y follow-up study. Am J Clin Nutr 101:173–183
182. Wolf I, Levanon-Cohen S, Bose S et al (2008) Klotho: a tumor suppressor and a modulator of the IGF-1 and FGF pathways in human breast cancer. Oncogene 27:7094–7105
183. Xie J, Yoon J, An SW et al (2015) Soluble Klotho protects against uremic cardiomyopathy independently of fibroblast growth factor 23 and phosphate. J Am Soc Nephrol 26:1150–1160
184. Xin L, Jiang TT, Kinder JM et al (2015) Infection susceptibility and immune senescence with advancing age replicated in accelerated aging Lmna(Dhe) mice. Aging Cell 14:1122–1126
185. Xing XB, Cai WB, Luo L et al (2013) The prognostic value of p16 Hypermethylation in cancer: a meta-analysis. PLoS One 8:e66587
186. Xu R, Wang F, Wu L et al (2013) A systematic review of hypermethylation of p16 gene in esophageal cancer. Cancer Biomark 13:215–226
187. Yamada S, Leaf EM, Chia JJ et al (2018) PiT-2, a type III sodium-dependent phosphate transporter, protects against vascular calcification in mice with chronic kidney disease fed a high-phosphate diet. Kidney Int 94:716–727
188. Yamada S, Tatsumoto N, Tokumoto M et al (2015) Phosphate binders prevent phosphate-induced cellular senescence of vascular smooth muscle cells and vascular calcification in a modified, adenine-based uremic rat model. Calcif Tissue Int 96:347–358
189. Yamamoto KT, Robinson-Cohen C, De Oliveira MC et al (2013) Dietary phosphorus is associated with greater left ventricular mass. Kidney Int 83: 707–714
190. Ye Z, Palazzo JP, Lin L et al (2013) Postoperative hyperphosphatemia significantly associates with adverse survival in colorectal cancer patients. J Gastroenterol Hepatol 28:1469–1475
191. Yoon KJ, Zhang D, Kim SJ et al (2019) Exercise-induced AMPK activation is involved in delay of skeletal muscle senescence. Biochem Biophys Res Commun 512:604–610
192. Yoshikawa R, Yamamoto H, Nakahashi O et al (2018) The age-related changes of dietary phosphate responsiveness in plasma 1,25-dihydroxyvitamin D levels and renal Cyp27b1 and Cyp24a1 gene expression is associated with renal alpha-Klotho gene expression in mice. J Clin Biochem Nutr 62:68–74
193. Yzydorczyk C, Li N, Chehade H et al (2017) Transient postnatal overfeeding causes liver stress-induced premature senescence in adult mice. Sci Rep 7:12911
194. Zhou L, Li Y, Zhou D et al (2013) Loss of Klotho contributes to kidney injury by derepression of Wnt/beta-catenin signaling. J Am Soc Nephrol 24:771–785
195. Zhu L, Stein LR, Kim D et al (2018) Klotho controls the brain-immune system interface in the choroid plexus. Proc Natl Acad Sci U S A 115:E11388–E11396

Phosphate Toxicity and Epithelial to Mesenchymal Transition

8

Eric Lewis, Faith Seltun, Mohammed S. Razzaque, and Ping He

Abstract

The underlying role of inadequate or excess intake of phosphate is evident in disease states, including metabolic, skeletal, cardiac, kidney and various cancers. Elevated phosphate levels can induce epithelial to mesenchymal transition (EMT) and cell death. EMT and associated lethal, metastatic or fibrinogenic responses are known to be underlying disease processes in fibrotic diseases and various solid tumors. Studies have shown EMT is regulated by induction of different signaling pathways, including TGF-β, RTK, SRC, Wnt and Notch signal transduction. However, cross-talk amongst these signaling pathways is less understood. We have shown that elevated phosphate levels enhanced EMT partially through activating ERK1/2 pathway, resulting in massive cell death. We thus proposed excess phosphate-mediated lethal EMT as one of the underlying mechanisms of phosphate-induced cytotoxicity, which could explain high phosphate-associated renal fibrosis and cancer metastasis in preclinical and clinical studies. This chapter provides the overview of EMT with the highlights of its regulation by various signaling pathways induced by phosphate toxicity. We further put lately reported lethal EMT in the context of phosphate toxicity with the intent to explain it to excessive phosphate-associated pathologies.

Keywords

Phosphate · Cytotoxicity · EMT · Apoptosis

E. Lewis · F. Seltun · P. He (✉)
Department of Biochemistry, Lake Erie College of Osteopathic Medicine, Erie, PA, USA
e-mail: ELewis27603@med.lecom.edu; FSeltun73884@med.lecom.edu; pinghe718@gmail.com

M. S. Razzaque
Department of Pathology, Lake Erie College of Osteopathic Medicine, Erie, PA, USA
e-mail: mrazzaque@lecom.edu; msr.nagasaki@gmail.com

8.1 Phosphate Toxicity

Phosphate (PO_4) is one of the most abundant nutrients in the body and is undeniably essential for the survival of all living organisms. In vivo, inorganic phosphate (Pi), in the form of *di*hydrogen phosphate (H_2PO_4) and *mono*hydrogen phosphate (HPO_4), is synthesized from organic PO_4 metabolism. Pi is involved in numerous essential biological processes, including the synthesis of DNA and RNA, the storage and transfer of energy in the form of ATP, regulation of cell metabolism and cell signaling via protein phosphorylation, and maintenance of cell membrane integrity as phospholipids [53]. Pi also plays a crucial role in skeletogenesis by forming, developing, and maintaining skeleton and dentin in vertebrates [9]. In the body, sufficient phosphorus intake is

M. S. Razzaque (ed.), *Phosphate Metabolism*, Advances in Experimental Medicine and Biology 1362,
https://doi.org/10.1007/978-3-030-91623-7_8

essential for maintaining musculoskeletal functions and beyond. Insufficient intake of dietary phosphorus results in malnutrition, leading to a deficiency in skeletal mineralization and subsequent development of rickets. In contrast, excess Pi may have harmful consequences by inducing pathological calcification [25], oxidative stress secondary to mitochondrial dysfunction [21], and dysregulated signal transduction [34]. These consequences may present as various disorders, including gingivitis [28], dental decay [27], heart disease [10], impaired fertility [64], diabetes [50], kidney disease [51], premature aging [61], and cancer [7].

Recent studies have demonstrated the ability of extracellular Pi to manipulate skeletal and extraskeletal cells. In skeletal cells, elevated Pi levels can alter osteoblast [5] and osteoclast [42] differentiation, as well as vascular calcification [25]. Extraskeletally, Pi has an essential role in cell proliferation [66] and growth [11, 16]. Several studies have provided evidence to suggest that high levels of Pi may play a role in promoting tumorigenesis by altering various metabolic pathways. For instance, high levels of serum phosphate have been found to positively correlate with the development of lung, thyroid, pancreas, and bone cancer in men, and esophageal, non-melanoma skin, and lung cancer in women [82]. In animal models, mice fed with high phosphate diets showed an increased risk of cancer growth than those with low-phosphate diets [7]. A human study found that high phosphate intake in men resulted in not only an increased overall risk of prostate cancer, but an increased risk of advanced-stage or lethal prostate cancer [81]. A possible explanation for this may be due to the excessive expression and activity of phosphate cotransporters on tumor cells and their ability to store more inorganic phosphate in comparison to normal cells [46]. In fact, tumor cells of cancer patients have been found to store nearly double the amount of phosphorus seen in normal cells [6, 23]. This excess intracellular phosphorus leads to increased metabolic activity in tumor cells, the induction of growth-promoting cell signaling, chromosome instability, and create a potential microenvironment for metastasis [7].

Studies have also shown the response of cells to elevated cytotoxic levels of Pi. Mitochondrial oxidative stress in endothelial cells resulted in cell death, possibly by apoptosis and tissue damage [21]. Experimental studies have highlighted the role of Pi-induced endothelial injury as the possible initial event in hyperphosphatemia-associated diseases, including chronic kidney disease, cardiovascular disease, and various cancers. Recent studies have convincingly shown excess Pi-induced ER stress, epithelial-mesenchymal transition (EMT), and cell death [34]. High Pi stimulates Transforming growth factor-β (TGF-β) signaling via the Raf/Mitogen-activated protein kinase (MEK)/Extracellular signal-regulated kinases (ERK) pathways by decreasing E-Cadherin expression in HEK293 cells and increased phosphorylated Smad2 and Snail in both HEK293 and HeLa cells; the resulting effects being EMT to lethal EMT. We believe that lethal EMT exerts a pathologic role in phosphate toxicity-mediated tissue injuries and organ damages. One of the goals of this chapter is to explain how elevated Pi may induce different pathways to initiate EMT and related cellular pathologies.

8.2 EMT Overview

EMT is defined as the loss of epithelial cell polarity and adhesion, gain of invasive and migratory properties, and resulting phenotype conversion to mesenchymal cells. The expression of N-cadherin and vimentin (mesenchymal markers), and loss of E-cadherin (epithelial markers) are the hallmarks of EMT. Classification falls into 3 groups, which is based upon the biological process it plays a role in. Type 1 EMT includes embryogenesis and gastrulation, while type 2 EMT is associated with wound healing and fibrosis. Type 3 EMT is involved with cancer cells invasiveness and metastasis. Figure 8.1 demonstrates a few of the pathways that induce EMT, including TGF-β, Wnt/β-catenin, Notch, proto-oncogene c-Src (Src) and Ras-Mitogen-activated Protein Kinase (MAPK) [26]. These mechanisms then lead to the

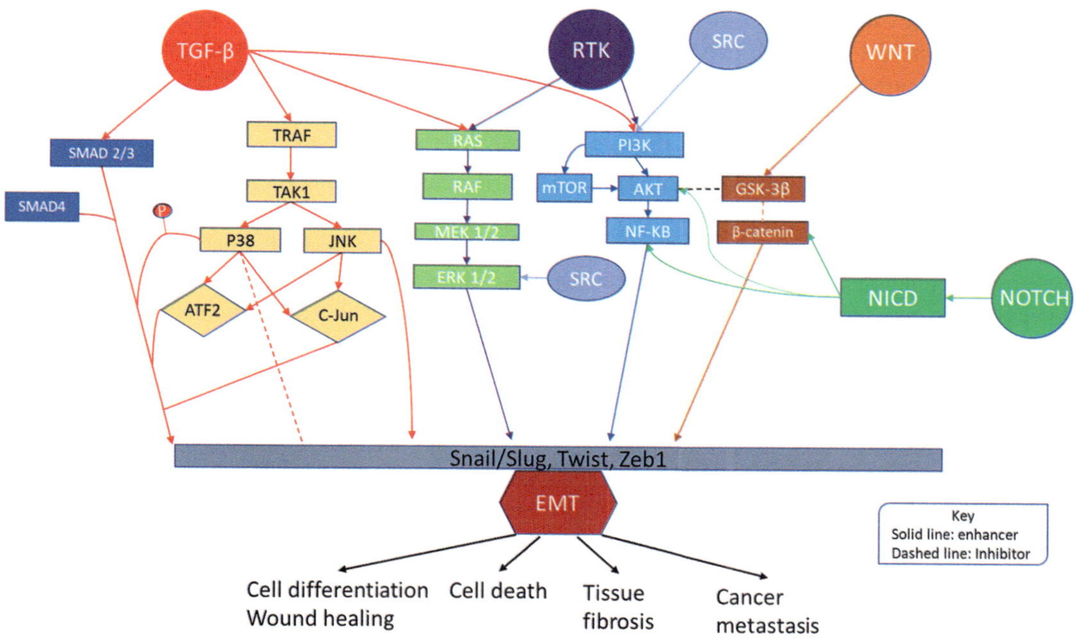

Fig. 8.1 Regulation of EMT

synthesis of transcription factors, including zinc finger proteins Snail 1 and 2 (Slug), basic helix-loop-helix factors Zinc finger homeodomain proteins 1 and 2 (ZEB1/2), and Twist 1, 2 and 3. While EMT has been indicated as a normal process occurring in embryonic differentiation of cells and wound healing, aberrant stimulation of EMT has been identified in metastasis of cancer, as well as tissue fibrosis. The considerable cross-talk that occurs between the different pathways remains less explained. Thus, we will firstly review the pathways involved, downstream effects and major regulators of these pathways.

8.2.1 TGFβ-Induced EMT

TGF-β plays roles in cell proliferation, differentiation and apoptosis [35]. TGF-β has been identified as one of the master regulators of the EMT process [45, 73]. The activation of the TGF-β pathway occurs through several different proteases and integrins, as well as situational activators such as reactive oxygen species or pH-dependent environments. One of the resulting downstream effects of TGF-β signaling is EMT. TGF-β-mediated EMT is either a Smad-dependent or a Smad-independent phenomenon [20, 26, 45].

8.2.1.1 Smad-Dependent EMT

Activation of TGF-β and subsequently TGF-β type I receptor (TBRI) results in Smad2 and Smad3 activation via the phosphorylation of serine residues on Smad. The activated Smad2/3 then forms a trimeric complex with Smad4, which is translocated to the nucleus and binds transcription factors [41, 85]. The transcription of Snail1/2, and Twist1 proteins is then upregulated, leading to the repression of epithelial markers, such as E-cadherin, claudins and occludin. Snail and Twist1 are also involved in the activation of mesenchymal gene expressions such as fibronectin, vitronectin and N-cadherin. Activation of ZEB proteins occurs through an alternative mechanism, resulting from the interaction with Smad-3. However, the downstream effects of ZEB1 and ZEB2 activation can repress epithelial markers and the expression of mesenchymal markers [83]. TGF-β activated Smad was also identified as an enhancer of the MAPK pathway [3, 52] (Fig. 8.1). Specifically, the Smad2/

3-smad4 complex induced by TGF-β upregulates transcription of LEF-1, which can form a complex with β-catenin, and allow the gene transcription that promotes EMT [52]. The significance of Smad-dependent, TGF-β induced EMT was demonstrated in the study performed by Thacker et al., in which cervical cancer cells were treated with curcumin and Emodin. The treatment caused the down regulation of TGF-β signaling via the decreased expression of TGF-β receptor II, Smad3 and Smad 4 and also counterbalanced the tumorigenic effects of TGF- β via the induction of G_2/M phase arrest of cell cycle [75].

8.2.1.2 Smad-Independent EMT

While Smad-dependent TGF-β signaling has been identified as a central mechanism in EMT, Smad-independent pathways have also been shown to play a role in the genesis of EMT. Studies have identified Smad-independent roles in TGF-β-induced EMT for MAP kinases and Phosphoinositide 3- kinase (PI3K)-Protein kinase B (AKT) pathway, as well as other signaling pathways [31, 41].

MAP kinases represent a family of proteins that play a role in different aspects of cell signaling, including gene expression, cell proliferation and cell cycle regulation. Within this family, TGF-β signals ERK, p38 MAPK and c-Jun N-terminal kinases (JNK). ERK 1/2 proteins have been identified as a potent activator of EMT. Stimulation of the TBRI leads to the activation of Raf-Ras, which then activates MEK and is followed by MEK activation of ERK. The effects of activated ERK leads to downregulation of E-cadherin, resulting in a gain of motility and invasiveness. Studies have shown that the ERK-mediated step is important in TGF-β induced EMT [86]. Another study looking at this relationship showed the inhibition of MEK-ERK signaling activities prevented the induction of TGF-β- mediated EMT [8]. Together, these findings highlight the importance of the MAPK-ERK pathways in the evolvement of EMT.

JNK and p38 are modulated by TGF-β signaling through a common mediator, TRAF 4/6. The interaction of TRAF with TGF-βR1 leads to TGF-β activated kinase 1 (TAK1) [79, 84]. The downstream effects of p38 and JNK activation signaling were associated with proliferation, differentiation, and apoptosis. Thus, multiple studies have focused on the role of activated p38 and JNK in EMT. The role of activated JNK is linked to increased cell invasion and motility of cells, though its role in apoptosis is not yet conclusive. Studies have proposed both pro-apoptotic and pro-survival roles of JNK [79], which can be determined by the duration of JNK activation. Persistent JNK stimulation led to apoptosis [13], while short-term JNK activation by growth factors promotes cell survival and proliferation [71]. Activated JNK signaling was also shown to result in the stabilization of Twist1 and enhancement of Smad-mediated effects [20]. The downstream effects of p38 MAPK have been identified in both downregulation of E-cadherin and induction of EMT. The phosphorylation of Smad proteins by p38 can cause downregulation of E-cadherin. Conversely, p38 repression of Snail1 and upregulation of Twist1, was also reported. The study claimed the role of p38 in maintaining the expression of E-cadherin [48]. TAK1 is a necessary step regarding the activity of the Smad pathway. Activated JNK and p38, downstream of effects of TAK1, are required for the activation of c-JUN and Activating transcription factor 2 (ATF2). The JNK-c-JUN was identified as a mediator of TGF-β-induced Smad transcription, while p38-ATF2 complex cooperates with Smads in EMT [24]. The interactions of TGF-β1, MAPK, and Smads promoted EMT in human malignant keratinocyte cell lines, with ERK and p38 enhancing Smad 2/3 transcription [18]. However, in a study on the role of Galectin-1 (Gal-1) in human ovarian cancer cells, it was shown that Gal-1 enhanced EMT through the activation of the MAPK JNK/p38 pathway [89]. These findings show cooperative and isolated signaling mechanisms involvling TGF-β, Smads, and MAPKs, and suggest that further studies are needed to identify the molecular mechanisms of EMT in various pathological states.

The role of *PI3K/AKT* has also been identified as another key regulator of EMT, but the specifics behind this pathway continue to be further investigated. The study showed PI3K-AKT was a necessary pathway for EMT [3]. Further studies looked at the specific role AKT plays in the signaling process. Long-term activation of AKT resulted in cells with enhanced motility and decreased adhesion, features typical to EMT. These features associated with activated Akt have been linked to the downstream effects of increased expression of Snail1 and repression of E-cadherin [30]. It is thought that TGF-β signaling activates the PI3K pathway by its own receptors or indirectly through Epidermal growth factor (EGF) and Platelet-derived growth factor (PDGF) receptors [26]. Qian et al., found that TGF-β1 stimulation of PI3K/AKT-induced phosphorylation of FOXO3a to be an underlying mechanism of Bleomycin-induced pulmonary fibrosis and potential therapeutic target [63].

8.2.2 Non-TGFβ-Induced EMT Pathways

While the role of TGF-β in EMT is well defined and considered a major part, studies have shown that TGF- β independent mechanisms can also lead to EMT.

The *Src family* of proteins (non-receptor tyrosine kinase) serve as a regulating point of different pathways. Specifically, an increase in Src kinase activity has been shown to promote EMT activity. The studies have highlighted such possibilities by demonstrating that Src regulates E-cadherin, and by increasing or suppressing Src activity resulted in a corresponding induction or suppression of EMT [55].

Wnt signaling is actively involved in tumorigenesis. With a number of those Wnt-mediated tumors displaying activated EMT factors. Wnt acts on GSK-3β, which acts as an inhibitor of β-catenin destruction complex. This inhibition prevents degradation of β-catenin, which accumulates and is translocated to the nucleus. β-catenin then acts as transcriptional inducer of EMT, increasing the expression of Snail/Slug, Twist1/2, and ZEB1/2 [39]. The combined treatment of curcumin and emodin showed effects on the Wnt/β-catenin in cervical cancer, HeLa cells; TGF-β activated the Wnt/β-catenin pathway in HeLa cells but combination therapy with curcumin and emodin downregulated Wnt signaling via the inhibition of β-catenin. [75].

Notch signaling has been identified to play a direct and/or indirect role in EMT signaling. Notch protein acts as a paracrine substance, regulating the actions of cells locally. The role of Notch involves the downstream effects of Notch Intracellular domain (NICD), which is released by Notch receptor interaction with Delta-like 1/3/4 (Dll) or Jagged 1/2 (JAG) interaction. NICD then binds C protein binding factor 1/Suppressor of Lag 1(CSL) and acts to regulate the expression of proteins such as Nuclear factor kappa-B (NF-kB), p21/27, AKT or the β-catenin destruction complex [20, 80]. The binding of JAG2 was also indicated as an indirect inducer of EMT by inhibiting miRNAs. Of note, Notch signaling also plays a role in TGF-β induced EMT. Snail-Notch interaction was identified as a β-catenin activator and E-cadherin repressor [80].

8.2.3 EMT and Diseases

The knowledge gained to this point has implicated the pathological role of EMT in different disease states. Studies conducted thus far have looked explicitly at the role EMT plays in cancer and fibrosis.

8.2.3.1 Cancer Metastasis

As mentioned, EMT has been found to have a significant role in cancer development and metastasis, which is predominantly driven by TGF-β induced EMT. During the late stages of tumor development, cancer cells become resistant to TGF- β induced cytostasis and TGF- β subsequently functions as a tumor promoter by inducing EMT [33]. Activation of EMT can result in decreased cellular adhesion and tight junctions, allowing for cancer cell dissemination, which is pivotal to early metastasis [33]. Mouse models have demonstrated that activation of EMT through the TGF- β signaling pathway with hyperactivity of the Raf/MAPK pathway results in the invasion of cancer cells to distant organs

[38]. The TGF- β pathway has been found to induce metastasis in various cancers, including breast, gastric, lung, and prostate [58]. Interestingly, new evidence also suggests that EMT may play a role in resistance against cancer therapy in a process driven by EMT transcription factors (EMT-TFs) [77]. *In vitro* and xenograft mouse studies have been able to identify increased expression of EMT-TFs on chemoresistant cells [77]. For example, a study performed by Zhu et al. was able to demonstrate a significant positive correlation between expression of TWIST1 (a helix-loophelix EMT-TF), and poor post-operative, and post-chemotherapeutic survival in colorectal cancer patients [88].

8.2.3.2 Fibrosis

One of the main functions of EMT includes wound healing, or type II EMT. The physiological role of EMT was indicated in the re-epithelialization of cutaneous wounds by kertinocytes and post-ovulation in ovarian surface epithelium [1, 2]. Tissue is further repaired through the generation of scar tissue, mostly via myofibroblasts. Studies to this point have identified EMT as one of the driving mechanisms behind myofibroblastic activities [72]. However, over activities of myofibroblasts results in pathologic tissue scarring or fibrosis through EMT and subsequent accumulation of collagen proteins. With that in mind, studies have further looked at the potential association of EMT and fibrosis of different organs. Studies have shown that TGF-β induced loss of epithelial morphology in epicardial cells [15]. EMT is also an important event in the development of pulmonary fibrosis, as the alterations of epithelial to mesenchymal phenotype in alveolar epithelial cells have been documented in the genesis of pulmonary fibrosis [43]. In a study looking at hepatic fibrosis, TGF-β-dependent activation of Smad 2/3 was indicated as the mechanism that mediated phenotype change in hepatocytes [40]. Numerous other studies have also highlighted the role of EMT in cardiac, pulmonary and hepatic fibrosis, as well as renal fibrosis [12, 29, 32, 49, 70]. A partial-EMT model was also identified as the driver of keratinocyte phenotype switch and resulting fibrotic skin lesions in scleroderma [59].

8.2.3.3 Therapeutics by Targeting EMT

Using the information known on EMT signaling pathways, studies have been performed to find potential therapeutic targets for EMT-associated changes. One such study by Zhou et al. looked at the effects of Thalidomide (THL) on EMT in alveolar cells during pulmonary fibrosis. Their results showed that THL decreased expression of the mesenchymal phenotype by inhibiting several signal transducers in TGF-β signaling pathways, which included p38, JNK, ERK, AKT, GSK3β and smads 2/3 [87]. The role of EMT in cancer was indicated to induce not only metastasis (as reviewed at Sect. 8.2.3.1), but also aid in escape of immune response through immunosuppression [44]. The role of Programmed death-1 (PD-1) on EMT cells in non-small cell lung cancer (NSCLC) was investigated in regards to immune evasion, showing the use of AKT pathway inhibitor (LY294002), ERK pathway inhibitor (PD98059) and TAK1 pathway inhibitor (5Z-7) could effectively inhibit the characteristic phenotype expressed in EMT cells [47]. These studies highlight the prospects of reducing EMT by therapeutic manipulation of various signaling pathways.

8.3 Lethal EMT

Studies have described the overexpression of Snail/slug, zinc finger E-box-binding homeobox (Zeb), and Twist1, as well as the decrease in E-cadherin and regulation of β-catenin as major factors in EMT. As shown by Padmanaban et al. E-cadherin loss has been shown to result in increased apoptosis of cells. Cells displaying loss of E-cadherin exhibited nuclear localization of Smad 2/3 [62]. This is consistent with other findings indicating the importance of both TGF-β and E-cadherin in EMT process. Their study further defined the relationship by showing E-cadherin loss triggers TGF-β signaling dependent ROS, that resulted in in apoptosis of the cells. Recent studies have shown that TGF-β signaling induced fibrosis and cell injury are partly regulated through lethal EMT [17, 73]. Studies have demonstrated the role of EMT in tumor suppression by remodeling the transcription factor landscape and converting SOX4 into a promoter of EMT-mediated cell death [17]. Another study found a potential role of RREB1 in TGF-β induced lethal EMT, by coordinated activation of RAS and TGF-β signaling cascade [73]. Studies have also shown KRAS mutation combined with TGF-β stimulation had a 30-fold increase in Snail1 and Zeb1 expression, decreased E-cadherin, and increased organoid dissociation and apoptosis.

8.4 High Pi-Mediated EMT

Our recent studies showed that abnormally high Pi could markedly enhance EMT by up-regulating the expression of Snail and Vimentin, and repression of E-Cadherin (Fig. 8.2) [34].

Pharmacologically (by the chemical inhibitor of Pi transporters) or genetically (by siRNA knockdown of Pi transporters) blocking cellular Pi transport resulted in prevention of high Pi-mediated EMT, indicating excess Pi's primary effect on the initiation and progression of EMT. Interestingly, elevated Pi-induced EMT was not predominantly mediated by TGF-β signaling but partially regulated by ERK1/2 signaling (Fig. 8.3). Elevated Pi also triggered massive cell death via activating extrinsic and intrinsic apoptotic pathways. Our cell-based studies have also shown that high Pi-mediated lethal EMT was partly mediated by reducing Snail mitigated Pi-triggered apoptosis. This finding is based on an acute high Pi treatment (24–48 h) model. In a chronic high-Pi stress model, where cells are exposed to moderately high levels (above 1–10 mM) of Pi for a longer duration (>3 days), high Pi-mediated EMT may have differential effects, such as increased cell mobility and fibrosis. This speculation is further echoed by high-Pi-related cancer metastasis and renal fibrosis in human and experiment studies.

It is believed that Pi-driven metastases could be the consequence of an adaptation of cancer cells to meet their phosphorus needs for rapid growth [19]. A higher concentration of interstitial Pi was detected in metastatic tumors compared to the non-invasive ones [6], suggesting a requirement of elevated Pi for the rapid cell growth, invasion and migration. Alpha-klotho functions in pair with FGF23 and reduces renal reabsorption of phosphate [76]. Elevated phosphate levels have been linked to decreased klotho expression in mouse genetic studies [56, 65]. Of relevance, klotho has been considered a tumor suppressor with universally depressed in breast cancer, pancreatic cancer, ovarian cancer, lung cancer, colorectal cancer, and melanoma [67]. In mice, secreted klotho protein administration could suppress renal fibrosis and cancer metastasis by inhibiting TGF-β, Wnt and IGF-1 signaling and delaying EMT process [22]. The causal links between Pi overload, hyperphosphatemia and the progression of chronic kidney disease (CKD) have been reported in human [60, 69, 78,

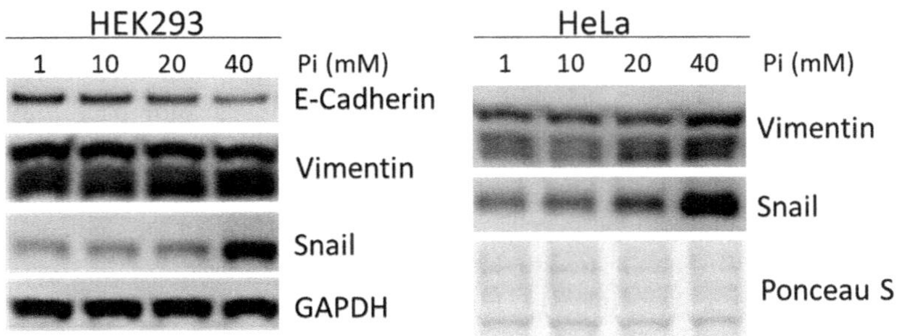

Fig. 8.2 High Pi (24 h treatment) enhanced EMT in HEK293 and HeLa cells by Westen blot analysis

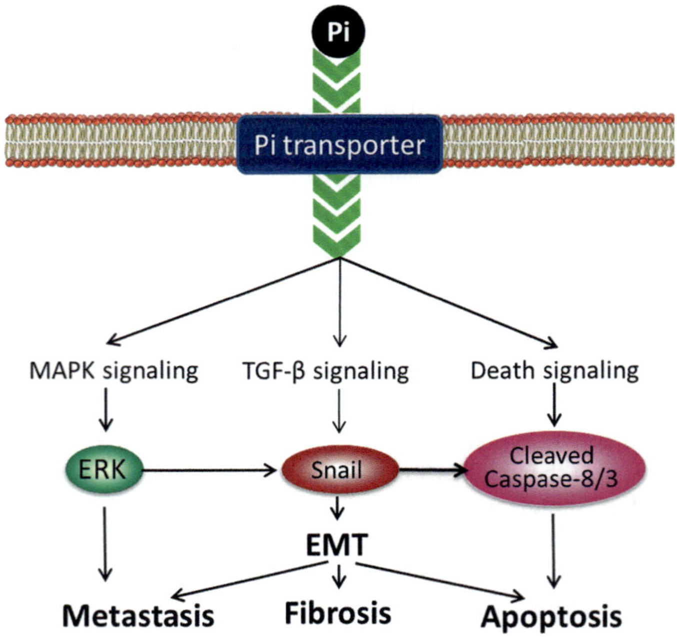

Fig. 8.3 High Pi-mediated EMT

90] and animal [37, 54, 57, 68] studies. Interstitial fibrosis is the main morphological alteration that eventually progresses to CKD. *In vitro*, elevated Pi could induce myofibroblast activation, promote cell proliferation and augment the synthesis of interstitial matrix protein, such as type I collagen [74] and fibronectin [14]. As mentioned, EMT is one of the causal events for fibrosis. Hence, high Pi-related renal fibrosis is partly mediated by the induction of EMT [90].

8.5 Open Questions and Future Directions

Excess Pi-mediated lethal EMT is dispensable of TGF-β signaling. Beyond the MAPK pathway, it will be clinically rewarding to discover other Pi-mediated EMT regulators by applying proteomic or phosphoproteomic analysis. Pi-mediated EMT may result in differential effects, such as lethal, metastatic, or fibrogenic. *In vitro* assays (such as invasion and migration assays) and animal models (genetically or dietary-induced high Pi) are desired to demonstrate these Pi-related phenotypes of EMT. Finally, it will be critical to determine the factors involved in Pi-mediated EMT's differential effects. High Pi-induced changes in the microenvironment may be one of the key driving factors.

8.6 Concluding Remarks

EMT plays essential roles in biological (organogenesis and wound healing) and pathological (tumorigenesis and fibrogenesis) cellar and tissue events. It can be regulated by canonical TGF-β signaling and noncanonical pathways (such as RTK, PI3K/AKT, Wnt, Notch signaling). Recently, TGF-β induced EMT has shown to have a higher potential of cell death, indicating a novel function of EMT as a lethality mediator. Dietary phosphate overload directly causes severe cytotoxicity (cell stress and apoptosis), which is partly linked to elevated Pi-mediated EMT. Further insights into the molecular details of Pi-medicated lethal EMT will assist us to understand the pathology of Pi toxicity, and likely to offer novel therapeutic targets for the clinical management of phosphate toxicity-induced tissue and organ damages.

Acknowledgments Eric Lewis and Faith Seltun are Osteopathic Medical Students at the Lake Erie College of Osteopathic Medicine, Erie (USA).

References

1. Ahmed N, Maines-Bandiera S, Quinn MA, Unger WG, Dedhar S, Auersperg N (2006) Molecular pathways regulating EGF-induced epithelio-mesenchymal transition in human ovarian surface epithelium. Am J Physiol Cell Physiol 290(6):C1532–C1542. https://doi.org/10.1152/ajpcell.00478.2005
2. Arnoux V, Nassour M, L'Helgoualc'h A, Hipskind RA, Savagner P (2008) Erk5 controls slug expression and keratinocyte activation during wound healing. Mol Biol Cell 19(11):4738–4749. https://doi.org/10.1091/mbc.e07-10-1078
3. Bakin AV, Rinehart C, Tomlinson AK, Arteaga CL (2002) p38 mitogen-activated protein kinase is required for TGFbeta-mediated fibroblastic transdifferentiation and cell migration. J Cell Sci 115(Pt 15): 3193–3206, Retrieved from https://www.ncbi.nlm.nih.gov/pubmed/12118074
4. Bakin AV, Tomlinson AK, Bhowmick NA, Moses HL, Arteaga CL (2000) Phosphatidylinositol 3-kinase function is required for transforming growth factor beta-mediated epithelial to mesenchymal transition and cell migration. J Biol Chem 275(47): 36803–36810. https://doi.org/10.1074/jbc.M005912200
5. Beck GR (2003) Inorganic phosphate as a signaling molecule in osteoblast differentiation. J Cell Biochem 90(2):234–243. https://doi.org/10.1002/jcb.10622
6. Bobko AA, Eubank TD, Driesschaert B, Dhimitruka I, Evans J, Mohammad R et al (2017) Interstitial inorganic phosphate as a tumor microenvironment marker for tumor progression. Sci Rep 7:41233. https://doi.org/10.1038/srep41233
7. Brown RB, Razzaque MS (2018) Phosphate toxicity and tumorigenesis. Biochim Biophys Acta Rev Cancer 1869(2):303–309. https://doi.org/10.1016/j.bbcan.2018.04.007
8. Buonato JM, Lazzara MJ (2014) ERK1/2 blockade prevents epithelial-mesenchymal transition in lung cancer cells and promotes their sensitivity to EGFR inhibition. Cancer Res 74(1):309–319. https://doi.org/10.1158/0008-5472.CAN-12-4721
9. Camalier CE, Yi M, Yu LR, Hood BL, Conrads KA, Lee YJ et al (2013) An integrated understanding of the physiological response to elevated extracellular phosphate. J Cell Physiol 228(7):1536–1550. https://doi.org/10.1002/jcp.24312
10. Cancela AL, Santos RD, Titan SM, Goldenstein PT, Rochitte CE, Lemos PA et al (2012) Phosphorus is associated with coronary artery disease in patients with preserved renal function. PLoS One 7(5):e36883. https://doi.org/10.1371/journal.pone.0036883
11. Chang SH, Yu KN, Lee YS, An GH, Beck GR, Colburn NH et al (2006) Elevated inorganic phosphate stimulates Akt-ERK1/2-Mnk1 signaling in human lung cells. Am J Respir Cell Mol Biol 35(5): 528–539. https://doi.org/10.1165/rcmb.2005-0477OC
12. Chapman HA (2011) Epithelial-mesenchymal interactions in pulmonary fibrosis. Annu Rev Physiol 73:413–435. https://doi.org/10.1146/annurev-physiol-012110-142225
13. Chen YR, Wang X, Templeton D, Davis RJ, Tan TH (1996) The role of c-Jun N-terminal kinase (JNK) in apoptosis induced by ultraviolet C and gamma radiation. Duration of JNK activation may determine cell death and proliferation. J Biol Chem 271(50): 31929–31936. https://doi.org/10.1074/jbc.271.50.31929
14. Chen Z, Chen D, McCarthy TL, Centrella M, Zhang Y, Moeckel GW (2012) Inorganic phosphate stimulates fibronectin expression in renal fibroblasts. Cell Physiol Biochem 30(1):151–159. https://doi.org/10.1159/000339054
15. Compton LA, Potash DA, Mundell NA, Barnett JV (2006) Transforming growth factor-beta induces loss of epithelial character and smooth muscle cell differentiation in epicardial cells. Dev Dyn 235(1):82–93. https://doi.org/10.1002/dvdy.20629
16. Conrads KA, Yi M, Simpson KA, Lucas DA, Camalier CE, Yu LR et al (2005) A combined proteome and microarray investigation of inorganic phosphate-induced pre-osteoblast cells. Mol Cell Proteomics 4(9):1284–1296. https://doi.org/10.1074/mcp.M500082-MCP200
17. David CJ, Huang YH, Chen M, Su J, Zou Y, Bardeesy N et al (2016) TGF-beta tumor suppression through a lethal EMT. Cell 164(5):1015–1030. https://doi.org/10.1016/j.cell.2016.01.009
18. Davies M, Robinson M, Smith E, Huntley S, Prime S, Paterson I (2005) Induction of an epithelial to mesenchymal transition in human immortal and malignant keratinocytes by TGF-beta1 involves MAPK, Smad and AP-1 signalling pathways. J Cell Biochem 95(5): 918–931. https://doi.org/10.1002/jcb.20458
19. de Carvalho CC, Caramujo MJ (2012) Tumour metastasis as an adaptation of tumour cells to fulfil their phosphorus requirements. Med Hypotheses 78(5):664–667. https://doi.org/10.1016/j.mehy.2012.02.006
20. Derynck R, Muthusamy BP, Saeteurn KY (2014) Signaling pathway cooperation in TGF-β-induced epithelial-mesenchymal transition. Curr Opin Cell Biol 31:56–66. https://doi.org/10.1016/j.ceb.2014.09.001
21. Di Marco GS, Hausberg M, Hillebrand U, Rustemeyer P, Wittkowski W, Lang D, Pavenstädt H (2008) Increased inorganic phosphate induces human endothelial cell apoptosis in vitro. Am J Physiol Renal Physiol 294(6):F1381–F1387. https://doi.org/10.1152/ajprenal.00003.2008
22. Doi S, Zou Y, Togao O, Pastor JV, John GB, Wang L et al (2011) Klotho inhibits transforming growth factor-beta1 (TGF-beta1) signaling and suppresses renal fibrosis and cancer metastasis in mice. J Biol Chem 286(10):8655–8665. https://doi.org/10.1074/jbc.M110.174037

23. Elser JJ, Kyle MM, Smith MS, Nagy JD (2007) Biological stoichiometry in human cancer. PLoS One 2(10):e1028. https://doi.org/10.1371/journal.pone.0001028
24. Gardner A, Fisher AJ, Richter C, Johnson GE, Moisey EJ, Brodlie M et al (2012) The critical role of TAK1 in accentuated epithelial to mesenchymal transition in obliterative bronchiolitis after lung transplantation. Am J Pathol 180(6):2293–2308. https://doi.org/10.1016/j.ajpath.2012.02.022
25. Giachelli CM (2009) The emerging role of phosphate in vascular calcification. Kidney Int 75(9):890–897. https://doi.org/10.1038/ki.2008.644
26. Gonzalez DM, Medici D (2014) Signaling mechanisms of the epithelial-mesenchymal transition. Sci Signal 7(344):re8. https://doi.org/10.1126/scisignal.2005189
27. Goodson JM, Shi P, Mumena CH, Haq A, Razzaque MS (2017) Dietary phosphorus burden increases cariogenesis independent of vitamin D uptake. J Steroid Biochem Mol Biol 167:33–38. https://doi.org/10.1016/j.jsbmb.2016.10.006
28. Goodson JM, Shi P, Razzaque MS (2019) Dietary phosphorus enhances inflammatory response: a study of human gingivitis. J Steroid Biochem Mol Biol 188: 166–171. https://doi.org/10.1016/j.jsbmb.2019.01.023
29. Grande MT, Sánchez-Laorden B, López-Blau C, De Frutos CA, Boutet A, Arévalo M et al (2015) Snail1-induced partial epithelial-to-mesenchymal transition drives renal fibrosis in mice and can be targeted to reverse established disease. Nat Med 21(9):989–997. https://doi.org/10.1038/nm.3901
30. Grille SJ, Bellacosa A, Upson J, Klein-Szanto AJ, van Roy F, Lee-Kwon W et al (2003) The protein kinase Akt induces epithelial mesenchymal transition and promotes enhanced motility and invasiveness of squamous cell carcinoma lines. Cancer Res 63(9): 2172–2178, Retrieved from https://www.ncbi.nlm.nih.gov/pubmed/12727836
31. Gui T, Sun Y, Shimokado A, Muragaki Y (2012) The roles of mitogen-activated protein kinase pathways in TGF-β-induced epithelial-mesenchymal transition. J Signal Transduct 2012:289243. https://doi.org/10.1155/2012/289243
32. Guo F, Yi X, Li M, Fu J, Li S (2017) Snail1 is positively correlated with atrial fibrosis in patients with atrial fibrillation and rheumatic heart disease. Exp Ther Med 14(5):4231–4237. https://doi.org/10.3892/etm.2017.5084
33. Hao Y, Baker D, Ten Dijke P (2019) TGF-β-mediated epithelial-mesenchymal transition and cancer metastasis. Int J Mol Sci 20(11). https://doi.org/10.3390/ijms20112767
34. He P, Mann-Collura O, Fling J, Edara N, Hetz R, Razzaque MS (2020) High phosphate actively induces cytotoxicity by rewiring pro-survival and pro-apoptotic signaling networks in HEK293 and HeLa cells. FASEB J. https://doi.org/10.1096/fj.202000799RR
35. Heldin CH, Landström M, Moustakas A (2009) Mechanism of TGF-beta signaling to growth arrest, apoptosis, and epithelial-mesenchymal transition. Curr Opin Cell Biol 21(2):166–176. https://doi.org/10.1016/j.ceb.2009.01.021
36. Hetz R, Beeler E, Janoczkin A, Kiers S, Li L, Willard BB, Razzaque MS, He P (2021) Excessive inorganic phosphate burden perturbed intracellular signaling: quantitative proteomics and phosphoproteomics analyses. Front. Nutr. 8:765391. https://doi.org/10.3389/fnut.2021.765391.
37. Ibels LS, Alfrey AC, Haut L, Huffer WE (1978) Preservation of function in experimental renal disease by dietary restriction of phosphate. N Engl J Med 298(3): 122–126. https://doi.org/10.1056/NEJM197801192980302
38. Janda, E., Lehmann, K., Killisch, I., Jechlinger, M., Herzig, M., Downward, J., . . . Grünert, S. (2002). Ras and TGF[beta] cooperatively regulate epithelial cell plasticity and metastasis: dissection of Ras signaling pathways. J Cell Biol, 156(2), 299–313. doi:https://doi.org/10.1083/jcb.200109037
39. Jiang YG, Luo Y, He DL, Li X, Zhang LL, Peng T et al (2007) Role of Wnt/beta-catenin signaling pathway in epithelial-mesenchymal transition of human prostate cancer induced by hypoxia-inducible factor-1alpha. Int J Urol 14(11):1034–1039. https://doi.org/10.1111/j.1442-2042.2007.01866.x
40. Kaimori A, Potter J, Kaimori JY, Wang C, Mezey E, Koteish A (2007) Transforming growth factor-beta1 induces an epithelial-to-mesenchymal transition state in mouse hepatocytes in vitro. J Biol Chem 282(30): 22089–22101. https://doi.org/10.1074/jbc.M700998200
41. Kamato D, Burch ML, Piva TJ, Rezaei HB, Rostam MA, Xu S et al (2013) Transforming growth factor-β signalling: role and consequences of Smad linker region phosphorylation. Cell Signal 25(10): 2017–2024. https://doi.org/10.1016/j.cellsig.2013.06.001
42. Kanatani M, Sugimoto T, Kano J, Kanzawa M, Chihara K (2003) Effect of high phosphate concentration on osteoclast differentiation as well as bone-resorbing activity. J Cell Physiol 196(1):180–189. https://doi.org/10.1002/jcp.10270
43. Kim KK, Kugler MC, Wolters PJ, Robillard L, Galvez MG, Brumwell AN et al (2006) Alveolar epithelial cell mesenchymal transition develops in vivo during pulmonary fibrosis and is regulated by the extracellular matrix. Proc Natl Acad Sci U S A 103(35): 13180–13185. https://doi.org/10.1073/pnas.0605669103
44. Kudo-Saito C, Shirako H, Takeuchi T, Kawakami Y (2009) Cancer metastasis is accelerated through immunosuppression during snail-induced EMT of cancer cells. Cancer Cell 15(3):195–206. https://doi.org/10.1016/j.ccr.2009.01.023
45. Lamouille S, Xu J, Derynck R (2014) Molecular mechanisms of epithelial-mesenchymal transition.

Nat Rev Mol Cell Biol 15(3):178–196. https://doi.org/10.1038/nrm3758
46. Levan K, Mehryar M, Mateoiu C, Albertsson P, Bäck T, Sundfeldt K (2017) Immunohistochemical evaluation of epithelial ovarian carcinomas identifies three different expression patterns of the MX35 antigen, NaPi2b. BMC Cancer 17(1):303. https://doi.org/10.1186/s12885-017-3289-2
47. Li F, Zhu T, Yue Y, Zhu X, Wang J, Liang L (2018) Preliminary mechanisms of regulating PD-L1 expression in non-small cell lung cancer during the EMT process. Oncol Rep 40(2):775–782. https://doi.org/10.3892/or.2018.6474
48. Lin S, Yu L, Ni Y, He L, Weng X, Lu X, Zhang C (2020) Fibroblast growth factor 21 attenuates diabetes-induced renal fibrosis by negatively regulating TGF-β-p53-Smad2/3-mediated epithelial-to-mesenchymal transition via activation of AKT. Diabetes Metab J 44(1):158–172. https://doi.org/10.4093/dmj.2018.0235
49. Lin Y, Mallen-St Clair J, Wang G, Luo J, Palma-Diaz F, Lai C et al (2016) p38 MAPK mediates epithelial-mesenchymal transition by regulating p38IP and snail in head and neck squamous cell carcinoma. Oral Oncol 60:81–89. https://doi.org/10.1016/j.oraloncology.2016.06.010
50. Mancini FR, Affret A, Dow C, Balkau B, Clavel-Chapelon F, Bonnet F et al (2018) High dietary phosphorus intake is associated with an increased risk of type 2 diabetes in the large prospective E3N cohort study. Clin Nutr 37(5):1625–1630. https://doi.org/10.1016/j.clnu.2017.07.025
51. Marks J, Debnam ES, Unwin RJ (2013) The role of the gastrointestinal tract in phosphate homeostasis in health and chronic kidney disease. Curr Opin Nephrol Hypertens 22(4):481–487. https://doi.org/10.1097/MNH.0b013e3283621310
52. Medici D, Hay ED, Goodenough DA (2006) Cooperation between snail and LEF-1 transcription factors is essential for TGF-beta1-induced epithelial-mesenchymal transition. Mol Biol Cell 17(4):1871–1879. https://doi.org/10.1091/mbc.e05-08-0767
53. Michigami T, Kawai M, Yamazaki M, Ozono K (2018) Phosphate as a signaling molecule and its sensing mechanism. Physiol Rev 98(4):2317–2348. https://doi.org/10.1152/physrev.00022.2017
54. Nagano N, Miyata S, Obana S, Kobayashi N, Fukushima N, Burke SK, Wada M (2003) Sevelamer hydrochloride, a phosphate binder, protects against deterioration of renal function in rats with progressive chronic renal insufficiency. Nephrol Dial Transplant 18(10):2014–2023. https://doi.org/10.1093/ndt/gfg309
55. Nagathihalli NS, Merchant NB (2012) Src-mediated regulation of E-cadherin and EMT in pancreatic cancer. Front Biosci (Landmark Ed) 17:2059–2069. https://doi.org/10.2741/4037
56. Nakatani T, Sarraj B, Ohnishi M, Densmore MJ, Taguchi T, Goetz R et al (2009) In vivo genetic evidence for klotho-dependent, fibroblast growth factor 23 (Fgf23) -mediated regulation of systemic phosphate homeostasis. FASEB J 23(2):433–441. https://doi.org/10.1096/fj.08-114397
57. Neves KR, Graciolli FG, dos Reis LM, Pasqualucci CA, Moyses RM, Jorgetti V (2004) Adverse effects of hyperphosphatemia on myocardial hypertrophy, renal function, and bone in rats with renal failure. Kidney Int 66(6):2237–2244. https://doi.org/10.1111/j.1523-1755.2004.66013.x
58. Nguyen DX, Bos PD, Massagué J (2009) Metastasis: from dissemination to organ-specific colonization. Nat Rev Cancer 9(4):274–284. https://doi.org/10.1038/nrc2622
59. Nikitorowicz-Buniak J, Denton CP, Abraham D, Stratton R (2015) Partially evoked epithelial-mesenchymal transition (EMT) is associated with increased TGFβ signaling within Lesional scleroderma skin. PLoS One 10(7):e0134092. https://doi.org/10.1371/journal.pone.0134092
60. O'Seaghdha CM, Hwang SJ, Muntner P, Melamed ML, Fox CS (2011) Serum phosphorus predicts incident chronic kidney disease and end-stage renal disease. Nephrol Dial Transplant 26(9):2885–2890. https://doi.org/10.1093/ndt/gfq808
61. Ohnishi M, Razzaque MS (2010) Dietary and genetic evidence for phosphate toxicity accelerating mammalian aging. FASEB J 24(9):3562–3571. https://doi.org/10.1096/fj.09-152488
62. Padmanaban V, Krol I, Suhail Y, Szczerba BM, Aceto N, Bader JS, Ewald AJ (2019) E-cadherin is required for metastasis in multiple models of breast cancer. Nature 573(7774):439–444. https://doi.org/10.1038/s41586-019-1526-3
63. Qian W, Cai X, Qian Q, Zhang W, Wang D (2018) Astragaloside IV modulates TGF-β1-dependent epithelial-mesenchymal transition in bleomycin-induced pulmonary fibrosis. J Cell Mol Med 22(9):4354–4365. https://doi.org/10.1111/jcmm.13725
64. Razzaque MS (2011) Phosphate toxicity: new insights into an old problem. Clin Sci (Lond) 120(3):91–97. https://doi.org/10.1042/CS20100377
65. Razzaque MS (2012) The role of Klotho in energy metabolism. Nat Rev Endocrinol 8(10):579–587. https://doi.org/10.1038/nrendo.2012.75
66. Roussanne MC, Lieberherr M, Souberbielle JC, Sarfati E, Drüeke T, Bourdeau A (2001) Human parathyroid cell proliferation in response to calcium, NPS R-467, calcitriol and phosphate. Eur J Clin Investig 31(7):610–616. https://doi.org/10.1046/j.1365-2362.2001.00809.x
67. Rubinek T, Wolf I (2016) The role of alpha-Klotho as a universal tumor suppressor. Vitam Horm 101:197–214. https://doi.org/10.1016/bs.vh.2016.03.001
68. Sakaguchi Y, Hamano T, Matsui I, Oka T, Yamaguchi S, Kubota K et al (2019) Low magnesium diet aggravates phosphate-induced kidney injury. Nephrol Dial Transplant 34(8):1310–1319. https://doi.org/10.1093/ndt/gfy358

69. Schwarz S, Trivedi BK, Kalantar-Zadeh K, Kovesdy CP (2006) Association of disorders in mineral metabolism with progression of chronic kidney disease. Clin J Am Soc Nephrol 1(4):825–831. https://doi.org/10.2215/CJN.02101205
70. Sicklick JK, Choi SS, Bustamante M, McCall SJ, Pérez EH, Huang J et al (2006) Evidence for epithelial-mesenchymal transitions in adult liver cells. Am J Physiol Gastrointest Liver Physiol 291(4): G575–G583. https://doi.org/10.1152/ajpgi.00102.2006
71. Sluss HK, Barrett T, Derijard B, Davis RJ (1994) Signal transduction by tumor necrosis factor mediated by JNK protein kinases. Mol Cell Biol 14(12): 8376–8384. https://doi.org/10.1128/mcb.14.12.8376
72. Stone RC, Pastar I, Ojeh N, Chen V, Liu S, Garzon KI, Tomic-Canic M (2016) Epithelial-mesenchymal transition in tissue repair and fibrosis. Cell Tissue Res 365(3):495–506. https://doi.org/10.1007/s00441-016-2464-0
73. Su J, Morgani SM, David CJ, Wang Q, Er EE, Huang YH et al (2020) TGF-β orchestrates fibrogenic and developmental EMTs via the RAS effector RREB1. Nature 577(7791):566–571. https://doi.org/10.1038/s41586-019-1897-5
74. Tan X, Xu X, Zeisberg EM, Zeisberg M (2016) High inorganic phosphate causes DNMT1 phosphorylation and subsequent fibrotic fibroblast activation. Biochem Biophys Res Commun 472(3):459–464. https://doi.org/10.1016/j.bbrc.2016.01.077
75. Thacker PC, Karunagaran D (2015) Curcumin and emodin down-regulate TGF-β signaling pathway in human cervical cancer cells. PLoS One 10(3): e0120045. https://doi.org/10.1371/journal.pone.0120045
76. Urakawa I, Yamazaki Y, Shimada T, Iijima K, Hasegawa H, Okawa K et al (2006) Klotho converts canonical FGF receptor into a specific receptor for FGF23. Nature 444(7120):770–774. https://doi.org/10.1038/nature05315
77. van Staalduinen J, Baker D, Ten Dijke P, van Dam H (2018) Epithelial-mesenchymal-transition-inducing transcription factors: new targets for tackling chemoresistance in cancer? Oncogene 37(48): 6195–6211. https://doi.org/10.1038/s41388-018-0378-x
78. Voormolen N, Noordzij M, Grootendorst DC, Beetz I, Sijpkens YW, van Manen JG et al (2007) High plasma phosphate as a risk factor for decline in renal function and mortality in pre-dialysis patients. Nephrol Dial Transplant 22(10):2909–2916. https://doi.org/10.1093/ndt/gfm286
79. Wang J, Kuiatse I, Lee AV, Pan J, Giuliano A, Cui X (2010a) Sustained c-Jun-NH2-kinase activity promotes epithelial-mesenchymal transition, invasion, and survival of breast cancer cells by regulating extracellular signal-regulated kinase activation. Mol Cancer Res 8(2):266–277. https://doi.org/10.1158/1541-7786.MCR-09-0221
80. Wang Z, Li Y, Kong D, Sarkar FH (2010b) The role of Notch signaling pathway in epithelial-mesenchymal transition (EMT) during development and tumor aggressiveness. Curr Drug Targets 11(6):745–751. https://doi.org/10.2174/138945010791170860
81. Wilson KM, Shui IM, Mucci LA, Giovannucci E (2015) Calcium and phosphorus intake and prostate cancer risk: a 24-y follow-up study. Am J Clin Nutr 101(1):173–183. https://doi.org/10.3945/ajcn.114.088716
82. Wulaningsih W, Michaelsson K, Garmo H, Hammar N, Jungner I, Walldius G et al (2013) Inorganic phosphate and the risk of cancer in the Swedish AMORIS study. BMC Cancer 13:257. https://doi.org/10.1186/1471-2407-13-257
83. Xu J, Lamouille S, Derynck R (2009) TGF-beta-induced epithelial to mesenchymal transition. Cell Res 19(2):156–172. https://doi.org/10.1038/cr.2009.5
84. Yamashita M, Fatyol K, Jin C, Wang X, Liu Z, Zhang YE (2008) TRAF6 mediates Smad-independent activation of JNK and p38 by TGF-beta. Mol Cell 31(6): 918–924. https://doi.org/10.1016/j.molcel.2008.09.002
85. Yoshida K, Murata M, Yamaguchi T, Matsuzaki K, Okazaki K (2016) Reversible human TGF-β signal shifting between tumor suppression and fibro-carcinogenesis: implications of Smad Phospho-isoforms for hepatic epithelial-mesenchymal transitions. J Clin Med 5(1). https://doi.org/10.3390/jcm5010007
86. Zavadil J, Bitzer M, Liang D, Yang YC, Massimi A, Kneitz S et al (2001) Genetic programs of epithelial cell plasticity directed by transforming growth factor-beta. Proc Natl Acad Sci U S A 98(12):6686–6691. https://doi.org/10.1073/pnas.111614398
87. Zhou XL, Xu P, Chen HH, Zhao Y, Shen J, Jiang C et al (2017) Thalidomide inhibits TGF-β1-induced epithelial to mesenchymal transition in alveolar epithelial cells via Smad-dependent and Smad-independent signaling pathways. Sci Rep 7(1):14727. https://doi.org/10.1038/s41598-017-15239-2
88. Zhu DJ, Chen XW, Zhang WJ, Wang JZ, Ouyang MZ, Zhong Q, Liu CC (2015) Twist1 is a potential prognostic marker for colorectal cancer and associated with chemoresistance. Am J Cancer Res 5(6):2000–2011. Retrieved from https://www.ncbi.nlm.nih.gov/pubmed/26269759
89. Zhu J, Zheng Y, Zhang H, Liu Y, Sun H, Zhang P (2019) Galectin-1 induces metastasis and epithelial-mesenchymal transition (EMT) in human ovarian cancer cells via activation of the MAPK JNK/p38 signalling pathway. Am J Transl Res 11(6):3862–3878, Retrieved from https://www.ncbi.nlm.nih.gov/pubmed/31312395
90. Zoccali C, Ruggenenti P, Perna A, Leonardis D, Tripepi R, Tripepi G et al (2011) Phosphate may promote CKD progression and attenuate renoprotective effect of ACE inhibition. J Am Soc Nephrol 22(10):1923–1930. https://doi.org/10.1681/ASN.2011020175

Phosphate and Endothelial Function: How Sensing of Elevated Inorganic Phosphate Concentration Generates Signals in Endothelial Cells

9

Nima Abbasian, Alan Bevington, and Dylan Burger

Abstract

Present in all cells, inorganic phosphate (Pi) is involved in regulating a wide range of fundamental cellular processes including energy homeostasis; nucleotide, nucleic acid and phospholipid metabolism; and signalling through protein phosphorylation events. However, at excess concentrations, Pi is known to exert adverse effects on cells, particularly on endothelial cells. This review gives a brief overview of the functional effects of elevated extracellular Pi concentration on mammalian cells and tissues *in vitro* and *in vivo*. We then address the cardiovascular effects of elevated extracellular Pi concentration *in vitro* and *in vivo*, emphasising that effects have been reported *in vivo* even within the top end of normal range for plasma [Pi]. Cardiovascular sites of action of Pi are then considered, with a focus on the role of soluble Pi in endothelial dysfunction. The regulation of intracellular Pi concentration by Pi transporter proteins in mammalian cells is described, followed by consideration in detail of how changes in Pi concentration are sensed in mammalian cells and how these trigger functional effects in endothelial cells.

Keywords

Phosphate · Microvesicles · Angiogenesis · Endothelial-mesenchymal transition · Signaling

N. Abbasian (✉)
Department of Respiratory Sciences, University of Leicester, Leicester, UK

School of Life and Medical Sciences, University of Hertfordshire, Hertfordshire, UK
e-mail: abbasian.n.174@gmail.com

A. Bevington
Department of Respiratory Sciences, University of Leicester, Leicester, UK
e-mail: 57rrd1955@gmail.com

D. Burger
Kidney Research Centre, The Ottawa Hospital Research Institute, Department of Cellular and Molecular Medicine, University of Ottawa, Ottawa, ON, Canada
e-mail: dburger@uottawa.ca

9.1 Introduction

Comprising about 1% of an individual's body weight, phosphorus is one of the most abundant minerals in the human body. Predominantly found in conjunction with calcium (hydroxyapatite), or in cells as organic phosphates in carbohydrates, proteins, nucleic acids and lipids, only a very small part (~0.1%) of phosphorus present in extracellular fluids occurs in the form of inorganic phosphate (Pi). Plasma Pi (comprising dihydrogen orthophosphate $H_2PO_4^-$ and monohydrogen orthophosphate HPO_4^{2-} in equilibrium at a normal physiological pH) ranges

M. S. Razzaque (ed.), *Phosphate Metabolism*, Advances in Experimental Medicine and Biology 1362,
https://doi.org/10.1007/978-3-030-91623-7_9

between 0.8 and 1.5 mM, but this range can be exceeded transiently following ingestion of a phosphate enriched diet [1], or chronically under certain pathological conditions such as chronic kidney disease (CKD) [2]. It should also be noted that other pathophysiological disorders, such as hypoxia/ischemia [3] and chemotherapy in cancer patients resulting in tumor lysis syndrome [4, 5], may lead to a large-scale Pi generation from cytosolic organophosphorus metabolites, such as ATP and phosphocreatine, resulting in localised and systemic hyperphosphatemia. Even though Pi plays a pivotal role in all cells, in processes as disparate as energy homeostasis; nucleotide, nucleic acid and phospholipid biosynthesis and signalling through protein phosphorylation events; excess concentrations of extracellular Pi are known to exert pathological effects on cells including cells of the vasculature (both the endothelial cells [1, 6–8] and the underlying smooth muscle cells [9–11]) and can also disturb bone and mineral metabolism in humans [12].

Commonly in CKD patients, hyperphosphatemia drives soft-tissue calcification [9, 10, 13, 14], through insoluble calcium phosphate deposition, osteogenic transformation in vascular smooth muscle [9, 10, 13, 15] and increased vascular stiffness [16]. This chapter will focus principally on the endothelial effects of elevated *soluble* extracellular Pi in the concentration range ~1 mM to ~2.5 mM (i.e. from normal human plasma Pi concentration to the elevated concentrations that transiently follow ingestion of a dietary Pi load or that may occur chronically in disorders such as CKD). Interesting and important effects also occur outside this range, but will not be discussed here as they may be qualitatively different – for example because of ATP depletion as a consequence of severe Pi depletion (<<1 mM Pi) or because of the substantial complexing of ionised Ca^{2+} by Pi ions that occurs at very high Pi concentrations >>2.5 mM [17].

A further reason for focussing here on the extracellular Pi concentration range 1–2.5 mM is that, in recent years, serum or plasma Pi concentrations even within the top end of normal range have been shown to be associated with poor cardiovascular outcomes [18–23]. In a population-based study consisting of 8953 participants with normal kidney function, high serum Pi levels greater than 1.3 mM were shown to be associated with elevated low-grade albuminuria [19] which *per se* increases the risk of CV events and mortality [24]. In another population-based cohort study consisting of 13,340 subjects, higher serum Pi levels were shown to be associated with increased CV risk (i.e. carotid intima-media thickness (cIMT)) independent of participants' eGFR, hypercholesterolemia, diabetes, age, sex, and hypertension) [21]. Additionally, in a study on patients with coronary disease but no record of overt hyperphosphataemia, it has been shown that higher Pi levels well within the normal range are associated with greater all-cause mortality and adverse cardiovascular outcomes [22]. Foley, el al., 2009, demonstrated that there is a link between higher serum Pi levels and increased coronary atherosclerosis in a large population of young adults with no concomitant kidney disease and overt hyperphosphataemia [23]. Notably, there may be sex differences in the relationship between serum Pi and cardiovascular mortality with clear associations in men but less consistent associations in women [25, 26].

Taken together, these data imply that high Pi levels within the normal range (e.g. >1.3 < 1.5 mM) may nevertheless accelerate the onset and development of CV morbidity in the general population with or without kidney disease. Indeed in 2012, Ellam and colleagues reviewed Pi as a "new cholesterol" meaning that (by analogy with LDL cholesterol), intervention to manage Pi even at levels not regarded as particularly high, may benefit the general population and prevent development of atherosclerotic vascular mortalities [18]. It has therefore been suggested that modern diets containing large amounts of canned products and prepared food which are rich in Pi and/or preservatives containing Pi, may be a suitable target for intervention to manage serum Pi levels with possible vascular benefits even in the general population [18].

Harmful effects of Pi *in vivo* have been confirmed by specific manipulation of phosphate status in experimental animals. Hyperphosphatemic Klotho knockout mice demonstrated features of premature aging and a reduced life span. Klotho and NaPi-IIa double knockout mice (in which deletion of NaPi-IIa transporters of the SLC34 gene family favoured Pi excretion), reversed these pathological features. However, feeding these double-knockout animals with high dietary Pi supplements restored premature aging features, indicating that premature aging in these animals is predominantly a manifestation of Pi toxicity [27]. Furthermore, Yamada, et al. demonstrated that on feeding adenine-induced CKD rats with high (1.2%) dietary Pi for two months (and manifesting hyperphosphatemia; Table 2 in [14]), serum and tissue levels of TNF-α were significantly increased. In this study the apparent Pi toxicity involved premature aging phenotypes, vascular calcification, malnutrition, and mortality without any effect on kidney function [14]. All of the observed Pi-induced changes were blunted after feeding CKD rats on the 1.2% phosphorus diet with 6% lanthanum carbonate as a Pi binder [14].

These emerging effects on CV risks of higher serum Pi levels within the normal range might be attributable to indirect cardiovascular effects of phosphate-responsive hormones (i.e. phosphatonins) rather than to direct action by Pi on the cells of the cardiovascular system. Such indirect endocrine effects of Pi may include (but are not restricted to) inhibition of 1,25-dihydroxyvitamin D synthesis [28] and increased secretion of PTH [29] and FGF23 both of which are considered as predictors of cardiovascular mortality in populations with or without kidney disease [18]. Nevertheless, there is also mounting evidence of direct toxic effects of excess soluble Pi on mammalian cells. These might include impaired cell signalling, increased cell death, impaired fertility, renal fibrosis, osteoblastic transformation of smooth muscle cells, premature aging, angiogenesis, carcinogenesis, tumour progression, endothelial–mesenchymal transition, generation of procoagulant endothelial microvesicles (MVs) and enhanced systemic inflammation and malnutrition [6, 7, 10, 11, 16, 27, 30–35]. The implications of such effects for a direct role of Pi in endothelial dysfunction are therefore discussed in the next section. (The more specific question of how a relatively modest increase in extracellular Pi concentration could generate a biologically significant intracellular signal in endothelial cells is discussed later in Sect. 9.5).

9.2 Pi and Endothelial Function – What Pi Does to Endothelium *In Vitro* and *In Vivo*

9.2.1 Effects on Vasodilation

Important early evidence of direct endothelial effects of Pi came from work by Shuto et al. who showed that ingestion of a high phosphate diet impaired flow-mediated dilation (FMD) of the brachial artery at 2 h after a meal [1]. Nishi et al. also reported that excessive dietary phosphate intake can acutely (within 1–4 h) impair endothelial function in healthy people as determined by decreased FMD [36]. Levac et al. showed that an acute oral phosphate load resulted in impaired FMD even in the absence of measurable changes to plasma phosphate levels [37]. A possible mechanism for this effect was proposed by Di Marco et al. who demonstrated that in cultured human coronary artery endothelial cells, exposure to high phosphate media directly results in stiffened endothelial cells which could, in turn, reduce mechanosensing and nitric oxide generation [16]. Consistent with this, studies on cultured human umbilical vein endothelial cells (HUVECs) show inhibition of endothelial nitric oxide synthase (eNOS) expression/activity in response to Pi [8]. In addition to effects on endothelium-dependent vasorelaxation, Six et al. have also shown that a high phosphate diet increases phenylephrine-induced vasoconstriction in mice [38]. Thus effects of Pi on vascular function likely involve both endothelial cells and vascular smooth muscle.

9.2.2 Oxidative Stress and Cell Survival

Oxidative stress is a condition whereby the production of reactive oxygen species (ROS) exceeds antioxidant defenses [39]. A further potentially important mechanism by which a Pi-load could induce dysfunction in endothelial cells is by triggering oxidative stress, resulting in apoptosis. It has been reported that an extracellular Pi-load induces apoptosis in HUVECs [8] and triggers mitochondrial oxidative stress and subsequently cellular apoptosis in the human endothelial cell line EA.hy926 *in vitro* [7]. However, work elsewhere using the same cell line failed to detect these effects [6]. It has also been reported that a high extracellular Pi concentration induces autophagy in endothelial cells via the inhibition of Akt/mTOR signalling, a process that the authors suggested may serve as a protective mechanism that shields endothelial cells from high Pi-induced apoptosis [40]. A further distinct response was reported by Olmos et al. who demonstrated that a high extracellular Pi concentration induces oxidative stress and cellular senescence in endothelial cells via upregulation of endothelin-1 (ET-1) [41]. Thus several Pi-mediated effects have been described in endothelial cells. The reason for discrepant, sometimes conflicting results from different laboratories is currently unknown. While some may arise from differences in the origin of the endothelial cells studied, this does not explain the difference reported between [7] and [6], possibly indicating technical differences. Technical variables that may merit future investigation include the precise timing of the addition of the Pi load to the culture medium, which could affect precipitation of biologically active factors such as calcium phosphate nanocrystals [42]; and the presence or absence of trace impurities in the added Pi load, for example arsenate and arsenite which are structurally and chemically similar to Pi and may promote oxidative stress in endothelial cells [43]. These suggest that there is a need in future studies to take such variables into consideration to avoid results variation between laboratories.

9.2.3 Microvesicles

Membrane-derived microvesicles (MVs) are sub-micron (<1 μm diameter) vesicles which are shed from plasma membrane in response to apoptosis or cellular activation (notably in platelets, endothelial cells, and leukocytes) [6, 7] (Reviewed in [44]). It has been demonstrated *in vitro* that an extracellular Pi-load consistently leads to liberation of endothelial MVs in human coronary artery endothelial cells [16], in human EA.hy926 cells [6] and in HUVECs (Supplemental data in [6]). This has also recently been demonstrated in response to hyperphosphatemia in the rat partial nephrectomy model of CKD *in vivo* [35]. It is worth noting that, once released from endothelial cells, MVs may exert effects elsewhere in the vasculature as they are potently procoagulant [6], and carry miRNA cargo [45], capable of delivering miRNAs to other cells, including feedback effects on the endothelium [46–48]. This Pi-induced release of MVs may partly explain the high concentration of pro-coagulant endothelial MVs reported in circulation in hyperphosphatemic CKD patients *in vivo* [49]. Similarly this has shown that MVs induce reactive oxygen species production and cell cycle arrest in cultured endothelial cells and this may contribute to Pi-induced effects on oxidative stress and senescence [50]. Shedding of MVs in response to Pi may also be sufficient to deplete the cell-surface expression of biologically important endothelial proteins. In this regard, Di Marco et al. showed in human coronary artery endothelial cells [16] that a higher Pi milieu induces a decrease in cellular Annexin II protein level which these authors attributed to shedding of

Annexin II in endothelial MVs rather than reduced synthesis or increased degradation of the Annexin II protein within the cells.

9.2.4 Angiogenesis

New blood vessel development (angiogenesis) is a complex and highly regulated process which involves the proliferation, migration, and re-modelling of endothelial cells from pre-existing blood vessels. Pro-angiogenic genes such as forkhead box protein C2 (FOXC2), osteopontin, and VEGF-α have been shown to be influenced by an elevated Pi concentration [9, 10, 13, 14]. However, the precise effect of Pi is dependent upon the origin of the endothelial cells that are being studied. Even though an increase in extracellular Pi concentration has been reported to *promote* the angiogenic potential of cancer cells through a mechanism requiring FOXC2 and osteopontin (OPN, a secreted cytokine like factor) [51], hyperphosphataemia has also been reported to *impair* endothelial cells' angiogenic competence by inducing alteration in the structure and functionality of endothelial cells and thus contributing to endothelial dysfunction [16]. It should be noted that the effect of higher extracellular Pi in promoting the angiogenic potential of cancer cells reported by Lin and colleagues [51] differs from the reported Pi-induced impairment of endothelial cells' angiogenic potency [16]. In [51] it was the conditioned medium from Pi-loaded cancer cells that promoted angiogenic markers in HUVECs: however in [16] the reported effect of Pi on impairment of angiogenetic potency (cell migration and tube formation) was the consequence of direct effect of Pi-loaded medium on human coronary artery endothelial cells (HCAECs) and EA.hy926 human endothelial cells. It has been suggested that the down-regulation of Annexin II that occurs under these conditions (as described in Sect. 9.2.3 above) is an important contributor to this angiogenic impairment in view of Annexin II's angiogenesis-promoting effects [16].

9.2.5 Endothelial-Mesenchymal Transition

Endothelial–mesenchymal transition (EndoMT) [52, 53] is a biological mechanism involving endothelial cell transformation into mesenchymal cells in which endothelial cells lose their specific morphology and markers and obtain myofibroblast-like features [52]. EndoMT is distinct from epithelial-mesenchymal transition which is a fundamental part of metazoan embryogenesis and characterises the structural development of organs [54]. EndoMT has been shown to be associated with the loss of common endothelial cell antigenic markers such as CD31, CD144, and von Willebrand factor (vWF) and resulting instead in acquisition of the expression of mesenchymal marker proteins for example smooth muscle actin (α-SMA), vimentin, fibronectin, calponin and fibroblast-specific protein-1 (FSP1) [52, 53]. A wide range of signalling pathways have been associated with the initiation and progression of EndoMT during both development and disease conditions. Transforming growth factor β (TGF-β) signalling is a potent inducer of EndoMT [55]. The activation of TGF-β signalling results in EndoMT by inducing accumulation of nuclear transcription factor complexes (SMADs) that induce the expression of the mesenchymal transcription factors (e.g. Snail, Twist, and Slug) which initiate EndoMT [55, 56]. Whether the TGF-β-induced EndoMT can be directly affected by intracellular rises in Pi in endothelial cells remains unclear. Such interaction between TGF-β and Pi merits further investigation however in view of the induction of EndoMT by Pi discussed below, and reports that TGF-β increases expression of SLC20A1 (GLVR1) Pi transporters in chondrogenic ATDC5 cells [57].

A chronic Pi load applied to coronary endothelial cells has been reported to induce EndoMT even in the absence of added TGF-β [53]. Tan and colleagues reported that applying an elevated extracellular Pi concentration results in an increased Pi influx which results in phosphorylation and activation of the DNA methyltransferase (DNMT1), which is recruited to the RAS protein activator like 1 (RASAL1) promoter by histone deacetylase 2 (HDAC2) and hence induces RASAL1 promoter CpG island hypermethylation which has been implicated in pathological EndoMT in cardiac pathologies such as Calcific Aortic Valve Disease and cardiac fibrosis [53, 58]. However, such promotion of EndoMT cannot be assumed to occur in all endothelial cells. Wang, et al. demonstrated that in retinal microvascular endothelial cells autophagy can induce a reduction in Smad3 phosphorylation and prevent its association with Smad4 to be translocated to the nucleus and in this way inhibits the transcription of specific genes such as snail (the master regulator of EndoMT), slug and twist [59]. As discussed in Sect. 9.2.2 above, a higher extracellular Pi concentration can inhibit Akt/mTOR signalling in endothelial cells and subsequently induces protective autophagy in the cells [40]. It remains to be determined whether the direct effect of Pi on the induction of EndoMT or the indirect suppressive effect through the induction of autophagy dominates under these conditions.

9.3 How Pi Concentration Is Regulated in Mammalian Cells

Pi is a major intracellular anion in mammalian cells. The negative resting membrane potential inside the cells tends to repel this negatively charged anion from the cytosol into extracellular fluid. To overcome this tendency, and therefore maintain a cytosolic Pi concentration of the order of millimolar inside the cells, mammalian cells express active Pi transporters in their plasma membrane which accumulate Pi anions in the cytosol against the electrical gradient of the membrane potential [60]. In humans, transporter proteins (solute carriers, SLCs) are encoded by genes which are classified in 65 SLC gene families [61]. Of these, three gene families – SLC20, SLC34 (and possibly SLC17) encode active Pi transporters (i.e. "pumps") that are expressed in the plasma membrane and pump Pi into the cytosol. These Na^+-dependent transporters perform secondary active co-transport of Pi anions and Na^+ into the cell, thus using the electrochemical gradient of Na^+ that exists across the plasma membrane to drive Pi into the cell. Besides these plasma membrane Na^+-dependent Pi transporters, there are other (non – Na^+-dependent) Pi transporters that are expressed in intracellular compartments, for instance in mitochondria (where Pi transport is coupled to H^+ and dicarboxylate ions) and the endoplasmic reticulum (where transport is coupled to Ca^{2+}-ATPase) [60, 62, 63]. However, the role of these intracellular Pi transporters in the regulation of intracellular Pi concentration is less clear.

Even though the influx of Pi through active Na^+-linked Pi transporters into mammalian cells clearly plays a role in the regulation of the intracellular Pi concentration, it is not obvious at first sight how changes in the influx of Pi through these transporters could alter the intracellular Pi concentration (and hence exert functional effects on the cell) when the extracellular Pi concentration is varied in the range 1–2.5 mM as discussed in Sects. 9.1 and 9.2 above. An apparent problem is that these transporters, for example the ubiquitously expressed transporters of the SLC20 gene family (Pit1/SLC20A1 and Pit2/SLC20A2), have a transport Michaelis constant (K_M) for Pi which is << 1 mM i.e. the transporters are saturated with Pi at physiological extracellular Pi concentration, and are operating near their V_{MAX} in the Pi concentration range 1–2.5 mM. It might be expected therefore that the intracellular Pi concentration would be unaffected by raising the extracellular concentration from 1 mM to 2.5 mM. However, mathematical modelling of intracellular Pi concentration [63] suggests that, at steady-state, the response of the intracellular Pi concentration to the extracellular concentration could in principle depend upon a <u>*combination*</u> of influx of Pi through active Na^+-linked symporters balanced by efflux of Pi through a passive "back-leak" transporter. This "pump-leak" model is shown schematically in Fig. 9.1. According to this model, at concentrations of extracellular Pi > > the K_M of the influx transporter for Pi, the behaviour of the corresponding steady-state intracellular Pi concentration would depend on the relative magnitude of the K_M for Pi on the influx

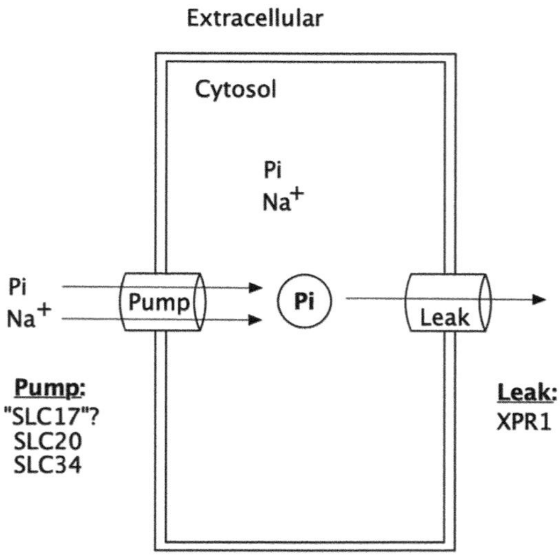

Fig. 9.1 Schematic representation of a "Pump-Leak" theoretical model for the regulation of intracellular Pi concentration in cells [63]. At steady state, active Na^+-linked influx of Pi into the cell through transporters (such as those of the SLC20 and SLC34 families) is balanced by passive efflux of Pi through a "leak" transporter such as XPR1

transporter and the K_M for Pi on the "back-leak" transporter. If the K_M of the influx transporters performing Na^+-dependent active pumping of Pi into the cell is significantly less than the K_M of the "leak" transporters, then the intracellular Pi concentration should show little change when extracellular Pi is increased. In contrast, if the K_M of the "pump" is similar to that of the "leak" transporters, an increase in extracellular Pi concentration over the range ~1–2.5 mM is predicted to lead to a commensurate increase in the intracellular concentration [63].

The relatively recent identification of a member of the SLC53 gene family (SLC-53A1 or XPR1), initially identified as the cell surface receptor for xenotropic and polytropic murine leukemia retrovirus (X/P-MLV) [64], as an efflux transporter [65] has now provided a plausible molecular basis for the "pump-leak" model (Fig. 9.1). Furthermore, the discovery of the responsiveness of the efflux transport activity of XPR1 to ambient Pi concentration sensed via InsP6 kinase/InsP8 signalling [66] (as described in Sect. 9.4 below) has revealed a further level of molecular control over the regulation of intracellular Pi concentration.

9.4 How Changes in Pi Concentration Are Sensed in Mammalian Cells

Even though mechanisms for the sensing of Pi by bacteria [67], yeast [68, 69] and plants [70] have been described in detail, corresponding explanations for how changes in Pi availability are sensed in mammalian cells have only recently emerged. Bon and colleagues have demonstrated a direct sensing mechanism for extracellular Pi concentration by SLC20 transporters through transport-independent signalling to ERK [71]. Binding of extracellular Pi to SLC20 transporters in the plasma membrane (i.e. a heterodimer of PiT-1/SLC20A1 and PiT-2/SLC20A2) in murine pre-osteoblastic MC3T3-E1 cells stimulated Pi-dependent ERK1/2 phosphorylation and subsequently resulted in up-regulation of gene expression for the mineralization inhibitors matrix Gla protein and OPN. It is worth noting that this effect of Pi occurred in the concentration range 1–10 mM even though (as pointed out in Sect. 9.3) the transport K_M of these transporters for Pi is <<1 mM. The transport-independence of this signalling effect through PiT1/PiT2 hetero-dimers apparently overcomes this K_M problem through binding of Pi to putative "Pi-sensing" sites in the heterodimer which are distinct from the amino acid residues responsible for Pi transport.

In a more recent study, Wilson, et al. reported intracellular sensing of ambient Pi concentration by inositol hexakisphosphate kinase (IP6K) signalling to XPR1 [66]. Using HCT116 cells (a human colon cancer cell line), which expresses PiT1/SLC20A1 but no other inwardly directed Pi-transporters, the authors demonstrated that deletion of IP6K1/2 (proposed to be involved in regulation of mammalian intracellular Pi concentration) results in an increase in free measurable intracellular Pi through a regulatory effect on the plasma membrane Pi exporter (XPR1). It is interesting to note that this novel and important role of

IP6K1/2 in Pi-sensing in HCT116 cells may be distinct from the biological role of IP6 kinases in other cell types. IP6K1 also regulates polyphosphate (PolyP) levels in eukaryotic cells [72], and it has recently been shown that high extracellular Pi concentration *in vitro* increases platelet PolyP content in a Pi-transport and IP6K dependent manner [73]. In contrast the HCT116 cells that show the IP6K1/2-dependent Pi-sensing and regulatory effect possesses no detectable PolyP [66].

While the two mechanisms above are clearly important in sensing Pi in mammalian cells, they have not yet been investigated in endothelial cells. A further Pi sensing mechanism has however been demonstrated in EA.hy926 human endothelial cells in which it was shown that raising extracellular Pi concentration can increase the intracellular Pi concentration by transport through active Na^+-linked PiT1/SLC20A1 transporters, thus allowing Pi to inhibit intracellular phosphoprotein phosphatases and subsequently induce a global increase in both protein Tyr and protein Ser/Thr phosphorylation [6, 74, 75].

9.5 Amplification of Pi Signals in Endothelial Cells

A number of phosphoprotein phosphatases expressed in mammalian cells are directly inhibited by physiological Pi concentrations [6, 74–78], potentially acting as sensors of intracellular Pi concentration which are directly capable of affecting intracellular signalling through protein phosphorylation. There have been no recent studies of the enzymology of this process. However, older literature (much of it pre-dating detailed molecular characterisation of the enzymes and the genes that encode them), contains reports of inhibition of protein tyrosine phosphatases and protein serine/threonine phosphatases by physiologically relevant Pi concentrations *in vitro* [6, 74, 75, 77]. By itself this is of limited value in explaining effects of Pi on endothelium, especially the observation (noted in Sect. 9.1) that even modest increases in plasma Pi concentration *in vivo* are associated with cardiovascular effects. This implies the existence not just of a Pi sensor but also some form of amplification. Such a mechanism has now been described in EA.hy926 human endothelial cells [74]. The phosphoprotein phosphatase PP2A, that accounts for the majority of protein Ser/Thr phosphatase activity in eukaryotic cells [79], is directly inhibited by Pi acting on the enzyme protein [74]. A substrate for PP2A in these cells is the phosphorylated (activated) form of the regulatory nonreceptor tyrosine kinase Src [80] (Fig. 9.2). Dephosphorylation of Src's carboxy-terminal Tyr-530 and autophosphorylation on Tyr-419 activates the Src kinase [81]. Consequently inhibition of PP2A by Pi increases phosphoactivation of Src. Src can then further inhibit PP2A by inhibitory phosphorylation of the PP2A-C catalytic sub-unit of the phosphatase [74]. Together the mutual effects of PP2A-C and Src on one another may allow the initial direct inhibitory effect of Pi ions on PP2A-C to be amplified (Fig. 9.2).

9.6 How Pi Signals Trigger Functional Effects in Endothelial Cells

In principle the Pi/PP2A/Src regulatory cycle described above has multiple targets in the control of cell signalling: PP2A is a ubiquitously expressed phosphatase which is involved in the regulation of a wide range of signals and responses, including neural development; Akt, NF-kB and MAPK signalling; apoptosis, and cell cycle progression [79], whereas Src plays a crucial role in cellular processes as diverse as cell proliferation, cell survival and drug resistance [80, 82, 83] and angiogenesis [84]. In EA.hy926 human endothelial cells a potentially important target of this Pi/PP2A/Src cycle is death-associated protein kinase 1 (DAPK-1) [74] (Fig. 9.2). Wang, el al. demonstrated that the activation of Src results in inhibitory phosphorylation of DAPK-1 [85]. DAPK-1 is a 160 kDa, serine/threonine protein kinase that is involved in a number of cellular processes including tumor suppressive function, apoptosis and autophagy

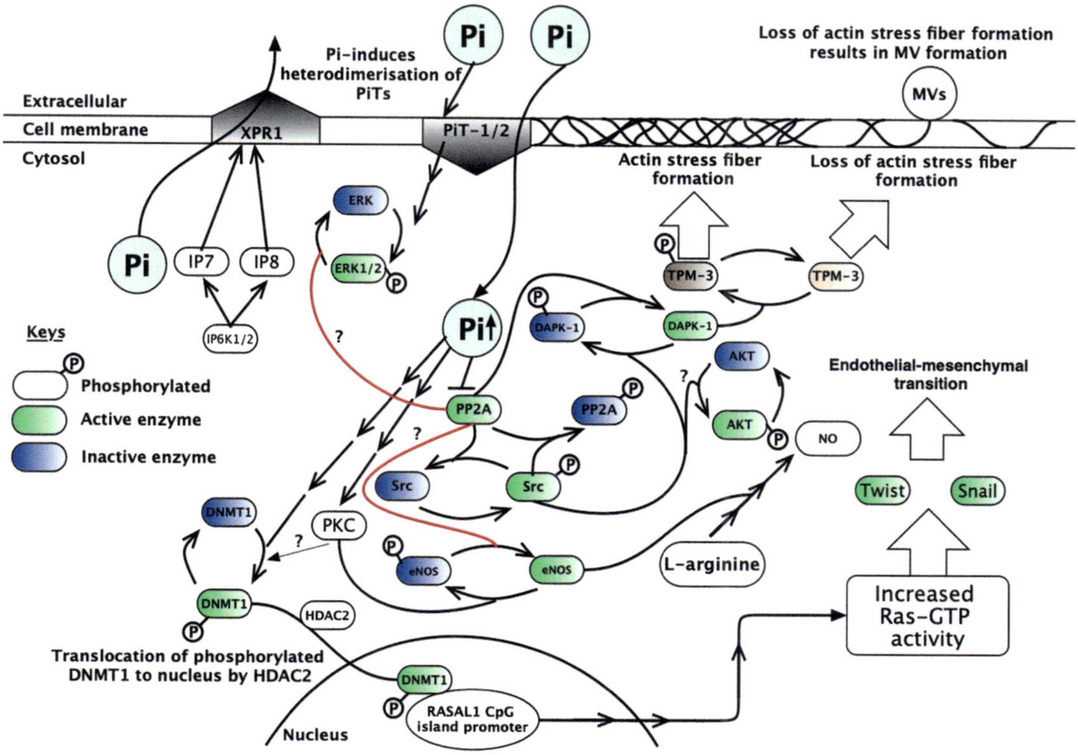

Fig. 9.2 **Pi-sensing mechanisms and proposed interplay between different pathways which elicit biological responses to elevated extracellular Pi concentration in mammalian cells.** The involvement of the pathways, molecules and effector enzymes shown here is based on data presented in [1, 53, 66, 71, 74]. Abbreviations: DNMT1, DNA methyltransferase; HDAC2, histone deacetylase 2; RASAL1, RAS protein activator like 1

[85, 86]. The kinase activity of this enzyme has been previously shown to be reduced by phosphorylation [85, 87], and subsequent proteasomal degradation of the phosphorylated DAPK-1 protein reduces the kinase activity of this enzyme even further. Houle, et al. reported that DAPK-1 phosphorylates cytoskeletal regulatory protein Tropomyosin [88], and this phosphorylation has been shown to protect the integrity of the cell membrane by inducing actin-stress fibre formation. Conversely, hypophosphorylation of -Tropomyosin-3 (TPM-3) (which has been shown to be triggered by Pi-induced PP2A inhibition [74]), results in the loss of actin stress fibre formation and an associated increase in membrane blebbing and MV generation [74] (Fig. 9.2). Thus in endothelial cells Pi-induced signalling through PP2A/Src and DAPK-1 culminates in cytoskeleton disruption and generation of pro-coagulant microvesicle (MVs) as outlined in Sect. 9.2.3 above.

9.7 Directions for Future Work

In addition to Pi signalling through the PP2A/Src/DAPK-1/TPM-3 pathway described above, it will be of interest in future work to investigate whether the Pi/PP2A/Src cycle plays any role in the other functional effects of Pi loading that were described above in Sect. 9.2, for example the Pi-induced inhibition of eNOS in Sect. 9.2.1. Previous work [1] demonstrated that this occurs as a result of inhibitory phosphorylation at Thr-497 of eNOS via protein kinase C (PKC) (Fig. 9.2). PP2A is a key determinant of eNOS dephosphorylation and enzyme activity [89], suggesting that a possible contributor to inhibitory phosphorylation of eNOS induced by Pi may be the inhibition by Pi of PP2A. However, a possible countervailing effect may arise from the accompanying Src activation induced by Pi [74] because (at least in EA.hy926 endothelial cells and HUVECs [90]) Src activation results in eNOS activation via PI3K/Akt.

Further work is also needed to clarify whether Pi affects protein phosphorylation simply by inhibiting phosphoprotein phosphatases. Firstly, even for a given phosphatase, this may vary depending on the phosphoprotein substrate. Even though the activity of two phosphoprotein phosphatases (PPP-I and PPP-II, EC 3.1.3.16) from rabbit liver has been shown to be inhibited by Pi and by pyrophosphate (PPi) while using casein as a substrate, Pi was found to activate these enzymes when a histone was the substrate [91]. Furthermore, in addition to rapid direct inhibitory effects of Pi on phosphoprotein phosphatases, countervailing compensatory effects may also occur. In EA.hy926 cells prolonged Pi-loading led to compensatory upregulation of low molecular weight protein tyrosine phosphatase (LMW-PTP)) [6].

More generally, as at least three distinct Pi sensing mechanisms have now been described in different cell types (i.e. SLC20/PiT heterodimer signalling to ERK; IP6K signalling to XPR1; and PP2A/Src signalling (Fig. 9.2), an important priority for future work is to determine whether all 3 mechanisms are expressed together in endothelial cells and, if so, how these three pathways interact. For example, in addition to activating Akt as discussed above, in cerebral ischemia [92] Src also activates ERK, through decreasing PP2A activity, suggesting potential interaction between Pi signalling through the SLC20/PiT heterodimer and PP2A/Src pathways.

Finally, even though it has been shown that high serum Pi (or a Pi-dependent hormonal response derived from it) is sufficient to induce a marked increase in circulating pro-coagulant MVs *in vivo* [35] (indicative of endothelium dysfunction), it must be acknowledged that much of the knowledge related to Pi-mediated effects on the endothelium stems from *in vitro* monoculture systems. However, we know that monoculture does not reflect the *in vivo* condition. In particular, we know that adjacent vascular smooth muscle cells are also responsive to Pi and could indirectly influence endothelial cells through paracrine mechanisms (and vice-versa). *In vitro* co-culture systems could aid in better understanding the interplay between smooth muscle cells and the endothelium in response to Pi.

9.8 Conclusions

The potent biological effects described in this review underline both the importance of Pi in normal cell function and in the pathological responses of endothelial cells to hyperphosphatemia. In particular, because of the almost universal role of protein phosphorylation signals in regulating cell function, the recent evidence that elevation of the intracellular Pi concentration profoundly affects protein phosphorylation, at least partly through its direct action on phosphoprotein phosphatases, has far-reaching implications both in endothelial cells and more widely throughout human cell biology.

References

1. Shuto E, Taketani Y, Tanaka R, Harada N, Isshiki M, Sato M, Nashiki K, Amo K, Yamamoto H, Higashi Y, Nakaya Y, Takeda E (2009) Dietary phosphorus acutely impairs endothelial function. JASN 20(7): 1504–1512
2. Moore LW, Nolte J, v, Gaber AO, Suki WN. (2015) Association of dietary phosphate and serum phosphorus concentration by levels of kidney function. Am J Clin Nutr 102(2):444–453
3. Rath DP, Bailey M, Zhang H, Jiang Z, Abduljalil AM, Weisbrode S, Hamlin RL, Robitaille PM (1995) 31P-nuclear magnetic resonance studies of chronic myocardial ischemia in the Yucatan micropig. J Clin Invest 95(1):151–157
4. Date K, Hall J, Greenman J, Maraveyas A, Madden LA (2013) Tumour and microparticle tissue factor expression and cancer thrombosis. Thromb Res 131(2):109–115
5. Cohen LF, Balow JE, Magrath IT, Poplack DG, Ziegler JL (1980) Acute tumor lysis syndrome. A

review of 37 patients with Burkitt's lymphoma. Am J Med 68(4):486–491
6. Abbasian N, Burton JO, Herbert KE, Tregunna B-E, Brown JR, Ghaderi-Najafabadi M, Brunskill NJ, Goodall AH, Bevington A (2015) Hyperphosphatemia, phosphoprotein phosphatases, and microparticle release in vascular endothelial cells. J Am Soc Nephrol 26(9):2152–2162
7. Di Marco GS, Hausberg M, Hillebrand U, Rustemeyer P, Wittkowski W, Lang D, Pavenstadt H (2008) Increased inorganic phosphate induces human endothelial cell apoptosis in vitro. Am J Physiol Renal Physiol 294(6):F1381–F1387
8. Peng A, Wu T, Zeng C, Rakheja D, Zhu J, Ye T, Hutcheson J, Vaziri ND, Liu Z, Mohan C, Zhou XJ (2011) Adverse effects of simulated hyper- and hypophosphatemia on endothelial cell function and viability. PLoS One 6(8):e23268
9. Giachelli CM, Jono S, Shioi A, Nishizawa Y, Mori K, Morii H (2001) Vascular calcification and inorganic phosphate. Am J Kidney Dis 38(4 Suppl 1):S34–S37
10. Jono S, McKee MD, Murry CE, Shioi A, Nishizawa Y, Mori K, Morii H, Giachelli CM (2000) Phosphate regulation of vascular smooth muscle cell calcification. Circ Res 87(7):E10–E17
11. Reynolds JL, Joannides AJ, Skepper JN, McNair R, Schurgers LJ, Proudfoot D, Jahnen-Dechent W, Weissberg PL, Shanahan CM (2004) Human vascular smooth muscle cells undergo vesicle-mediated calcification in response to changes in extracellular calcium and phosphate concentrations: a potential mechanism for accelerated vascular calcification in ESRD. JASN 15(11):2857–2867
12. Gutiérrez OM, Luzuriaga-McPherson A, Lin Y, Gilbert LC, Ha S-W, Beck GRJ (2015) Impact of phosphorus-based food additives on bone and mineral metabolism. J Clin Endocrinol Metab 100(11): 4264–4271
13. Shanahan CM, Crouthamel MH, Kapustin A, Giachelli CM (2011) Arterial calcification in chronic kidney disease: key roles for calcium and phosphate. Circ Res 109(6):697–711
14. Yamada S, Tokumoto M, Tatsumoto N, Taniguchi M, Noguchi H, Nakano T, Masutani K, Ooboshi H, Tsuruya K, Kitazono T (2014) Phosphate overload directly induces systemic inflammation and malnutrition as well as vascular calcification in uremia. Am J Physiol Renal Physiol 306(12):F1418–F1428
15. Lau WL, Pai A, Moe SM, Giachelli CM (2011) Direct effects of phosphate on vascular cell function. Adv Chronic Kidney Dis 18(2):105–112
16. Di Marco GS, Konig M, Stock C, Wiesinger A, Hillebrand U, Reiermann S, Reuter S, Amler S, Kohler G, Buck F, Fobker M, Kumpers P, Oberleithner H, Hausberg M, Lang D, Pavenstadt H, Brand M (2013) High phosphate directly affects endothelial function by downregulating annexin II. Kidney Int 83(2):213–222
17. Adler AJ, Ferran N, Berlyne GM (1985) Effect of inorganic phosphate on serum ionized calcium concentration in vitro: a reassessment of the "trade-off hypothesis". Kidney Int 28(6):932–935
18. Ellam TJ, Chico TJA (2012) Phosphate: the new cholesterol? The role of the phosphate axis in non-uremic vascular disease. Atherosclerosis 220(2):310–318
19. Lee H, Oh SW, Heo NJ, Chin HJ, Na KY, Kim S, Chae D-W (2012) Serum phosphorus as a predictor of low-grade albuminuria in a general population without evidence of chronic kidney disease. Nephrol Dial Trans 27(7):2799–2806
20. Gonzalez-Parra E, Tuñón J, Egido J, Ortiz A (2012) Phosphate: a stealthier killer than previously thought? Cardiovasc Pathol 21(5):372–381
21. Onufrak SJ, Bellasi A, Shaw LJ, Herzog CA, Cardarelli F, Wilson PW, Vaccarino V, Raggi P (2008) Phosphorus levels are associated with subclinical atherosclerosis in the general population. Atherosclerosis 199(2):424–431
22. Tonelli M, Sacks F, Pfeffer M, Gao Z, Curhan G (2005) Relation between serum phosphate level and cardiovascular event rate in people with coronary disease. Circulation 112(17):2627–2633
23. Foley RN, Collins AJ, Herzog CA, Ishani A, Kalra PA (2009) Serum phosphorus levels associate with coronary atherosclerosis in young adults. JASN 20(2):397–404
24. Danziger J (2008) Importance of low-grade albuminuria. Mayo Clin Proc 83(7):806–812
25. Campos-Obando N, Lahousse L, Brusselle G, Stricker BH, Hofman A, Franco OH, Uitterlinden AG, Zillikens MC (2018) Serum phosphate levels are related to all-cause, cardiovascular and COPD mortality in men. Eur J Epidemiol 33(9):859–871
26. Onufrak SJ, Bellasi A, Cardarelli F, Vaccarino V, Muntner P, Shaw LJ, Raggi P (2009) Investigation of gender heterogeneity in the associations of serum phosphorus with incident coronary artery disease and all-cause mortality. Am J Epidemiol 169(1):67–77
27. Ohnishi M, Razzaque MS (2010) Dietary and genetic evidence for phosphate toxicity accelerating mammalian aging. FASEB J 24(9):3562–3571
28. Portale AA, Halloran BP, Morris RCJ (1989) Physiologic regulation of the serum concentration of 1,25-dihydroxyvitamin D by phosphorus in normal men. J Clin Invest 83(5):1494–1499
29. Smogorzewski M, Zayed M, Zhang YB, Roe J, Massry SG (1993) Parathyroid hormone increases cytosolic calcium concentration in adult rat cardiac myocytes. Am J Phys 264(6 Pt 2):H1998–H2006
30. Razzaque MS (2011) Phosphate toxicity: new insights into an old problem. Clin Sci (London, England: 1979) 120(3):91–97
31. Brown RB, Razzaque MS (2018) Phosphate toxicity and tumorigenesis. Biochim Biophys Acta Rev Cancer 1869(2):303–309
32. Goodson JM, Shi P, Razzaque MS (2019) Dietary phosphorus enhances inflammatory response: a study

of human gingivitis. J Steroid Biochem Mol Biol 188: 166–171
33. Hetz R, Beeler E, Janoczkin A, Kiers S, Li L, Willard BB, Razzaque MS, He P (2021) Excessive inorganic phosphate burden perturbed intracellular signaling: quantitative proteomics and phosphoproteomics analyses. Front. Nutr. 8:765391. https://doi.org/10.3389/fnut.2021.765391.
34. He P, Mann-Collura O, Fling J, Edara N, Hetz R, Razzaque MS (2020) High phosphate actively induces cytotoxicity by rewiring pro-survival and pro-apoptotic signaling networks in HEK293 and HeLa cells. FASEB J 35:e20997
35. Abbasian N, Goodall AH, Burton JO, Bursnall D, Bevington A, Brunskill NJ (2020) Hyperphosphatemia drives procoagulant microvesicle generation in the rat partial nephrectomy model of CKD. J Clin Med 9(11): 3534
36. Nishi T, Shuto E, Ogawa M, Ohya M, Nakanishi M, Masuda M, Katsumoto M, Yamanaka-Okumura H, Sakai T, Takeda E, Sakaue H, Taketani Y (2015) Excessive dietary phosphorus intake impairs endothelial function in young healthy men: a time- and dose-dependent study. JMI 62(3–4):167–172
37. Levac BM, Adams MA, Pyke KE (2017 Dec) The impact of an acute oral phosphate load on endothelium dependent and independent brachial artery vasodilation in healthy males. Appl Physiol Nutr Metab 42(12):1307–1315
38. Six I, Maizel J, Barreto FC, Rangrez AY, Dupont S, Slama M, Tribouilloy C, Choukroun G, Mazière JC, Bode-Boeger S, Kielstein JT, Drüeke TB, Massy ZA (2012 Oct) Effects of phosphate on vascular function under normal conditions and influence of the uraemic state. Cardiovasc Res 96(1):130–139
39. Pizzino G, Irrera N, Cucinotta M, Pallio G, Mannino F, Arcoraci V, Squadrito F, Altavilla D, Bitto A (2017) Oxidative stress: harms and benefits for human health. Oxidative Med Cell Longev 2017:8416763
40. Hsu Y-J, Hsu S-C, Huang S-M, Lee H-S, Lin S-H, Tsai C-S, Shih C-C, Lin C-Y (2015) Hyperphosphatemia induces protective autophagy in endothelial cells through the inhibition of Akt/mTOR signaling. J Vasc Surg 62(1):210–221.e2
41. Olmos G, Martínez-Miguel P, Alcalde-Estevez E, Medrano D, Sosa P, Rodríguez-Mañas L, Naves-Diaz M, Rodríguez-Puyol D, Ruiz-Torres MP, López-Ongil S (2017) Hyperphosphatemia induces senescence in human endothelial cells by increasing endothelin-1 production. Aging Cell 16(6):1300–1312
42. Sage AP, Lu J, Tintut Y, Demer LL (2011 Feb) Hyperphosphatemia-induced nanocrystals upregulate the expression of bone morphogenetic protein-2 and osteopontin genes in mouse smooth muscle cells in vitro. Kidney Int 79(4):414–422
43. Ma Y, Ma Z, Yin S, Yan X, Wang J (2017 Jan) Arsenic and fluoride induce apoptosis, inflammation and oxidative stress in cultured human umbilical vein endothelial cells. Chemosphere 167:454–461
44. Batool S, Abbasian N, Burton JO, Stover C (2013) Microparticles and their roles in inflammation: a review. Open Immunol J 6(1):1–14
45. Abbasian N, Herbert KE, Pawluczyk I, Burton JO, Bevington A (2018) Vesicles bearing gifts: the functional importance of micro-RNA transfer in extracellular vesicles in chronic kidney disease. Am J Physiol Renal Physiology 315:F1430
46. Liang H-Z, Li S-F, Zhang F, Wu M-Y, Li C-L, Song J-X, Lee C, Chen H (2018 Nov) Effect of endothelial microparticles induced by hypoxia on migration and angiogenesis of human umbilical vein endothelial cells by delivering MicroRNA-19b. Chin Med J 131(22): 2726–2733
47. Zernecke A, Bidzhekov K, Noels H, Shagdarsuren E, Gan L, Denecke B, Hristov M, Köppel T, Jahantigh MN, Lutgens E, Wang S, Olson EN, Schober A, Weber C (2009) Delivery of microRNA-126 by apoptotic bodies induces CXCL12-dependent vascular protection. Sci Signal 2(100):ra81
48. Li C, Li S, Zhang F, Wu M, Liang H, Song J, Lee C, Chen H (2018 Jan) Endothelial microparticles-mediated transfer of microRNA-19b promotes atherosclerosis via activating perivascular adipose tissue inflammation in apoE(−/−) mice. Biochem Biophys Res Commun 495(2):1922–1929
49. Burton JO, Hamali HA, Singh R, Abbasian N, Parsons R, Patel AK, Goodall AH, Brunskill NJ (2013) Elevated levels of procoagulant plasma microvesicles in dialysis patients. PLoS One 8(8):1–11
50. Burger D, Kwart DG, Montezano AC, Read NC, Kennedy CRJ, Thompson CS, Touyz RM (2012 Jun) Microparticles induce cell cycle arrest through redox-sensitive processes in endothelial cells: implications in vascular senescence. J Am Heart Assoc 1(3):e001842
51. Lin Y, McKinnon KE, Ha SW, Beck GR Jr (2015) Inorganic phosphate induces cancer cell mediated angiogenesis dependent on forkhead box protein C2 (FOXC2) regulated osteopontin expression. Mol Carcinog 54(9):926–934
52. Pinto MT, Covas DT, Kashima S, Rodrigues CO (2016) Endothelial mesenchymal transition: comparative analysis of different induction methods. Biol Proced Online 18:10
53. Tan X, Xu X, Zeisberg M, Zeisberg EM (2016) DNMT1 and HDAC2 cooperate to facilitate aberrant promoter methylation in inorganic phosphate-induced endothelial-mesenchymal transition. PLoS One 11(1)
54. Lee JM, Dedhar S, Kalluri R, Thompson EW (2006 Mar) The epithelial-mesenchymal transition: new insights in signaling, development, and disease. J Cell Biol 172(7):973–981
55. van Meeteren LA, ten Dijke P (2011) Regulation of endothelial cell plasticity by TGF-β. Cell Tissue Res 347(1):177–186
56. Hulshoff MS, Xu X, Krenning G, Zeisberg EM (2018 Sep) Epigenetic regulation of endothelial-to-mesenchymal transition in chronic heart disease. Arterioscler Thromb Vasc Biol 38(9):1986–1996

57. Palmer G, Guicheux J, Bonjour JP, Caverzasio J (2000 Jun) Transforming growth factor-beta stimulates inorganic phosphate transport and expression of the type III phosphate transporter Glvr-1 in chondrogenic ATDC5 cells. Endocrinology 141(6):2236–2243
58. Xu X, Tan X, Tampe B, Nyamsuren G, Liu X, Maier LS, Sossalla S, Kalluri R, Zeisberg M, Hasenfuss G, Zeisberg EM (2015 Mar) Epigenetic balance of aberrant Rasal1 promoter methylation and hydroxymethylation regulates cardiac fibrosis. Cardiovasc Res 105(3): 279–291
59. Wang J, Feng Y, Wang Y, Xiang D, Zhang X, Yuan F (2017 Jun) Autophagy regulates endothelial-mesenchymal transition by decreasing the phosphorylation level of Smad3. Biochem Biophys Res Commun 487(3):740–747
60. Kemp GJ, Bevington A, Khodja D, Russell RG (1988 Apr) Net fluxes of orthophosphate across the plasma membrane in human red cells following alteration of pH and extracellular Pi concentration. Biochim Biophys Acta 969(2):148–157
61. SLC TABLES [Internet]. [cited 2020 Nov 1]. Available from: http://slc.bioparadigms.org/
62. Bevington A, Kemp GJ, Graham R, RUSSEL G. (1992) Phosphate-sensitive enzymes: a possible molecular basis for cellular disorders of phosphate metabolism. Clin Chem Enzymol Commun (Print) 4(4):235–257
63. Kemp GJ, Bevington A (1993) The regulation of intracellular orthophosphate concentration. J Theor Biol 161(1):77–94
64. Battini JL, Rasko JE, Miller AD (1999 Feb) A human cell-surface receptor for xenotropic and polytropic murine leukemia viruses: possible role in G protein-coupled signal transduction. Proc Natl Acad Sci U S A 96(4):1385–1390
65. Giovannini D, Touhami J, Charnet P, Sitbon M, Battini J-L (2013 Jun) Inorganic phosphate export by the retrovirus receptor XPR1 in metazoans. Cell Rep 3(6):1866–1873
66. Wilson MS, Jessen HJ, Saiardi A (2019 Jul) The inositol hexakisphosphate kinases IP6K1 and −2 regulate human cellular phosphate homeostasis, including XPR1-mediated phosphate export. J Biol Chem 294(30):11597–11608
67. McCleary WR (2017) Molecular mechanisms of phosphate homeostasis in Escherichia coli. In: Samie A (ed) Escherichia coli—recent advances on physiology, pathogenesis and biotechnological applications. InTechOpen, Rijeka, pp 333–357
68. Bergwitz C, Jüppner H (2011) Phosphate sensing. Adv Chronic Kidney Dis 18(2):132–144
69. Santos-Beneit F (2015) The pho regulon: a huge regulatory network in bacteria. Front Microbiol 6:402
70. Ticconi CA, Abel S (2004) Short on phosphate: plant surveillance and countermeasures. Trends Plant Sci 9(11):548–555
71. Bon N, Couasnay G, Bourgine A, Sourice S, Beck-Cormier S, Guicheux J, Beck L (2018 Feb) Phosphate (Pi)-regulated heterodimerization of the high-affinity sodium-dependent Pi transporters PiT1/Slc20a1 and PiT2/Slc20a2 underlies extracellular Pi sensing independently of Pi uptake. J Biol Chem 293(6):2102–2114
72. Ghosh S, Shukla D, Suman K, Lakshmi BJ, Manorama R, Kumar S, Bhandari R (2013) Inositol hexakisphosphate kinase 1 maintains hemostasis in mice by regulating platelet polyphosphate levels. Blood 122(8):1478–1486
73. Abbasian N, Harper MT (2020) High extracellular phosphate increases platelet polyphosphate content. Platelets 1–3:992–994
74. Abbasian N, Bevington A, Burton JO, Herbert KE, Goodall AH, Brunskill NJ (2020) Inorganic phosphate (Pi) signaling in endothelial cells: a molecular basis for generation of endothelial microvesicles in uraemic cardiovascular disease. Int J Mol Sci 21(19):6993
75. Buzalaf MA, Taga EM, Granjeiro JM, Ferreira C, v, Lourencao VA, Ortega MM, Poletto DW, Aoyama H. (1998) Kinetic characterization of bovine lung low-molecular-weight protein tyrosine phosphatase. Exp Lung Res 24(3):269–272
76. Szajerka G, Kwiatkowska J (1984) The effect of cortisol on rabbit red cell acid phosphatase isoenzymes. Mol Cell Biochem 59(1–2):183–186
77. Zhang ZY, VanEtten RL (1991) Pre-steady-state and steady-state kinetic analysis of the low molecular weight phosphotyrosyl protein phosphatase from bovine heart. J Biol Chem 266(3):1516–1525
78. Gil G, Calvet VE, Ferrer A, Hegardt FG (1982) Inactivation and reactivation of rat liver 3-hydroxy-3-methylglutaryl-CoA-reductase phosphatases: effect of phosphate, pyrophosphate and divalent cations. Hoppe-Seyler's Zeitschrift fur physiologische Chemie 363(10):1217–1224
79. Nematullah M, Hoda MN, Khan F (2018) Protein phosphatase 2A: a double-faced phosphatase of cellular system and its role in neurodegenerative disorders. Mol Neurobiol 55(2):1750–1761
80. Xu J, Xu Z, Zhou J-Y, Zhuang Z, Wang E, Boerner J, Wu GS (2013) Regulation of the Src-PP2A interaction in tumor necrosis factor (TNF)-related apoptosis-inducing ligand (TRAIL)-induced apoptosis. J Biol Chem 288(46):33263–33271
81. Bjorge JD, Jakymiw A, Fujita DJ (2000) Selected glimpses into the activation and function of Src kinase. Oncogene 19(49):5620–5635
82. Roskoski R Jr (2004) Src protein–tyrosine kinase structure and regulation. Biochem Biophys Res Commun 324(4):1155–1164
83. Wang Y, Liu J, Chen M, Du T, Duan C, Gao G, Yang H (2016) The novel mechanism of rotenone-induced α-synuclein phosphorylation via reduced protein phosphatase 2A activity. Int J Biochem Cell Biol 75:34–44
84. Li P, Chen D, Cui Y, Zhang W, Weng J, Yu L, Chen L, Chen Z, Su H, Yu S, Wu J, Huang Q, Guo X (2018) Src plays an important role in AGE-induced endothelial cell proliferation, migration, and Tubulogenesis. Front Physiol 9:765

85. Wang W-J, Kuo J-C, Ku W, Lee Y-R, Lin F-C, Chang Y-L, Lin Y-M, Chen C-H, Huang Y-P, Chiang M-J (2007) The tumor suppressor DAPK is reciprocally regulated by tyrosine kinase Src and phosphatase LAR. Mol Cell 27(5):701–716
86. Singh P, Ravanan P, Talwar P (2016) Death associated protein kinase 1 (DAPK1): a regulator of apoptosis and autophagy. Front Mol Neurosci 9:46
87. Jin Y, Blue EK, Gallagher PJ (2006) Control of death-associated protein kinase (DAPK) activity by phosphorylation and proteasomal degradation. J Biol Chem 281(51):39033–39040
88. Houle F, Poirier A, Dumaresq J, Huot J (2007 Oct) DAP kinase mediates the phosphorylation of tropomyosin-1 downstream of the ERK pathway, which regulates the formation of stress fibers in response to oxidative stress. J Cell Sci 120(Pt 20):3666–3677
89. Greif DM, Kou R, Michel T (2002 Dec) Site-specific dephosphorylation of endothelial nitric oxide synthase by protein phosphatase 2A: evidence for crosstalk between phosphorylation sites. Biochemistry 41(52): 15845–15853
90. Haynes MP, Li L, Sinha D, Russell KS, Hisamoto K, Baron R, Collinge M, Sessa WC, Bender JR (2003 Jan) Src kinase mediates phosphatidylinositol 3-kinase/Akt-dependent rapid endothelial nitric-oxide synthase activation by estrogen. J Biol Chem 278(4): 2118–2123
91. Khandelwal RL, Kasmani SA (1980) Studies on in activation and reactivation of homogeneous rabbit liver phosphoprotein phosphatases by inorganic pyorphosphate and divalent cations. Biochim Biophys Acta 613(1):95–105
92. Hu X, Wu X, Xu J, Zhou J, Han X, Guo J (2009 Jul) Src kinase up-regulates the ERK cascade through inactivation of protein phosphatase 2A following cerebral ischemia. BMC Neurosci 10:74

Common Dietary Sources of Natural and Artificial Phosphate in Food

10

Ken-ichi Miyamoto, Joanna Oh, and Mohammed S. Razzaque

Abstract

The Recommended Dietary Allowance (RDA) for phosphate in the U.S. is around 700 mg/day for adults. The majority of healthy adults consume almost double the amount of phosphate than the RDA. Lack of awareness, and easy access to phosphate-rich, inexpensive processed food may lead to dietary phosphate overload with adverse health effects, including cardiovascular diseases, kidney diseases and tumor formation. Nutritional education and better guidelines for reporting phosphate content on ingredient labels are necessary, so that consumers are able to make more informed choices about their diets and minimize phosphate consumption. Without regulatory measures, dietary phosphate toxicity is rapidly becoming a global health concern, and likely to put enormous physical and financial burden to the society.

Keywords

RDA · Adequate Intake · Processed food · CKD · Phosphate additives

10.1 Introduction

Phosphorus is a mineral found almost exclusively in the phosphorus-containing compound phosphate (PO_4). Throughout this chapter, the term phosphate will be used most typically to describe this element [9]. Chronic kidney disease (CKD) is increasingly recognized as a serious public health issue with the incidence rising worldwide [2, 3, 27]. CKD and declining renal function are associated with impaired phosphate and calcium homeostasis, hormonal imbalance and progressive pathology of the cardiovascular and skeletal systems [10]. Hyperphosphatemia is linked to vascular calcification in CKD and an independent risk factor for cardiovascular mortality in hemodialysis patients [18]. Furthermore, serum phosphate, even when within the normal range, has been associated with cardiovascular events and all-cause mortality, both in healthy subjects and in patients with CKD [8]. In these contexts, restriction of dietary phosphate is an important intervention in the prevention of CKD and cardiovascular diseases. However, estimation of the dietary intake of phosphate is complicated by inaccuracies in nutrient database due to varied use of phosphate-containing food additives

K.-i. Miyamoto (✉)
Department of Food Science and Human Nutrition, Faculty of Agriculture, Ryukoku University, Otsu-City, Japan
e-mail: kmiyamoto@agr.ryukoku.ac.jp

J. Oh · M. S. Razzaque
Department of Pathology, Lake Erie College of Osteopathic Medicine, Erie, PA, USA
e-mail: joh34422@med.lecom.edu; mrazzaque@lecom.edu

M. S. Razzaque (ed.), *Phosphate Metabolism*, Advances in Experimental Medicine and Biology 1362,
https://doi.org/10.1007/978-3-030-91623-7_10

[35]. In recent years, the amount of phosphate intake has increased worldwide, especially in countries with high consumption of processed food [9]. Unlike sodium and other food components, phosphate is usually not listed as an ingredient, making it difficult for patients to avoid phosphate-rich food and drinks [7]. This chapter will focus on common dietary sources of natural and artificial phosphate in food within the U.S. and Japan.

10.2 Recommended Dietary Allowance (RDA) in USA

In the United States, the average person consumes far more phosphate than the Recommended Dietary Allowance (RDA) (700 mg/d for adult), although very few individuals exceed the tolerable upper limit (4000 mg/d for adults <70 year old) [34]. Moore et al. evaluated the association of dietary phosphate intake and mortality by using a nationally representative sample of healthy participants in the NHANES III [20]. The estimates of usual phosphate intake are based on information from 24-h recall data, using a validated method, with second-day recalls from a subset of the surveyed population to assess intra-individual variation [20]. Twenty-four-hour recall data from the recently completed NHANES (2009–2010) were used to show the percentage of phosphate in various food categories. Milk and dairy had the highest percentages, followed by meat and poultry. The average phosphorus intake from both foods and supplements was 1,301 mg/24 hours for women and 1,744 mg/24 hours for men, as estimated by analyzing 2013–2014 NHANES data [20]. However, several lines of evidence suggest greater intakes of phosphate than those shown in the NHANES [12]. The evidence of underestimation came from comparisons of direct chemical analyses of foods with the available estimates from the nutrient database used in the NHANES [23].

10.3 Adequate Intake (AI) in Japan

In Japan, the Recommended Dietary Allowances (RDA) was first established in 1970, after which a revision was made every 5 years. The concept of Dietary Reference Intakes (DRIs) was introduced in the sixth revision of RDA (2000–2004). DRIs were established on a scientific basis, utilizing domestic and foreign research investigations and data that are available [32]. The Adequate Intake (AI) is a quantity that is sufficient to maintain a satisfactory nutritional status of a particular gender and age group. In general, the AI is determined based on epidemiological studies that estimate nutritional intake of healthy individuals. UL means tolerable upper intake level: AI of phosphate intake in adult male is 1000 mg/day, UL is 3000 mg/day. The AI values are higher than the RDA in the EU and US (~700 mg/day). These values are based on the report of the National Nutritional Survey in Japan (NNSJ), which is initiated to obtain factual information on the nutritional health, and actual food consumption and food requirements in Japan [39]. Food intake survey by weighed food records in three consecutive representative days were conducted by specially trained dietary interviewers. NNSJ showed that phosphate intake is 1000 ~1100mg /day in 2010 [21]. However, these intake values are not estimated dietary phosphate intake obtained from nutrient content database with direct chemical analyses.

10.4 Common Phosphate-Containing Food

The American and Japanese diet's phosphate content have been increasing due to the greater consumption of food processed with phosphate additives [4, 37, 38]. Two main sources of dietary phosphate are organic, which includes animal and vegetarian proteins, and inorganic, which are mainly food additives. In general, foods high in protein like meats, milk, eggs, and cereals are naturally high in phosphate and have represented the main source of dietary phosphate. In natural foods, phosphate is present both as inorganic ions and as a constituent of phosphoproteins, and also as membrane phospholipids, adenosine triphosphate, adenosine diphosphate, DNA, and RNA [14]. Protein-rich foods, such as meat, contain natural phosphate compounds (nucleotides, phospholipids, etc), along with naturally occurring orthophosphate. Besides the natural dietary phosphate present in food, phosphate is contained in functional food additives. Phosphate-containing additives work by sequestration of meat ions and dissociation of the actomyosin complex, bringing about an increase in water-holding capacity [1]. The direct identification of added phosphate in meat products is difficult. It represents the so-called hidden phosphate.

In contrast, plant phosphate is included in beans, cereals and nuts in the form of phytic acid or phytate [19]. In humans, the bioavailability of phosphate derived from plant foods is relatively low (less than 50%) [5]. Therefore, if phosphate content is "apparently" high in plants, the amount of phosphate actually absorbed from the intestinal tract may be less in vegetable proteins compared to animal proteins [19]. In addition, processing food by cooking, soaking, or fermenting is known to hydrolyze phytic acid. Therefore, a large amount of phosphate is likely to be bioavailable from cooked or fermented full-grain or legume foods than from raw or unprocessed foods [5].

Meat and milk products are important sources of dietary phosphate and protein. The use of phosphate additives is common both in processed cheese and meat products. Recently, Karp et al. reported that measuring in vitro digestible phosphate content of foods may reflect the absorbability of phosphate [15, 16] Measurements of in vitro digestible phosphate content of foods support that there is better phosphate absorbability in foods of animal origin than of plant origin [15, 16]. When healthy people ingested the same amount of phosphate from plant foods or animal foods, urinary excretion of phosphate was higher in the case of ingestion of meat [19]. Moe et al. demonstrated that meals having equivalent amounts of phosphate but of different sources (meats compared with grains) resulted in serum phosphate differences [19]. These data suggest the importance of bioaccessible phosphate content, not just total phosphate content in foods.

10.5 Inorganic Phosphate Additives

Inorganic Phosphate is the main component of most additives and is usually in the form of phosphoric acid, phosphates, or polyphosphates in processed foods. The food industry uses them in food processing for a variety of reasons (i.e, to extend conservation, enhance color, improve flavor, and retain moisture). No limit is given in regards to the amount of phosphate-containing preservatives. Currently, only a technological limit of 5 g/kg of food exists [7]. In the United States, the dietary phosphate burden from phosphate-containing preservatives has increased dramatically from an average of 470 mg/d in the 1990s to more than 1000 mg/d for a typical American diet in recent years [12]. In general, foods high in protein like meats, milk, eggs, and cereals are naturally high in inorganic phosphate and traditionally have represented the main source of dietary inorganic phosphate. However, this is changing as inorganic phosphate is currently being added to a large and increasing number of processed foods [36]. Preservatives are largely used in meat products (eg, chicken nuggets and hotdogs), processed cheese spreads, pasta, cooked and frozen dishes, puddings, sauces, bakery products, partially cooked and frozen foods, and soft drinks. As a result, depending on the food choices, additives may increase the inorganic phosphate intake by as much as 1.0 g/day and up to 100% of inorganic phosphate in processed foods may be absorbed [5].

The absorbability of phosphate may differ substantially among different plant foods. Despite high total phosphate content, legumes and seeds may be relatively poor phosphate sources, which could be used in the diets of patients with CKD [5]. In addition, most beverages contain little to no protein. Hence, any phosphate content is almost entirely from additives. As a consequence, patients who consumed beverages with a high phosphate content had serum phosphate levels that were quite high although their nutritional status may have been inferior [14]. In soft drinks and beer, all phosphate were digestible. Total phosphate content in soft drinks is from phosphoric acid, which is likely to degrade efficiently in the intestine. In beer, phosphate is from the grain used in preparing beer; in Finland, this is most often barley. On the basis of the results, it seems that during malting and other processing of grain when preparing beer, phosphate is efficiently released from the grain in a highly digestible form, resulting in even higher phosphate content than in soda beverages [16].

In Japan, although the risk of food additives attracted attention in the 1970s, there has been no study on phosphate additives [13]. However, there are many kinds of used food additives. Therefore, it is extremely difficult to show the content of food additives from the nation's foods precisely. In addition, Japanese eating habits vary, and even vary by generation. For example, elderly Japanese may enjoy traditional Japanese meals (fish and rice) whereas the younger generation likes Western dietary patterns of intake common to many Americans. Daily Intake of Food Additives in Japan was determined from 1976 to 2000 by the Market Basket Method [13]. For this method, researchers purchased about 250 foods reflecting the average eating habits in different parts of Japan and examined the quantity of additives included in the food. The study showed that the average phosphate additives, specifically polyphosphate and pyrophosphate, was 259 mg/day in 2013. However, a large-scale survey for phosphate additives in Japan has not been published.

10.6 Phosphate Intake and Bioaccessibility

As mentioned above, dietary phosphate assessment is complicated by inaccuracies in nutrient database due to the widespread and varied use of phosphate-containing food additives, as well as the regional and seasonal variation in naturally occurring phosphate (organic phosphate) in various crops [12]. Oenning et al. showed a significant 20–25% underestimation of phosphate content [23]. In addition, direct chemical analyses showed that a majority of chicken products (fresh frozen chicken products in Midwestern grocery stores) contained one or more phosphate-containing ingredients, contributing a mean of 84 mg phosphate/100 g serving (range = 12–165-mg/100 g serving). Sherman et al. reported similar observations in the content of phosphate in meat, poultry, and fish products [33]. The cumulative impact of added phosphate in many food categories can significantly raise actual phosphate intake well beyond food intake estimates with existing database information [5].

A major barrier to studying the relation between dietary phosphate and various health outcomes is the current inaccuracy in estimating dietary phosphate intake, especially in terms of bio-accessibility [12]. For the nutrient database, it is necessary for the values to account for the bioaccessibility of phosphate, not just total phosphate. Of relevance, phosphate is primarily absorbed from the intestinal tract where about 70% of the intake is believed to be used. The absorption mechanism of phosphate is in the intestinal tract, but the mechanism is complicated by factors that affect absorption, such as vitamin D [17]. As described above, phosphate forms (organic and inorganic phosphate) are also important in understanding the absorption of phosphate.

10.7 Food Additives and Health

Recent studies have revealed that high phosphate intake is associated with increased mortality in a healthy US population [6]. Increased dietary phosphate intake may be detrimental even in the absence of high serum phosphate concentrations [26]. Serum phosphate concentrations are effectively regulated by fibroblast growth factor 23 (FGF23), a hormone that increases urinary phosphate excretion [28, 29]. Although individuals with normal kidney function are able to regulate serum phosphate levels in a physiologic range by increasing FGF23 concentrations, long term effects may lead to left ventricular hypertrophy, heart failure, and mortality. More recently, studies showed that consumption of a diet rich in phosphate based food additives while stable for calcium for 1 week increased circulating FGF23, osteopontin and osteocalcin concentration relative to baseline values, and decreased mean sclerostin concentrations in healthy individuals [11]. Similar results were observed in animals fed diets with increased phosphate content; findings showed substantial decreases in bone mineral density (BMD) and structural indices. FGF23, osteopontin, and sclerostin are critically involved in regulation of bone and mineral metabolism and are associated with bone and cardiovascular disease [11]. These results indicate that high phosphate additives intake may have adverse effects in individuals with normal kidney function. In addition, Moore et al. showed that dietary intake from foods that are sources of inorganic phosphate have a greater impact on serum phosphate than do foods that are sources of mainly organic phosphate, which supports the importance of accounting for phosphate bioaccessibility in studies on dietary phosphate [20].

10.8 Conclusion and Perspective

Phosphate intake is an important issue of phosphate balance in CKD patients [22, 30, 31]. Phosphate toxicity can cause a wide range of organ damage, including renal injury, cardiovascular damage and accelerate aging processes (Fig. 10.1) [24, 25]. The information of phosphate content and phosphate type (organic and inorganic) in foods is not exactly observable. Phosphate in food additives is almost completely reabsorbed in the intestine. It is important to consider that inorganic phosphate is not only contained in natural food, but also inorganic phosphate added in processing food. In addition, since phosphate does not have mandatory labeling, accurate grasp of the phosphate content, which is present in many processed foods, is difficult to obtain. Even if the food is listed as having been prepared with phosphate salts by the United States Food and Drug Administration (FDA), the manufacturer is not obliged to display the phosphate content in food. To prevent excess phosphate intake above nutrient needs of healthy adults, the labeling of phosphate content and a database with accurate phosphate availability are required [9].

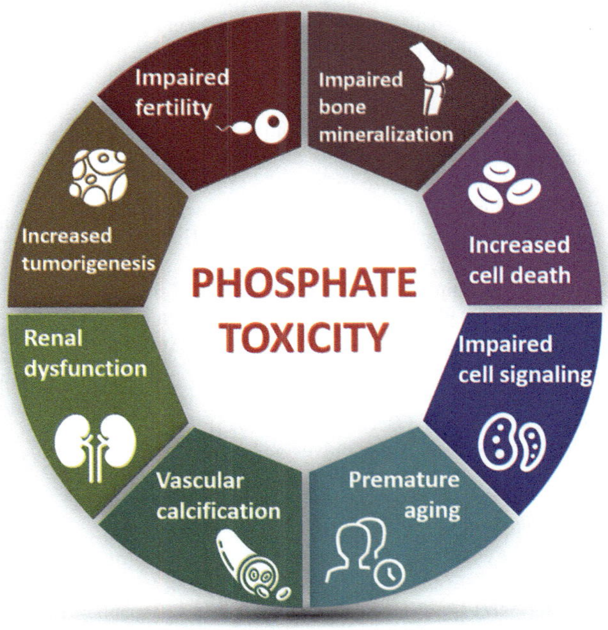

Fig. 10.1 Partial list of various pathologies induced by phosphate toxicity

Acknowledgement We want to express our sincere gratitude to Dr. Nuraly Akimbekov (Al-Farabi Kazakh National University, Kazakhstan) for his help in drawing the illustration. Joanna Oh is an Osteopathic Medical Student (OMS II) at the Lake Erie College of Osteopathic Medicine, Erie (USA).

References

1. Benini O, D'alessandro C, Gianfaldoni D, Cupisti A (2011) Extra-phosphate load from food additives in commonly eaten foods: a real and insidious danger for renal patients. J Ren Nutr 21:303–308
2. Block GA, Hulbert-Shearon TE, Levin NW, Port FK (1998) Association of serum phosphorus and calcium x phosphate product with mortality risk in chronic hemodialysis patients: a national study. Am J Kidney Dis 31:607–617
3. Block GA, Klassen PS, Lazarus JM, Ofsthun N, Lowrie EG, Chertow GM (2004) Mineral metabolism, mortality, and morbidity in maintenance hemodialysis. J Am Soc Nephrol 15:2208–2218
4. Calvo MS, Park YK (1996) Changing phosphorus content of the U.S. diet: potential for adverse effects on bone. J Nutr 126:1168S–1180S
5. Calvo MS, Uribarri J (2013) Contributions to total phosphorus intake: all sources considered. Semin Dial 26:54–61
6. Chang AR, Lazo M, Appel LJ, Gutierrez OM, Grams ME (2014) High dietary phosphorus intake is associated with all-cause mortality: results from NHANES III. Am J Clin Nutr 99:320–327
7. Cupisti A, Kalantar-Zadeh K (2013) Management of natural and added dietary phosphorus burden in kidney disease. Semin Nephrol 33:180–190
8. Dhingra R, Sullivan LM, Fox CS, Wang TJ, D'agostino RB Sr, Gaziano JM, Vasan RS (2007) Relations of serum phosphorus and calcium levels to the incidence of cardiovascular disease in the community. Arch Intern Med 167:879–885
9. Erem S, Razzaque MS (2018) Dietary phosphate toxicity: an emerging global health concern. Histochem Cell Biol 150:711–719
10. Foley RN, Collins AJ, Herzog CA, Ishani A, Kalra PA (2009) Serum phosphorus levels associate with coronary atherosclerosis in young adults. J Am Soc Nephrol 20:397–404
11. Gutierrez OM, Luzuriaga-Mcpherson A, Lin Y, Gilbert LC, Ha SW, Beck GR Jr (2015) Impact of phosphorus-based food additives on bone and mineral metabolism. J Clin Endocrinol Metab 100(11): 4264–4271. https://doi.org/10.1210/jc.2015-2279
12. Hill Gallant KM (2015) Studying dietary phosphorus intake: the challenge of when a gram is not a gram. Am J Clin Nutr 102:237–238
13. Ito Y (2007) Daily intake of food additives in Japan: determination of food additive residues in food-from 1976 to 2000 year by market basket method. FFI J 212: 815–838
14. Kalantar-Zadeh K, Gutekunst L, Mehrotra R, Kovesdy CP, Bross R, Shinaberger CS, Noori N, Hirschberg R, Benner D, Nissenson AR, Kopple JD (2010) Understanding sources of dietary phosphorus in the treatment of patients with chronic kidney disease. Clin J Am Soc Nephrol 5:519–530
15. Karp H, Ekholm P, Kemi V, Hirvonen T, Lamberg-Allardt C (2012a) Differences among total and in vitro digestible phosphorus content of meat and milk products. J Ren Nutr 22:344–349
16. Karp H, Ekholm P, Kemi V, Itkonen S, Hirvonen T, Narkki S, Lamberg-Allardt C (2012b) Differences among total and in vitro digestible phosphorus content of plant foods and beverages. J Ren Nutr 22:416–422
17. Lee GJ, Marks J (2015) Intestinal phosphate transport: a therapeutic target in chronic kidney disease and beyond? Pediatr Nephrol 30:363–371
18. Moe S, Drueke T, Cunningham J, Goodman W, Martin K, Olgaard K, Ott S, Sprague S, Lameire N, Eknoyan G, Kidney Disease: Improving Global, O (2006) Definition, evaluation, and classification of renal osteodystrophy: a position statement from Kidney Disease: Improving Global Outcomes (KDIGO). Kidney Int 69:1945–1953
19. Moe SM, Zidehsarai MP, Chambers MA, Jackman LA, Radcliffe JS, Trevino LL, Donahue SE, Asplin JR (2011) Vegetarian compared with meat dietary protein source and phosphorus homeostasis in chronic kidney disease. Clin J Am Soc Nephrol 6:257–264
20. U.S. Department of Agriculture, Agricultural Research Service. What We Eat in America, 2013–2014.
21. Nakamura Y, Okuda N, Turin TC, et al. (2010) Ministry of Health and Welfare. The National Nutrition Survey in Japan, 2010. Daiichi Shuppan, Tokyo
22. Nakatani S, Nakatani A, Mori K, Emoto M, Inaba M, Razzaque MS (2022) Fibroblast growth factor 23 as regulator of vitamin D metabolism. Adv Exp Med Biol. doi: 10.1007/978-3-030-91623-7_6
23. Oenning LL, Vogel J, Calvo MS (1988) Accuracy of methods estimating calcium and phosphorus intake in daily diets. J Am Diet Assoc 88:1076–1080
24. Ohnishi M, Nakatani T, Lanske B, Razzaque MS (2009) In vivo genetic evidence for suppressing vascular and soft-tissue calcification through the reduction of serum phosphate levels, even in the presence of high serum calcium and 1,25-dihydroxyvitamin d levels. Circ Cardiovasc Genet 2:583–590
25. Ohnishi M, Razzaque MS (2010) Dietary and genetic evidence for phosphate toxicity accelerating mammalian aging. FASEB J 24:3562–3571
26. Osuka S, Razzaque MS (2012) Can features of phosphate toxicity appear in normophosphatemia? J Bone Miner Metab 30:10–18
27. Palmer SC, Hayen A, Macaskill P, Pellegrini F, Craig JC, Elder GJ, Strippoli GF (2011) Serum levels of phosphorus, parathyroid hormone, and calcium and risks of death and cardiovascular disease in individuals

with chronic kidney disease: a systematic review and meta-analysis. JAMA 305:1119–1127
28. Razzaque MS (2009a) The FGF23-Klotho axis: endocrine regulation of phosphate homeostasis. Nat Rev Endocrinol 5:611–619
29. Razzaque MS (2009b) FGF23-mediated regulation of systemic phosphate homeostasis: is Klotho an essential player? Am J Physiol Renal Physiol 296:F470–F476
30. Razzaque MS (2022) Salivary phosphate as a biomarker for human diseases. FASEB Bioadv 4: 102–108. https://doi.org/10.1096/fba.2021-00104
31. Razzaque MS (2022) Phosphate metabolism: from physiology to toxicity. Adv Exp Med Biol 1362: 1–6. https://doi.org/10.1007/978-3-030-91623-7_1
32. Sasaki S (2008) Dietary Reference Intakes (DRIs) in Japan. Asia Pac J Clin Nutr 17(Suppl 2):420–444
33. Sherman RA, Mehta O (2009) Dietary phosphorus restriction in dialysis patients: potential impact of processed meat, poultry, and fish products as protein sources. Am J Kidney Dis 54:18–23
34. Sigrist MK, Chiarelli G, Lim L, Levin A (2009) Early initiation of phosphate lowering dietary therapy in non-dialysis chronic kidney disease: a critical review. J Ren Care 35(Suppl 1):71–78
35. Sullivan C, Sayre SS, Leon JB, Machekano R, Love TE, Porter D, Marbury M, Sehgal AR (2009) Effect of food additives on hyperphosphatemia among patients with end-stage renal disease: a randomized controlled trial. JAMA 301:629–635
36. Sullivan CM, Leon JB, Sehgal AR (2007) Phosphorus-containing food additives and the accuracy of nutrient databases: implications for renal patients. J Ren Nutr 17:350–354
37. Takeda E, Sakamoto K, Yokota K, Shinohara M, Taketani Y, Morita K, Yamamoto H, Miyamoto K, Shibayama M (2002) Phosphorus supply per capita from food in Japan between 1960 and 1995. J Nutr Sci Vitaminol (Tokyo) 48:102–108
38. Uribarri J, Calvo MS (2003) Hidden sources of phosphorus in the typical American diet: does it matter in nephrology? Semin Dial 16:186–188
39. Yoshiike N, Matsumura Y, Iwaya M, Sugiyama M, Yamaguchi M (1996) National nutrition survey in Japan. J Epidemiol 6:S189–S200

11 Phosphate Is a Cardiovascular Toxin

Maren Leifheit-Nestler, Isabel Vogt, Dieter Haffner, and Beatrice Richter

Abstract

Phosphate is essential for proper cell function by providing the fundamentals for DNA, cellular structure, signaling and energy production. The homeostasis of phosphate is regulated by the phosphaturic hormones fibroblast growth factor (FGF) 23 and parathyroid hormone (PTH). Recent studies indicate that phosphate induces phosphate sensing mechanisms via binding to surface receptors and phosphate cotransporters leading to feedback loops for additional regulation of serum phosphate concentrations as well as by phosphate itself. An imbalance to either side, enhances or reduces serum phosphate levels, respectively. The latter is associated with increased risk for cardiovascular diseases and mortality. Hyperphosphatemia is often due to impaired kidney function and linked to vascular disease, hypertension and left ventricular hypertrophy. In contrast, hypophosphatemia either due to reduced dietary intake or intestinal absorption of phosphate or hereditary or acquired renal phosphate wasting, may result in impaired energy metabolism and cardiac arrhythmias. Here, we review the effects and its underlying mechanisms of deregulated serum phosphate concentrations on the cardiovascular system. Finally, we summarize the current therapeutic approaches for both lowering serum phosphate levels and improvement of cardiovascular disease.

Keywords

Phosphate · Fibroblast growth factor 23 · Parathyroid hormone · Vascular disease · Left ventricular hypertrophy · Intervention

M. Leifheit-Nestler (✉) · I. Vogt · D. Haffner · B. Richter
Department of Pediatric Kidney, Liver and Metabolic Diseases, Pediatric Research Center, Hannover Medical School, Hannover, Germany
e-mail: leifheit-nestler.maren@mh-hannover.de; vogt.isabel@mh-hannover.de; haffner.dieter@mh-hannover.de; richter.beatrice@mh-hannover.de

11.1 Physiology of Phosphate Homeostasis

The element phosphorus is essential for normal cell function and occurs in combination with oxygen by generating phosphate. In the body, phosphate is found as structural element in teeth and bone in the form of hydroxyapatite and as phospholipids in cell membranes. Furthermore, DNA is a chain of PO_4^{3-}-containing molecules. Phosphate is involved in energy production as energy-carrying molecule adenosine triphosphate (ATP) and in the activation of enzymes, hormones and signaling cascades. In red blood cells, the salt 2,3-diphosphoglycerate controls the oxygen release to tissue by hemoglobin.

Serum phosphate levels are subject to tightly regulated homeostasis determined by dietary

M. S. Razzaque (ed.), *Phosphate Metabolism*, Advances in Experimental Medicine and Biology 1362,
https://doi.org/10.1007/978-3-030-91623-7_11

intake, intestinal absorption and renal glomerular filtration and tubular reabsorption. Active transport of phosphate is facilitated via type II sodium-dependent phosphate co-transporters NaPi-2b in the brush border membrane (BBM) of enterocytes and NaPi-2a and NaPi-2c in the BBM of renal proximal tubule cells. Type III sodium-coupled phosphate transporters, PiT1 and PiT2, initially discovered as retroviral receptors, are ubiquitously expressed in mammalian tissue [1, 2]. Low dietary phosphate upregulates the expression of PiT1 in the renal proximal tubule cell [3]. Both phosphaturic hormones parathyroid hormone (PTH) and fibroblast growth factor (FGF) 23 downregulate PiT2 in proximal tubules of the kidney [4, 5]. For now, the contribution of PiTs for renal and intestinal phosphate transport is not fully understood. The tubular reabsorption of phosphate is regulated by FGF23 and PTH, which regulate the renal production and secretion of active vitamin D ($1,25(OH)_2D_3$). The latter in turn enhances intestinal absorption of dietary phosphate in the jejunum and ileum by stimulating NaPi-2b-dependent phosphate transport. In the kidney, PTH decreases the abundance of NaPi-2a and NaPi-2c in the proximal tubule BBM leading to reduced tubular reabsorption of phosphate.

Normal serum phosphate levels in healthy adults are 2.5–4.5 mg/dL [6]. Hyperphosphatemia is defined as serum phosphate levels more than 4.5 mg/dL. For hypophosphatemia, it is distinguished between mild (2–2.5 mg/dL), moderate (1–2 mg/dL) and severe (<1 mg/dL).

11.1.1 FGF23

The main source of FGF23 is the bone where it is synthesized by osteocytes and osteoblasts due to different stimuli, among phosphate is one of them. However, the mechanisms how exactly phosphate affects FGF23 expression in bone are still unknown. FGF23`s main target organ are the kidneys where it binds to a complex of FGF receptor 1 (FGFR1) and its specific co-receptor klotho in renal proximal tubule cells [7]. The FGF23/FGFR1/klotho complex activates an intracellular cascade involving signal transduction via fibroblast growth factor receptor substrate (FRS) 2a and extracellular signal-regulated kinases (ERK) 1/2 affecting serum phosphate levels by two mechanisms. First, FGF23-mediated ERK1/2 activation inhibits both the expression of NaPi-2a and NaPi-2c and their abundance in the renal BBM. The decreased bioavailability of the transporters reduces renal phosphate reabsorption in the proximal tubules and thereby increases phosphate excretion in urine. Second, through the activation of the same pathway, FGF23 downregulates the expression of CYP27B1, encoding for 1α-hydroxylase, which converts prohormone 25-hydroxyvitamin D3 into the active form $1,25(OH)_2D_3$ [8], and upregulates the expression of CYP24A1, which encodes for the catabolic enzyme 24-hydroxylase. Reduced renal synthesis of active vitamin D leads to decreased serum $1,25(OH)_2D_3$ levels and thus, a low abundance of NaPi-2b in the gastrointestinal tract limiting phosphate uptake [9]. Both the suppression of renal phosphate reabsorption and reduction of intestinal phosphate absorption mediated by FGF23 lower the phosphate level in the serum.

11.1.2 PTH

PTH is the second phosphaturic hormone regulating phosphate homeostasis. High serum phosphate concentrations stimulate the secretion of PTH in the parathyroid glands that increases renal phosphate excretion. Thereby, PTH binds to the G-protein coupled PTH receptor 1 (PTHR1) that increases intracellular cyclic adenosine monophosphate (cAMP) levels and subsequently activates protein kinase A (PKA), protein kinase C (PKC) and ERK1/2 [10–12]. Activation of these signaling cascades leads to the internalization of NaPi-2a and NaPi-2c co-transporters from the BBM, resulting in renal phosphate wasting [13, 14]. To note, PTHR1 expression is found at both the apical and the basolateral membranes of renal proximal tubular cells [15]. In addition to maintaining phosphate homeostasis, PTH further interacts with FGF23 in the parathyroid gland.

Similarly to the kidney, FGF23 also binds to and activates the FGFR1/klotho complex in the parathyroid. The subsequently induced ERK1/2 signaling promotes the expression of the transcription factor early growth response 1 (Egr-1), which in turn decreases the transcription and secretion of PTH. *In vitro* and *in vivo* studies have shown that PTH induces the FGF23 expression in bone that initiates a negative feedback loop [16]. Thereby, FGF23 mRNA expression is increased via activation of intracellular PKA and WNT pathways by PTHR1. Furthermore, PTH affects bone mass by stimulating both bone resorption and bone formation. During chronic hyperparathyroidism, the catabolic actions of PTH reduce the bone mineral density causing an increased risk for fractures because of a net loss of bone mass [17]. During increased bone turn over, the liberation of calcium-phosphate product leads to an increase of serum phosphate levels and, over time, possibly to soft tissue calcifications. Interestingly, intermittent, low doses of PTH improve bone formation via stimulation of osteoanabolic WNT signaling [18, 19]. Due to the missing phosphaturic effect of PTH during hypoparathyroidism, patients display hyperphosphatemia [20]. Thus, hypoparathyroidism causes a reduction in bone remodeling with abnormalities in bone strength and microarchitecture as well as increased bone density [21].

11.1.3 Phosphate

Phosphate sensing has been extensively studied in bacteria and yeast (excellently reviewed in [22]). Since these unicellular organisms use types of phosphate transporters for phosphate sensing, it might be considered to hypothesize similar mechanisms for mammalian cells. *In vitro* data by Bon et al. support this idea. The authors show that by deleting PiT1 or PiT2, the phosphate-induced ERK1/2 activation as well as increase in gene expression of matrix Gla protein (MGP) and osteopontin were blunted [23]. Furthermore, overexpressing phosphate transport-deficient PiT mutants rescued the activation of ERK1/2. Even in the absence of phosphate transport activity, the heterodimerization of PiT1 and PiT2 was regulated by extracellular phosphate. These data suggest that the phosphate sensing function of PiT co-transporters does not depend on cellular phosphate uptake, but on phosphate binding. In a follow-up study, the authors addressed the question if PiT1 and PiT2 contribute to the regulation of FGF23 secretion from bone. *Ex vivo* results suggest that PiT1 deletion in long bones does not contribute to the phosphate-mediated FGF23 secretion [24]. In contrast, by using *ex vivo* organ culture of long bones from PiT2 knockout animals, the authors showed that PiT2 is involved in the phosphate-dependent secretion of FGF23. They conclude that at least in hyperphosphatemic situations PiT2 plays a role for phosphate-mediated sensing to induce osseous FGF23.

Recently, it was shown that unliganded FGFR1 serves as phosphate-sensing receptor regulating the production of FGF23 in bone [25]. *In vivo* and *in vitro* data show that phosphate does not directly increase the expression of FGF23, but the expression of the enzyme encoded by the N-acetylgalactosaminyltransferase (GALNT) 3 gene via induction of the ERK1/2 signaling pathway. FGF23 is posttranslationally modified via several steps. GALNT3 initiates the O-glycosylation of FGF23, which prevents its cleavage and increases the bioavailability of active FGF23 in the serum [26, 27]. The authors show that inhibition and deletion of FGFR1 in osteoblasts/osteocytes prevented the dietary phosphate-induced GALNT3 expression in bone and subsequent elevation of serum FGF23 levels [28]. Thus, high phosphate load activates ligand-independent FGFR1-mediated intracellular signaling events in bone cells resulting in increased serum FGF23 levels to regulate serum phosphate levels.

The calcium-sensing receptor (CaSR) is the main controller of PTH secretion [29]. Centeno et al. identified the CaSR in the parathyroid as phosphate sensor [30]. Here, the authors show that extracellular phosphate is a noncompetitive antagonist of CaSR resulting in phosphate-stimulated PTH secretion. Phosphate binds to phosphate-binding sites in the CaSR causing a conformational change of the receptor [30]. Phosphate-induced CaSR inhibition leads to its open, inactive form that permits increased PTH secretion. In turn, PTH stimulates renal phosphate excretion to normalize serum phosphate levels.

Taken together, most likely several molecules function as phosphate sensing receptors and interfere in the feedback of phosphate regulation. More studies are needed to clarify the involved players and signaling events.

11.2 Phosphate-Related Cardiovascular Disease

Impaired phosphate homeostasis can result in hyper- or hypophosphatemia causing multiple phosphate-related disorders, such as cardiovascular diseases (CVD). Hyperphosphatemia can develop as a result of increased phosphate intake, defects in bone mineralization, genetic disorders in genes of phosphaturic hormones or an impaired renal phosphate excretion [31]. Most studies, analyzing the cardiovascular (CV) risk of phosphate were obtained from chronic kidney disease (CKD) patients, because the progressive loss of kidney function leads to the development of hyperphosphatemia that is associated with vascular calcification, left ventricular hypertrophy (LVH) and increased CV mortality in CKD patients [32–36]. In the general population, adverse effects of high phosphate levels gained attention as the average dietary phosphate intake of 1200 mg greatly exceeds the recommended dietary allowance (RDA) of 700 mg in the Western population [37]. Epidemiologic studies show an association of high serum phosphate levels with an increased all-cause and CV mortality in the general population [37–39]. In subjects with preserved renal function, increased serum phosphate levels associate with an increased risk for vascular calcification even within normal phosphate range [40–43]. The higher risk of coronary artery calcification (CAC) was observed in young white and African-American adults (mean age: 25.2 years) [40] as well as in Korean participants with mean age of 40.8 [42] or 53.5 [43] years, respectively, suggesting an association of high phosphate with vascular calcification irrespective of age and ethnic origin. High phosphate levels enhance the ankle brachial index indicating that phosphate contributes to peripheral arterial stiffness [41]. In participants without pre-existing CVD or CKD, high serum phosphate and dietary phosphate intake correlate with an increased left ventricular (LV) mass and prevalence of LVH [44–46]. In a small cohort of 20 healthy young adults, controlled high dietary phosphate intake over 11 weeks increased systolic and diastolic blood pressure (BP) [47]. The underlying mechanisms of multiple hyperphosphatemia-related CV events and increased mortality are not completely understood and therefore focus of research.

11.2.1 Mechanisms of Phosphate-Induced Vascular Calcification

Phosphate-induced vascular calcification occurs predominantly in form of pathological calcium phosphate depositions in the medial layer of the arteries. Medial vascular calcification is a common health risk in CKD, diabetes and aging but also high serum phosphate levels in the general population are associated with CAC. Vascular smooth muscle cells (VSMCs) play a key role in the onset and progression of phosphate-induced vascular calcification. In high phosphate conditions, VSMCs undergo osteochondrogenic transdifferentiation promoting calcification by reducing calcification inhibitors, stimulating extracellular matrix (ECM) remodeling, apoptosis, senescence, and pro-inflammatory responses [48, 49].

VSMCs incorporate phosphate via PiT1 and PiT2 [50]. PiT1 also acts as phosphate sensor and mediates vascular calcification via activation of ERK1/2 signaling independent of phosphate uptake. In transport-deficient PiT1 mutants that cannot incorporate phosphate and wild type VSMCs, high phosphate levels promote ERK1/2 signaling and osteochondrogenic differentiation, which is attenuated by knockdown of PiT1 [51, 52]. In contrast, knockdown of PiT2 increased phosphate-induced calcification by lowering osteoprotegerin (OPG) in VSMCs suggesting a protective function of PiT2 in phosphate-mediated calcification [53] (Fig. 11.1). It is discussed that Toll-like receptor 4 (TLR4) might be involved in extracellular phosphate-sensing as it is upregulated under high phosphate conditions *in vitro* in VSCMs. Likewise, knockdown of TLR4 by siRNA reduces calcification and expression of inflammatory cytokines in VSMC [54].

11.2.1.1 Phosphate-Dependent Osteochondrogenic Transdifferentiation of VSMCs

High extracellular phosphate levels activate several intracellular pathways stimulating the transdifferentiation from contractile into osteochondrogenic VSMCs cells. The nuclear factor "'kappa-light-chain-enhancer" of activated B cells (NF-κB) is a key regulator of phosphate-induced osteochondrogenic transdifferentiation and subsequent vascular calcification (Fig. 11.1). Phosphate activates NF-κB signaling in VSMCs via serum- and glucocorticoid-inducible kinase (SGK1) [55], oxidative stress [56] or inflammatory pathways [57]. Activation of NF-κB promotes the expression of osteochondrogenic transcription factors, such as msh homeobox 2 (MSX2) and core-binding factor subunit α1 (CBFA1) and concurrently decreases the expression of the VSMC-specific marker SM22α [54, 55, 57]. The osteochondrogenic transcription factors further increase the expression of alkaline phosphatase (ALP) that promotes mineral deposition in VSMCs [57, 58]. Inhibition of NF-κB is sufficient to reduce calcification during high phosphate conditions in VSMCs *in vitro* and in a CKD mouse model *in vivo* [54, 59].

Furthermore, activation of WNT/β-Catenin signaling by phosphate contributes to osteochondrogenic transdifferentiation and vascular calcification (Fig. 11.1). In VSMCs, high phosphate induces β-Catenin activity and downstream MSX2, CBFA1 and PiT1 expression [60, 61]. Besides the induction of transdifferentiation, WNT signaling promotes calcification by stimulating the expression of the matrix metalloproteinases MMP2 and MMP9 in VSMCs [62]. Rodent CKD models fed with a high phosphate diet show an association of WNT/β-Catenin activation and aortic calcification [61, 63]. The lentiviral knockdown with short hair pin RNA of β-Catenin attenuates vascular calcification *in vivo* [61]. Likewise, inhibitors of WNT/β-Catenin signaling such as secreted frizzled-related proteins [64], Dickkopf-related protein 1 [65] or the knockdown of WNT8b [66] reduces phosphate-induced calcification in VSMCs.

Experimental studies suggest that interaction of WNT/β-Catenin and peroxisome proliferator-activated receptor gamma (PPARγ) signaling contribute to calcification during hyperphosphatemia [67]. In the differentiation of mesenchymal stem cells PPARγ and WNT act antagonistically. PPARγ activates adipocyte differentiation and suppresses osteoblast differentiation, whereas WNT/β-Catenin signaling promotes osteogenesis and inhibits PPARγ expression [68]. Hyperphosphatemia induces calcification by downregulating the expression of PPARγ and its downstream target klotho in VSMCs and rodent CKD. The PPARγ agonist Rosiglitazone stimulates the expression of klotho and reduces vascular calcification *in vitro* and *in vivo* [69, 70]. Other PPARγ agonists, Pioglitazone and Ginsenoside Rb 1, exert anti-calcific effects by inhibiting WNT/β-Catenin signaling [67, 71].

Phosphate-mediated downregulation of AKT signaling might as well contribute to vascular calcification in VSMCs [72, 73]. Upregulation of AKT signaling by the farnesyl transferase inhibitor 277 prevents osteochondrogenic differentiation and mineral deposition in VSMCs [73]. Extracellular acid loading stimulates AKT phosphorylation and could reduce phosphate-mediated apoptosis and calcification [72]

11.2.1.2 ECM Remodeling During Hyperphosphatemia

The ECM is a highly structured network composed of fibrous proteins and proteoglycans [74]. Remodeling of the ECM by phosphate creates an environment that fosters the development of vascular calcification in which elastin

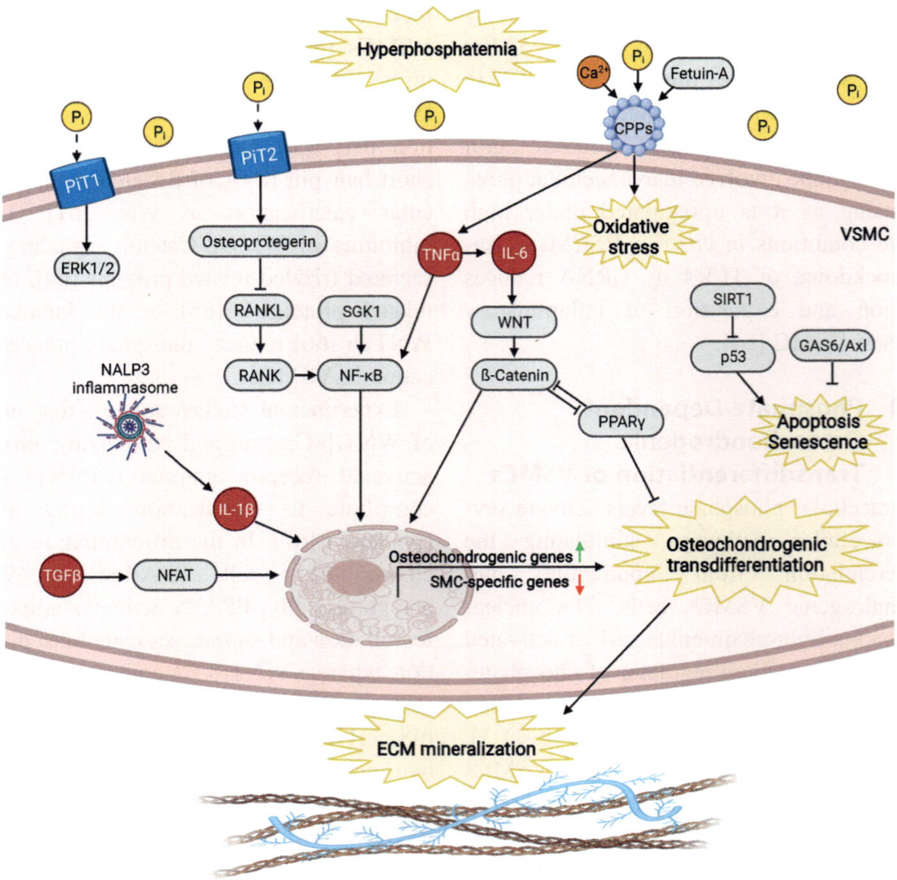

Fig. 11.1 Hyperphosphatemia-induced mechanisms in vascular smooth muscle cells contributing to vascular calcification. High extracellular phosphate (P_i) stimulates the production of inflammatory cytokines, oxidative stress, apoptosis, senescence and osteochondrogenic transdifferentiation in VSMCs, which together contribute to increased ECM mineralization. The simplified schematic illustration shows the main pathways involved in P_i-mediated VSMC dysfunction. P_i sensing and uptake via PiT1 promotes osteochondrogenic transdifferentiation, whereas PiT2-mediated increase in osteoprotegerin protects from calcification. P_i, TNFα and IL-6 promote osteochondrogenic transdifferentiation by upregulating the expression of osteochondrogenic genes and downregulating SMC-specific genes via NF-κB and WNT/β-Catenin signaling. Osteochondrogenic transdifferentiation is also stimulated by P_i-induced TGFβ/NFAT and NALP3/IL-1β activation. P_i stimulates apoptosis by downregulating GAS6/Axl and senescence by downregulating SIRT1 expression. Incorporated in CPPs, P_i induces oxidative stress and inflammation in VSMCs. (This figure was created with BioRender.com)

degradation via MMPs plays a crucial role. MMP2 and MMP9-knockout mice are rescued from calcium chloride induced aortic injury and neither develop elastin degradation nor calcification [75, 76]. Hyperphosphatemia predominantly triggers ECM remodeling and calcification in the medial layer of vasculature [77–79]. Phosphate-induced expression of MMP2, MMP9 and the cysteine protease cathepsin S promote degradation of extracellular matrix proteins and thereby calcification *in vitro* and *in vivo* [80–82]. Elastin-derived peptides only induce calcification under high phosphate conditions in VSMCs and in uremic mice suggesting that elastin-derived peptides are necessary but not sufficient to initiate calcification [76, 83]. In CKD patients on dialysis, increased MMP2 expression correlates with elastic fiber disorganization and calcification [84].

High phosphate levels are further associated with an increased collagen I (Col I) expression in clinical and experimental studies [85–88]. Col I promotes matrix vesicle-mediated mineralization and accelerates vascular calcification *in vitro* [88, 89]. Additionally, phosphate mediates collagen cross-linking by upregulating the enzymes lysyl hydroxylase 1 (PLOD1) and lysyl oxidase (LOX). Inhibition of collagen synthesis or cross-linking reduce phosphate-induced calcification in VSMCs [86, 90].

11.2.1.3 Phosphate-Induced Apoptosis, Senescence and Autophagy

Another mechanism how phosphate promotes calcification is apoptosis. In arteries of children with CKD, dialysis induces VSMCs apoptosis, followed by osteochondrogenic differentiation and subsequent severe calcification [91]. Further studies give evidence that VSMCs apoptosis occurs before the onset of calcification and inhibition of apoptosis with a caspase inhibitor can reduce calcification [92, 93]. In VSMCs, elevated phosphate levels induce apoptosis by downregulating the survival pathway mediated by growth arrest-specific gene 6 (Gas6) and its receptor Axl (Fig. 11.1). Statins could stabilize Gas6 mRNA and thereby have anti-apoptotic and anti-calcific effects in phosphate-induced calcification [94].

Senescence describes a permanent cell arrest that occurs in age or due to stress. In VSMCs and CKD rats, hyperphosphatemia increases the senescence-associated β-galactosidase activity that is associated with calcification. Phosphate mediates senescence via downregulation of Sirtuin 1 expression and subsequent activation of p53 and p21 [95–97] (Fig. 11.1).

Autophagy counteracts phosphate-induced vascular calcification by inhibiting osteochondrogenic transdifferentiation and matrix vesicle release [98–100]. In high phosphate diet fed mice and VSMCs under high phosphate culture conditions, autophagy is increased indicating an endogenous mechanism to protect from calcification. However, a recent study suggests that high phosphate inhibits VSMCs autophagy via upregulation of OGlcNAc transferase and subsequently the Hippo-YAP pathway promoting phosphate-induced calcification [101]. Although, the direct effect of phosphate on autophagy needs further investigation, induction of autophagy might be an interesting therapeutic strategy to reduce phosphate-induced calcification.

11.2.1.4 Regulation of Endogenous Calcification Inhibitors by Phosphate

Extracellular fluids are supersaturated with calcium and phosphate and therefore circulating calcification inhibitors are necessary to prevent spontaneous ectopic precipitation. Pyrophosphate, fetuin-A, vitamin K dependent MGP, osteopontin (OPN) and osteoprotegerin are endogenous calcification inhibitors [102, 103].

In end-stage kidney disease (ESKD), i.e., in patients on dialysis, reduced plasma pyrophosphate levels are associated with arterial calcification [104–106]. Experimental studies show that high dietary phosphate intake reinforces aortic calcification in mice lacking ectonucleotide pyrophosphatase phosphodiesterase, the enzyme that synthesizes extracellular pyrophosphate [107]. *In vitro*, physiological levels of pyrophosphate are sufficient to prevent calcium phosphate deposition in VSMCs under normal phosphate concentrations but not in hyperphosphatemia [108]. *In vivo*, daily injections of pyrophosphate could prevent aortic calcification in uremic rodent models [109, 110].

Fetuin-A is a liver-derived circulating glycoprotein that builds complexes with calcium and phosphate to prevent ectopic precipitation. Recently, it was shown that fetuin-A levels decrease with the progression of CKD [111]. Experimental studies in uremic rodents show that low fetuin-A levels promote the precipitation of calcium phosphate and high levels protect from calcification [112, 113]. In fetuin-A deficient CKD mice, additional challenge with a high phosphate diet induces more profound calcification than in wild type CKD mice

[114]. However, several clinical studies confirm a correlation of low circulating fetuin-A levels with increased calcification in dialysis patients [111, 115–117], whereas others did not find a correlation of fetuin-A or OPN with calcification [118].

MGP is another endogenous calcification inhibitor synthesized by VSMCs and chondrocytes [119]. High extracellular phosphate stimulates the expression of MGP by activating the ERK1/2-Fra-1 pathway *in vitro* [120–122] that might counteract phosphate-induced calcification. However, the activity of MGP is vitamin K dependent. It was shown in CKD and ESKD patients that dephosphorylated uncarboxylated MGP (dp-ucMGP), the inactive form of MGP, is associated with vitamin K deficiency and vascular calcification [123–125], indicating that MGP cannot exert its anti-calcific function in these patients due to the lack of vitamin K.

OPN is an inhibitor of hydroxyapatite crystal formation *in vivo* and *in vitro* that is induced by hyperphosphatemia [126–128]. Physiologically, OPN is not found in arteries but under pathological conditions, such as ESKD, it localizes to areas of calcification [129]. Several studies demonstrate that high levels of OPN protect from calcification, whereas OPN deficiency promotes calcification [130–132].

OPG, a soluble receptor of the TNF-receptor family, acts anti-calcific by inhibiting osteoclastogenesis. OPG binds to the Receptor Activator of NF-κB Ligand (RANKL) and thereby prevents its interaction with the RANK receptor. Inhibition of the RANK-RANKL interaction downregulates NF-κB signaling as key regulator of phosphate-induced osteochondrogenic differentiation and vascular calcification [133, 134]. Consequently, knockdown of OPG leads to medial calcification that is further enhanced by high dietary phosphate intake in mice [135]. In contrast, clinical studies show that elevated OPG levels are associated with increased vascular calcification and mortality in CKD patients [136–139]. However, increasing OPG levels might rather be a compensatory mechanism to inhibit progression of calcification instead of being the underlying cause for it.

11.2.1.5 Hyperphosphatemia-Associated Inflammatory Pathways

Pro-inflammatory proteins and cytokines are associated with vascular calcification in CKD patients [140–142] and promote vascular calcification in VSMCs *in vitro* [143–145]. In patients on chronic hemodialysis, increased C-reactive protein (CRP) levels associated with more severe vascular calcification [141]. Already in early CKD stages, medial calcification of the aorta is accompanied by vascular inflammation and osteochondrogenic differentiation of VSMCs [142]. Experimental studies reveal that high phosphate levels directly stimulate the expression of pro-inflammatory cytokines such as tumor necrosis factor α (TNFα), interleukin-6 (IL-6), IL-1β, IL-18 and transforming growth factor β (TGFβ) [146–151]. High cytokine levels activate several intracellular pathways that contribute to osteochondrogenic transdifferentiation and the development of vascular calcification (Fig. 11.1).

TNFα activates NF-κB signaling in VSMCs that promotes the osteochondrogenic transdifferentiation by upregulating the osteogenic transcription factor MSX2 and ALP activity [57]. In a diabetic mouse model, increased TNFα levels induce vascular calcification via bone morphogenetic protein 2 (BMP2), MSX2 and WNT signaling [152, 153]. Besides activating pro-calcific intracellular pathways, TNFα stimulates the expression of the pro-inflammatory cytokine IL-6 [154]. It is under debate whether IL-6 alone is sufficient to induce vascular calcification or just in an inflammatory environment. Sun et al. found that IL-6 induces calcification via BMP2 in VSMCs under otherwise normal non-calcific culture conditions [145]. In contrast, in the study of Deuell et al. TNFα, but not IL-6 alone, was sufficient to induce mineralization of VSMCs in high phosphate but not control media. However, co-incubation of TNFα-treated VSMCs with IL-6 further increased calcification suggesting a synergistic effect. TNFα and IL-6 also interfere with RANKL-NF-κB signaling. RANKL stimulates the expression of TNFα and IL-6 in macrophages and *vice versa*, neutralization of TNFα, IL-6 or both cytokines reduce RANKL-mediated calcification in co-cultures of macrophages and VSMCs [155]. Another study shows that neutralization of IL-6 further prevents RANKL-mediated expression of the osteogenic markers CBFA1 and BMP2 and the calcification inhibitor OPN [156]. In summary, both, TNFα and IL-6, interfere with NF-κB and WNT signaling to induce vascular calcification.

Furthermore, high phosphate activates the NALP3 inflammasome resulting in higher expression of IL-1β and calcification in VSMCs. Inhibition of the NALP3 inflammasome mitigates IL-1β secretion and calcification [144]. Phosphate also stimulates TGFβ expression in VSMCs [150, 157] and TGFβ promotes osteochondrogenic gene expression via upregulating the transcription factor NFAT5 and its downstream target SOX9 [157, 158]. Moreover, TGFβ contributes to remodeling of the ECM by induction of Col I, fibronectin, osteocalcin and OPN [150, 159].

11.2.1.6 The Role of Calciprotein Particles in Vascular Calcification

In calcium and phosphate supersaturated fluids, circulating calcification inhibitors are necessary to prevent spontaneous ectopic precipitation. Fetuin-A, MGP and GRP form primary calciprotein particles (CPP) with calcium and phosphate to enable efficient clearance of amorphous calcium phosphate clusters. The primary CPPs can undergo ripening into a highly structured more stable crystalline form called secondary CPPs [160, 161]. High dietary phosphate intake increases serum CPP levels in experimental as well as clinical studies [162, 163]. In serum of dialysis patients, predominantly primary CPPs occur, most likely, because the ripening process is time-consuming and primary CPPs are rapidly cleared in the liver by sinusoidal endothelial cells. Secondary CPPs are cleared by liver-resident Kupffer cells via class A scavenger receptor-mediated endocytosis [160, 164, 165]. Clinical studies give evidence that increased CPP levels are associated with aortic stiffness in pre-dialysis CKD patients and coronary atherosclerosis [166, 167].

The predisposition of serum to form crystals can be assessed by the T_{50} test *in vitro*. In this test, serum is supersaturated with calcium and phosphate to determine the necessary time to convert 50% of primary into secondary CPPs. Shorter T_{50} times represent a higher propensity of the serum to form crystals. In CKD patients, lower T_{50} times are associated with progression of vascular calcification, increased atherosclerotic CV disease events and CV and all-cause mortality [166, 168–171].

In vitro studies show that secondary but not primary CPPs cause calcification of VSMCs by activating inflammatory pathways and inducing oxidative stress (Fig. 11.1). Treatment of VSMCs with secondary CPPs stimulates TNFα and activates NF-κB signaling promoting osteochondrogenic transdifferentiation [148]. In macrophages, secondary CPPs also stimulate pro-inflammatory TNFα and IL-1β [172]. In CKD patients, there is evidence that CPPs directly induce pro-calcific pathways but at least the T_{50} test serves as a marker for the progression and appearance of CV complications.

11.2.2 Hyperphosphatemia-Associated Left Ventricular Hypertrophy

Elevated serum phosphate levels are associated with an increased LV mass and incidence of LVH in the general population [44–46]. In a large prospective study with 4005 healthy young adults, 4.5% develop LVH during a 5-year follow-up. The development of LVH correlates with increased baseline phosphate levels assessed at the beginning of the study [46]. Additionally, Dhingra et al. found a higher prevalence of incident heart failure and increased LV mass in subjects with high serum phosphate levels in a prospective study with 3300 participants [44]. Besides elevated serum phosphate levels, high dietary phosphate intake is associated with an increasing LV mass in a multi-ethnic cohort of 4494 healthy participants [45]. Hyperphosphatemia is also a common CV risk factor in CKD patients because of the insufficient renal phosphate clearance, and correlates with an increased LV mass [173, 174]. Nevertheless, these studies have limitations i.e., the serum FGF23 levels were not measured. FGF23 is a well-known risk factor and inducer for LVH [173–178]. Thus, it is not possible to discriminate whether cardiac hypertrophy is directly mediated by phosphate or phosphate-induced FGF23 elevation. Furthermore, vascular calcification, especially medial calcification, not only leads to vascular stiffness, but also to LVH and systolic hypertension [179]. The structural changes by calcium-phosphate deposition in the vasculature leading to CAC can be diagnosed and quantified by electron beam computer tomography. Functional changes are measured by pulse wave velocity indicating vascular stiffness (Fig. 11.2d–e).

So far, experimental studies could also not clarify whether phosphate directly causes LVH. In some studies, feeding of a 2% high phosphate diet to wild type mice for 12 weeks induce LVH and cardiac fibrosis [180–182], whereas in others the cardiac function is not altered [183, 184]. Grabner et al. suggest that a high

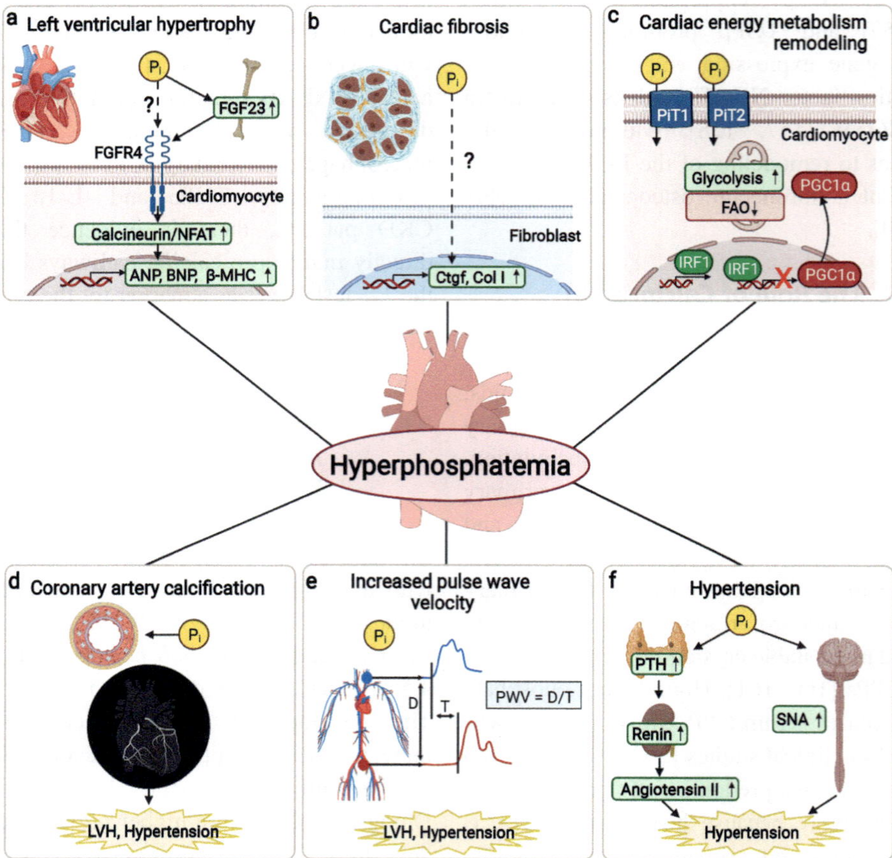

Fig. 11.2 Hyperphosphatemia-associated cardiac diseases. (**a**) P_i induces LVH via FGF23/FGFR4/calcineurin/NFAT signaling in cardiomyocytes or an unknown direct mechanism. (**b**) High P_i stimulates the expression of fibrotic genes Ctgf and Col I in cardiac fibroblasts via unknown mechanisms, (**c**) P_i promotes cardiac energy metabolism remodeling by entering cardiomyocytes via PiT1 and PiT2 to activate IRF1 expression that inhibits PGC1α. Downregulation of PGC1α reduces FAO and increases glycolysis. (**d**) Medial calcification, diagnosed by electron beam computer tomography and (**e**) increased pulse wave velocity, leads to LVH. (**f**) P_i triggers hypertension via PTH-mediated activation of the RAAS system and increased sympathetic nerve system (SNA) activity. (This figure was created with BioRender.com)

phosphate diet increases FGF23 and promotes LVH via FGFR4 [180] (Fig. 11.2a). This supports the hypothesis that phosphate mediates LVH only indirectly via FGF23. FGF23 promotes LVH by activating the PLCγ/calcineurin/NFAT pathway via FGFR4 [175]. *In vitro* studies in cardiac fibroblasts show pro-fibrotic effects of high phosphate that stimulates the expression of connective tissue growth factor and Col I but not Smad2/3 [181]. Thus, the contribution of phosphate for the development of cardiac fibrosis remains unsolved (Fig. 11.2b).

In the studies of Amann et al. and Neves et al., a 1.2% high phosphate diet for 8 weeks only promotes the development of LVH and cardiac fibrosis in uremic but not wild type rats [185, 186]. In CKD mice, feeding of a 2% high phosphate diet for 12 weeks further enhance LVH compared to a normal phosphate diet [187]. Huang et al. show that phosphate alone induces the hypertrophic markers ANP, BNP and β-MHC in the embryonic rat cardiomyocyte H9c2 cell line and in primary neonatal rat ventricular myocytes *in vitro*. Additionally, high phosphate disturbs the myocardial energy metabolism in H9c2 cells *in vitro* and in CKD mice *in vivo*. Phosphate enters cardiomyocytes via PiT1 and PiT2 and increases the expression of interferon regulatory factor 1 (IRF1) (Fig. 11.2c). IRF1 inhibits the expression of peroxisome proliferator-activated receptor gamma co-activator 1 alpha (PGC1α) and thereby, induces the switch from fatty acid oxidation to glucose dependent metabolism [187]. Together, these studies indicate that phosphate might only induce LVH under certain metabolic conditions. The source of phosphate and further composition of the diet, such as calcium or vitamin D content, could influence phosphate absorption and related effects. Inorganic phosphate is absorbed by over 90% in the intestine, organic phosphates only by 40–60% [188]. Therefore, phosphate composition of the diet plays a crucial role in the actual phosphate absorption and could influence phosphate-mediated effects. Furthermore, experimental studies give evidence that low vitamin D concentrations reduce high phosphate diet-induced LVH, but augment fibrosis. In contrast, high vitamin D concentrations promote high phosphate diet-induced cardiac hypertrophy and attenuate fibrosis [182]. Future studies need to investigate whether phosphate could cause LVH directly or indirectly and trigger cardiac remodeling only under certain metabolic conditions.

11.2.3 The Role of Hyperphosphatemia in Hypertension

Hyperphosphatemia is a potential risk factor for the development of hypertension [189]. In a large cohort of more than 9000 hypertensive participants, higher baseline serum phosphate levels are associated with increased systolic BP after 5-year follow-up. However, the poor BP control in patients with elevated baseline serum phosphate level does not correlate with the increased mortality in these patients [190]. A study in CKD patients reveals that serum phosphate levels only associate with higher systolic and diastolic BP in diabetic but not in non-diabetic CKD patients [191]. Also in the general population, high dietary phosphate intake for 6 weeks increases systolic and diastolic BP in young healthy adults that is not reversed by a concomitant vitamin D treatment for another 5 weeks. Reducing the high dietary phosphate load to a normal diet reverse the phosphate-induced elevation in BP after 2 months. Mohammad et al. suggest activation of the sympathetic nerve system as the underling mechanism for phosphate-induced hypertension and pulse rate elevation [47]. This study is limited by its small cohort size of 20 subjects, so the results have to be validated in larger cohorts.

However, experimental animal studies support the idea that phosphate stimulates the sympathetic nerve system (Fig. 11.2f). Likewise to the clinical study of Mohammad et al., feeding of a high phosphate diet activates the sympathetic nerve system and increases BP in healthy rats [192]. Independent of kidney function, a high phosphate diet also augments phenylephrine-induced vasoconstriction in aortic rings from mice *ex vivo* [193]. Another mechanism of phosphate-induced hypertension might be the activation of the renin angiotensin aldosterone system. High dietary phosphate intake increases renin expression via PTH and subsequently angiotensin II to induce hypertension in healthy rats [194]. The effect of hyperphosphatemia on hypertension and its underlying mechanisms has to be confirmed in larger clinical trials to develop suitable therapeutic strategies.

11.3 Hypophosphatemia and CVD Risk

The mechanisms causing the development of hypophosphatemia are decreased intestinal absorption, shift of extracellular phosphate into cells and urinary phosphate loss. A reduced phosphate intake can be caused by severe malnutrition and in patients with symptoms that affect their food intake, including vitamin D deficiency and alcohol [6, 195]. Thereby, vitamin D deficiency is associated with the development of CVD risk factors, e.g., hypertension [196]. This is supported by the results found in vitamin D receptor null mice developing high BP, cardiac hypertrophy and fibrosis [197, 198]. Furthermore, it is known for a long time that chronic heavy drinking is a cause for hypertension, arrhythmias and heart failure, thus, increases the risk for death and CVD [199]. Hypophosphatemia caused by redistribution of extracellular phosphate into cells does not mean a depletion of total body phosphate. The mechanisms causing internal redistribution can be respiratory alkalosis, catecholamines, administration of hormones and other drugs (e.g., insulin), alcohol and rapid cell proliferation.

Hereditary causes of hypophosphatemia are X-linked hypophosphatemia (XLH), autosomal dominant hypophosphatemic rickets (ADHR) and autosomal recessive hypophosphatemic rickets (ARDR) [200]. Furthermore, abundant expression of FGF23 in certain tumors can cause renal phosphate-wasting, so called tumor-induced osteomalacia (TIO).

Heart failure patients have an incidence of about 13% for hypophosphatemia [201]. Data investigating the association of hypophosphatemia with cardiomyopathy in regards for mortality are rare. Rozentryt et al. report that among their cohort with 722 heart failure patients and a 2-year follow-up, even the adjusted analysis for mortality was similar for patients with hypophosphatemia compared with patients having normal values of serum phosphate [202]. Hypophosphatemia is associated with an impairment of energy metabolism and the induction of arrhythmias [203]. Several studies discuss cardiac arrest as severe symptom of hypophosphatemia with the possibility leading to death [204–206]. Interestingly, several studies indicate a reversibility of cardiac impairments due to phosphate repletion [207, 208]. O'Connor et al. were the first showing that repletion of serum phosphate levels by potassium phosphate solution in seven patients with severe hypophosphatemia improved myocardial stroke work [207]. Returning serum phosphate levels to normal significantly increased the mean LV stroke work as determined by thermodilution technique. Zazzo et al. reported improved cardiac performance after normalizing serum phosphate concentrations in patients with moderate or severe hypophosphatemia by giving intravenous glucose phosphate during a short time [208]. By measuring cardiac output via thermodilution they demonstrate increased cardiac index, systolic index and LV stroke volume index after phosphate load. In contrast, several other cases show no evidence for an association between phosphate depletion and cardiac dysfunction at all [209, 210]. Thereby, Rasmussen et al. investigated the effect of moderate phosphate depletion (minimum serum phosphate levels of 1.46 mg/dL) induced by glucose infusion after surgery for myocardial performance [209]. The authors found no effect on stroke volume and cardiac output as well as unaltered mean arterial pressure concluding glucose-induced hypophosphatemia lacks affecting cardiac performance. Interestingly, Davis et al. show that only in cases of severe hypophosphatemia (0.9 ± 0.15 mg/dL), but not in patients with moderate hypophosphatemia (1.4 ± 0.11 mg/dL), the LV performance improved by corrections of serum phosphate levels

[210]. Thus, they hypothesized that the conflicting results are caused by the differences in serum phosphate levels among the different studies. To note, none of the hypophosphatemic patients in this cohort presented with clinical signs of congestive heart failure.

XLH is caused by inactivating mutations in the phosphate-regulating gene with homologies to endopeptidases on the X chromosome (PHEX) causing high serum FGF23 concentrations [211, 212]. For patients with XLH, the presence of CV complications is still under debate and varies in its characteristics. Alon et al. describe the presence of hypertension with the need for treatment in 8 out of 41 pediatric XLH patients [213]. Echocardiography revealed that two out of the eight hypertensive patients had LVH. Interestingly, all eight children with hypertension had hyperparathyroidism. Nehgme et al. investigated 13 patients with XLH, and despite no CV symptoms and normal heart structure as well as LV function, determined by cardiac ultrasonography, seven children presented LVH [214]. This was independent of the parathyroid status. During a maximal exercise stress test, all XLH patients with LVH had an abnormal diastolic BP response, but there was no significant correlation between these two parameters. Furthermore, cardiac calcification was not observed in these patients. More recent studies by Nakamura et al. (22 adult XLH patients) and Hernández-Frías et al. (24 pediatric patients) support the findings that characteristic symptoms of XLH patients can be the development of both hypertension and LVH [215, 216]. Only Hernández-Frías et al. reported the serum FGF23 concentrations in their XLH patient cohort (278.18 ± 294.45 pg/mL) and found no correlation with the observed echocardiographic parameters in the XLH patient cohort. Other studies give contradictory results in regards of CVD development in XLH patients [25, 217, 218]. Takashi et al. found no presence of LVH and FGF23 correlation with LVH-related parameters determined by echocardiography and electrocardiography in adult patients with FGF23-related hypophosphatemic rickets/osteomalacia [25]. In a cohort of pediatric patient with X-linked hypophosphatemic rickets, no evidence for LV dysfunction could be revealed by Doppler echocardiography [218]. Likewise, a study in adult patients with clinically significant hypophosphatemia, Vered Z et al. did observe normal LV function demonstrated using M-mode echocardiography and radionuclide ventriculography [217]. Interestingly, three patients displayed right ventricular dysfunction despite normal BP. Again, discrepancies might be explained by the initial serum phosphate concentration in the respective cohorts (Vered I et al.: lowest was 1.5 mg/dL; Vered Z et al. range between 1 and 2 mg/dL).

In line with clinical observations, investigations in the Hyp mouse, the animal model of XLH, show conflicting results too. In several studies, no signs of pathological remodeling were observed determined by histological analyses, gene expression analyses and echocardiography [219, 220]. Additionally, Liu et al. reported that tail cuff measurements did not reveal changes in systolic BP in male Hyp mice compared to wild type littermates [220]. In contrast, others showed that Hyp mice present LVH and hypertension [221, 222]. In another mouse model of XLH, the $Phex^{C733RMhda}$, bearing an amino acid substitution that causes mutations in the coding region for the large extracellular catalytic domain of the PHEX protein, despite high serum FGF23 levels in 27 weeks old mice, no LVH or altered cardiac function was present [223]. Importantly, the BP was not changed between the genotypes. Hypophosphatemic rickets in ARHR are caused by inactivating mutations in the gene dentin matrix protein 1 (DMP1) that encodes for a non-collagenous bone matrix protein in osteoblasts and osteocytes. This mutation leads to elevated serum FGF23 levels in patients and in animal models with DMP1 mutation, which causes the observed hypophosphatemia [224–226]. Wacker et al. showed that Dmp1 null mouse did not present an indication of cardiac hypertrophy, despite the increase of the heart weight to body weight ratio [200]. The authors did not observe the induction of classic pro-hypertrophic genes in cardiac tissue and they found no difference between wild type and Dmp1 null mice in *ex vivo* whole heart contractility test. The conflicting results of the different publications urge for the need of more studies to have a better understanding for the relationship between hypophosphatemia and cardiac complications.

11.4 Therapeutics Interventions to Prevent Phosphate-Dependent CVDs

Both, low and high phosphate levels may have severe effects on cardiac function. Thus, the detection of phosphate imbalance and its treatment is of high importance. Understanding the mechanisms of phosphate homeostasis shows that serum phosphate balance depends on an interplay between bone, kidneys and intestinal tract regulated by a tight endocrine system. Since phosphate, FGF23, PTH, $1{,}25(OH)_2D_3$ and calcium have a tight relation/interconnection to each other, changing one parameter influences all, making a therapeutic intervention rather complicated.

11.4.1 Treatment of Hypophosphatemia

Severe symptoms of phosphate depletion, such as cardiac arrhythmia and muscle weakness, were seen when serum phosphate levels are below 1 mg/dL [210]. However, most patients with hypophosphatemia are asymptomatic, but require the correction of the underlying cause, such as magnesium deficiency [195], which mostly normalizes serum phosphate levels. Oral phosphate supplementation is favorized for mild and moderate hypophosphatemia. Hereby, sodium phosphate tablets, potassium phosphate tablets or skim/cow milk are available phosphate sources. Corrections of the phosphate imbalance led to a reversibility of the myocardial dysfunction as shown in feeding experiments with dogs and rats [227, 228]. Phosphate depletion in dogs caused a decline in stroke volume and peak blood flow velocity as well as a decrease in maximum LV time rate change of pressure [227]. The values returned to control values after dietary phosphate repletion improving cardiac performance.

Intravenous supplementation of phosphate is used for patients with severe hypophosphatemia and in cases when oral supplementation is not possible. After a successful increase of serum phosphate concentration to >1 mg/dL, the intravenous administration can be switched to oral supplementation. Repletion of serum phosphate levels by intravenous administration of potassium phosphate solution or glucose phosphate in patients with moderate or severe hypophosphatemia improves cardiac performance [207, 208]. Thereby, normalizing serum phosphate concentrations increases mean LV stroke work, cardiac index, systolic index and LV stroke volume index after phosphate load. Furthermore, in early stages of sepsis low serum phosphate concentrations are associated with a high incidence for cardiac arrhythmias in patients without previous cardiac pathological conditions [229]. It has been shown that intravenous phosphate replacement in adult septic patients with hypophosphatemia, but without pre-existing cardiac condition, is associated with a significant reduction of incidence of arrhythmia (38% vs. 64% in non-treated group, $p = 0.04$) [203]. Hypophosphatemia is also a common problem after cardiac surgery [230]. Treatment of hypophosphatemia of a patient, who developed cardiac failure after cardiac surgery, by giving intravenous phosphate resulted in an improved cardiac index, indicating the consideration of preoperative serum phosphate measurements and postoperative management of hypophosphatemia [231]. In the case report of Frustaci and colleagues, the patient's characteristics were severe hypophosphatemia and impaired LV function measured by echocardiography [232]. Correction of serum phosphate levels by administration of potassium phosphate improved the myocardial contractility (ejection fraction increased from 40% to 72%).

All in all, ATP is essential for cardiac function, phosphate depletion will decrease cardiomyocyte function leading to cardiac failure, thus, management of hypophosphatemia is crucial for improving patient health. Hence, phosphate supplementation is not predictable, serum phosphate levels should be tightly controlled to avoid hyperphosphatemia or other electrolyte imbalances.

11.4.2 Treatment of Hyperphosphatemia

Increased serum phosphate levels are associated with a higher risk for CVD in the general population [233, 234]. Even more serious is hyperphosphatemia for patients with impaired kidney function. Thereby, the prevention and intervention of hyperphosphatemia is essential.

11.4.2.1 Phosphate Restricted Diet

The KDIGO (Kidney disease Improving Global Outcomes) guidelines suggest restricted dietary phosphate intake to attenuate hyperphosphatemia combined with other treatments for CKD patients [235]. In uremic rats, it has been shown that low dietary phosphate restriction reduces the perivascular fibrosis and cardiomyocyte size compared to rats on high phosphate diet [236]. Additionally, the mortality was reduced in rats under phosphate restriction. However, despite the reduction of serum FGF23 and phosphate levels in non-dialysis patients on a very low phosphate diet [237], more studies are needed to investigate the association between dietary phosphate uptake and improvement for general and CV outcome in the CKD cohort as well the general population.

11.4.2.2 Phosphate Binders Targeting Intestinal Phosphate Absorption

Phosphate binders prevent the intestinal phosphate absorption by binding phosphate, forming a nonabsorbable complex that is excreted in the feces. Different types of phosphate binders are approved to treat hyperphosphatemia in advanced stages of CKD, including calcium-containing phosphate binders (calcium acetate, calcium carbonate) and non-calcium-containing binders (sevelamer carbonate, lanthanum carbonate, ferric citrate). KDIGO guidelines suggest oral phosphate binder administration for patients with CKD stages 2–4 when serum phosphate levels are >4.6 mg/dL [238]. Calcium-containing phosphate binders appear to have pronounced adverse effects on vascular calcification and can cause hypercalcemia in patients requiring dialysis. The study of Block et al. compared lanthanum carbonate, sevelamer carbonate and calcium acetate in a randomized, double blind, placebo-controlled trial within a cohort of moderate to advanced CKD patients [239]. The authors show treatment with phosphate binders resulted in the progression of CAC and abdominal aortic calcification. Interestingly, Russo et al. studied the combined effect of low phosphate diet (<800 mg/d) with calcium carbonate or sevelamer in 90 pre-dialysis CKD patients with stable serum phosphate concentration [240]. This treatment combination reduced the progression of CAC, even in the calcium carbonate group, suggesting a relevance for avoiding CAC in patients not requiring dialysis. Despite beneficial effects of the phosphate binders sevelamer and ferric citrate on serum FGF23 levels [238, 241, 242] observations on cardiac function and performance were not objective of these studies. In a single-blinded, placebo-controlled, 3-month-study, Isakova et al. studied the effect of phosphate restricted diet and lanthanum in respect to serum FGF23 and cardiac function [243]. The study included 39 patients with CKD stages 3 or 4 with normal serum phosphate levels. The authors report that the combination of 900 mg phosphate-restricted diet plus lanthanum decreased serum FGF23 levels. However, echocardiography revealed no changes in ejection fraction or LV wall thickness between groups.

Ex vivo studies on aortic rings show that sevelamer treatment improved the CKD-induced endothelial dysfunction, aortic systolic expansion rate and pulse wave velocity [193]. In a murine model of chronic renal failure (CRF), Maizel et al. investigated the effects of sevelamer on CVD [244]. This model is characterized by the absence of hypertension and aortic calcification and by 6 weeks after surgery, CV abnormalities, such as LVH, diastolic dysfunction and aortic stiffness, have been developed in these mice. Six weeks after initiation of renal failure, CRF and sham-operated mice received regular chow or supplemented with 3% sevelamer for 14 weeks. Sevelamer reduced serum phosphate levels and improved aortic systolic expansion rate, pulse wave velocity and diastolic function, while LVH was not altered in CRF mice after 8 weeks of treatment. However, after 14 weeks of treatment, sevelamer prevented the progression of LVH in CRF mice. Furthermore, the multiple regression analysis revealed that serum phosphate concentrations, but not serum FGF23 levels, are associated with the LV diastolic function and mass in this mouse model. These results indicate that the phosphate binder sevelamer has beneficial effects on CV complications associated with renal impairment by lowering serum phosphate levels.

The iron-based intestinal phosphate binder ferric citrate hydrate further has positive effects on iron-related parameters, which in turn helps to normalized serum FGF23 levels [245]. Since iron deficiency stimulates FGF23 transcription and occurs frequently in CKD patients, correction of CKD-related anemia via increased hemoglobin, ferritin and transferrin saturation causes reduced serum FGF23 levels [246, 247]. Interestingly, an experimental study in Col4a3 knockout mice, a model for progressive CKD, showed that early initiation of ferric citrate slowed CKD progression, improved cardiac function and overall prolonged the survival of these mice [248]. Col4a3 knockout mice develop CKD-related anemia, have elevated serum phosphate concentrations and serum FGF23 levels rise by 6 weeks of age. In this study by Francis and colleagues, administration of ferric citrate rescued the iron deficiency and anemia regardless of the CKD stage. When starting the treatment in early CKD, but not in late CKD, both serum phosphate and FGF23 levels could be reduced. Furthermore, knockout mice on control chow developed hypertension and had significant lower ejection fraction, stroke volume and cardiac output compared to wild type mice on control diet. Feeding of ferric citrate mitigated the observed systolic dysfunction and lowered the BP in Col4a3 knockout mice. Summarizing that the combination of corrected serum phosphate levels, serum FGF23 levels and parameters of the iron metabolism might be beneficial for heart and kidney function.

Taken together, phosphate binders are well tolerated, efficacious, without safety concerns and commonly used to treat hyperphosphatemia among renal failure patients requiring dialysis. However, experimental and clinical studies about the impact of phosphate binders for CVD outcome in the CKD and in the general population are missing.

11.4.2.3 Drugs Targeting the Phosphate Transporters in the Intestine

Active intestinal phosphate absorption is mediated by the transporter sodium/hydrogen ion-exchange isoform 3 (NHE3) and novel minimally systemic NHE3 compounds have been developed. Tenapanor (previously referred to as RDX5791 or AZD1722) acts locally in the gut, inhibits NHE3 and therefore intestinal sodium and phosphate absorption. Spencer et al. showed that inhibition of intestinal sodium absorption by tenapanor has protective effects on sodium-driven cardiac and renal damage in rodents with CKD [249]. In a follow-up study, the authors showed that the inhibited sodium absorption in the intestine is accompanied by an inhibition of dietary phosphate uptake [250]. They show that administration of tenapanor increased fecal phosphate excretion and reduced urinary phosphate excretion accompanied by reduced ectopic calcification in CKD rats. In addition to the reduced hyperphosphatemia, CKD rats treated with tenapanor displayed reduced serum FGF23 levels and reduced cardiac hypertrophy. However, clinical trials investigating the benefits of tenapanor on cardiac function for patients displaying hyperphosphatemia are not available.

11.4.2.4 Drugs Targeting the Renal Phosphate Transporters

Studies in NaPi-2a knockout mice demonstrated that this cotransporter implements for roughly 70% of the renal phosphate reabsorption via the BBM of the proximal tubule [251]. The development of specific inhibitors against NaPi-2a might be useful to manage normal phosphate balance. A compound inhibiting NaPi-2a, named BAY767, showed in preclinical testing induction of phosphaturia and reduction of vascular calcification [252]. Of course, the value of this promising approach to treat hyperphosphatemia and associated CV dysfunctions must be tested in humans. Here, it should be considered that others hypothesized that NaPi-2c, not NaPi-2a, has a greater role for renal phosphate reabsorption in man [253]. As a second scenario, it is also possible that NaPi-2c expression increases for compensating the loss of NaPi-2a [254]. Maybe a combined approach of inhibiting both intestinal absorption and renal reabsorption could be beneficial to treat hyperphosphatemia and consequently reduce the risk for CV events and mortality.

11.4.2.5 Magnesium

A promising approach to prevent vascular calcification seems to be the administration of magnesium supplements, which are either inorganic (like magnesium chloride, magnesium oxide) or organic (magnesium citrate or magnesium glutamate) salts. *In vitro* studies showed that phosphate-induced calcification in VSMCs was reduced by treatment with magnesium [122, 255]. In bovine VSMCs, magnesium inhibited the beta-glycerophosphate-induced cell calcification, whereby, magnesium prevented the formation of secondary CCPs, which is a driving factor for vascular calcification [256]. In a register-based cohort with Japanese hemodialysis patients, increasing serum magnesium levels attenuate the risk for CV mortality associated with serum phosphate concentrations [257]. In rats with adenine diet-induced kidney disease, low-dose (375 mg/kg) and high-dose (750 mg/kg) of magnesium citrate significantly decreased serum phosphate levels compared to the untreated adenine diet alone [258]. The adenine diet-induced extensive vascular calcification was inhibited by magnesium citrate in a dose-dependent manner suggesting a protective role of magnesium citrate in calcification. The authors demonstrate that the underlying protective mechanism of magnesium is an increase of alpha smooth muscle actin and a decrease of pro-chondrogenic runt-related transcription factor-2 expression in aortas. This indicates magnesium inhibits the cellular transformation of VSMCs into osteoblast-like cells. Interestingly, Diaz-Tocados et al. reported that increasing dietary magnesium supplementation from 0.1% to 0.6% reversed established vascular calcification in 5/6 nephrectomized rats [259]. The serum phosphate levels in uremic rats were reduced by dietary magnesium, but the levels were still significantly higher compared to sham-operated animals. In addition, increased dietary magnesium reduced the BP in nephrectomized rats compared to controls. Furthermore, the beneficial action of magnesium was not restricted to oral application only, because intraperitoneally administered magnesium decreased vascular calcification too. Kaesler and colleagues reported reduced vascular calcification in heart aorta and kidney after treating 5/6 nephrectomized mice with either magnesium carbonate alone or the combination of magnesium carbonate plus nicotinamide [260]. The treatment significantly lowered serum FGF23 levels, but had only a slight effect on serum phosphate concentrations. Despite these promising results for attenuating calcification, single treatment of either magnesium carbonate or nicotinamide alone induced renal NaPi-2b expression and increased intestinal NaPi-2b expression, which in turn might enhance phosphate absorption and reabsorption. In summary, a combined strategy of magnesium carbonate and nicotinamide or another therapy reducing serum phosphate levels might be suitable to treat vascular calcification and control phosphate levels.

In an open-label, randomized, controlled trial in patients with CKD stages 3–4 presenting risk factors for calcification, orally administered magnesium oxide was effective to slow the progression of CAC [261]. Over the 2-year trial, the serum phosphate levels were unchanged by magnesium administration (3.5 mg/dL at baseline and 2 years later). Contrary to the results found for CAC, progression of thoracic aorta calcification was not suppressed by magnesium oxide. The authors speculate that thoracic aorta is resistant to uremic-induced calcification, because of the cellular origin of the VSMCs resulting in a less likelihood to undergo osteoblastic transdifferentiation. Furthermore, is has been reported that higher dialysate magnesium prolongs the conversation of primary CCP into secondary CPP in ESKD patients [262]. The conversation time, measured by T_{50} test, indicates the risk for vascular calcification, whereby a longer time means a lower risk. Additionally, serum phosphate levels were reduced by higher dialysate magnesium.

Despite the promising results from experimental and human studies, more investigations are need to clarify the role of magnesium for reducing the risk of CVD development and/ or progression.

11.5 Conclusions

Serum phosphate levels are regulated by FGF23, PTH and phosphate itself and are linked to the development of CVD. Hyperphosphatemia is a risk factor for the initiation of vascular calcification, LVH, fibrosis and hypertension. The main process of phosphate-induced vascular calcification is the osteochondrogenic transdifferentiation of VSMCs resulting in ECM remodeling, apoptosis, senescence and pro-inflammatory responses. In contrast, the mechanisms underlying phosphate-induced LVH, fibrosis and hypertension are poorly understood. Hypophosphatemic patients have a higher risk to develop cardiac arrhythmias. Keeping serum phosphate levels within the normal range is crucial for the CV outcome of patients especially suffering from renal failure. Despite several therapeutic approaches to normalize serum phosphate concentrations, only few studies present data on the impact of the treatment for associated CVD. Thereby, calcium free phosphate binders, such as sevelamer, combined with a second approach to lower serum levels of phosphate show promising results. Furthermore, ferric citrate potentially acts on two routes to prevent CVD. First, it functions as a phosphate binder reducing serum phosphate, and second, it helps to overcome iron deficiency and subsequently reducing serum FGF23 levels, a risk factor for LVH. Strategies to lower the intestinal absorption or renal reabsorption of phosphate via blocking the specific phosphate cotransporters are another important intervention. However, it must be demonstrated that this approach results in prevention or improvement of CVD. The most promising intervention for hyperphosphatemia-related CVD for now seems to be normalizing magnesium concentrations in the serum. Experimental studies indicate that dietary magnesium supplementation is beneficial for CV calcification. Recent clinical studies do also point in this direction. However, for all the discussed interventions to lower serum phosphate levels, the value for improvements of CVD, such as reducing the mortality risk, slow down the progression, prevent disease induction, reverse the disease outcome, are still not completely known. Summarizing the current knowledge of phosphate-induced CVD and therapeutic options reveals the urgent need for more experimental and clinical investigations addressing underlying mechanisms and especially CV outcomes of therapeutic strategies in patients suffering from out-off range serum phosphate levels.

References

1. Kavanaugh MP, Miller DG, Zhang W et al (1994) Cell-surface receptors for gibbon ape leukemia virus and amphotropic murine retrovirus are inducible sodium-dependent phosphate symporters. Proc Natl Acad Sci U S A 91(15):7071–7075. https://doi.org/10.1073/pnas.91.15.7071
2. Olah Z, Lehel C, Anderson WB et al (1994) The cellular receptor for gibbon ape leukemia virus is a novel high affinity sodium-dependent phosphate transporter. J Biol Chem 269(41):25426–25431. https://doi.org/10.1016/S0021-9258(18)47267-5
3. Villa-Bellosta R, Ravera S, Sorribas V et al (2009) The Na+-Pi cotransporter PiT-2 (SLC20A2) is expressed in the apical membrane of rat renal proximal tubules and regulated by dietary Pi. Am J Physiol Ren Physiol 296(4):F691–F699. https://doi.org/10.1152/ajprenal.90623.2008
4. Tomoe Y, Segawa H, Shiozawa K et al (2010) Phosphaturic action of fibroblast growth factor 23 in Npt2 null mice. Am J Physiol Ren Physiol 298(6): F1341–F1350. https://doi.org/10.1152/ajprenal.00375.2009
5. Picard N, Capuano P, Stange G et al (2010) Acute parathyroid hormone differentially regulates renal brush border membrane phosphate cotransporters.

Pflugers Arch 460(3):677–687. https://doi.org/10.1007/s00424-010-0841-1
6. Christopoulou EC, Filippatos TD, Megapanou E et al (2017) Phosphate imbalance in patients with heart failure. Heart Fail Rev 22(3):349–356. https://doi.org/10.1007/s10741-017-9615-6
7. Urakawa I, Yamazaki Y, Shimada T et al (2006) Klotho converts canonical FGF receptor into a specific receptor for FGF23. Nature 444(7120):770–774. https://doi.org/10.1038/nature05315
8. Shimada T, Hasegawa H, Yamazaki Y et al (2004) FGF-23 is a potent regulator of vitamin D metabolism and phosphate homeostasis. J Bone Miner Res 19(3): 429–435. https://doi.org/10.1359/JBMR.0301264
9. Marks J, Srai SK, Biber J et al (2006) Intestinal phosphate absorption and the effect of vitamin D: a comparison of rats with mice. Exp Physiol 91(3): 531–537. https://doi.org/10.1113/expphysiol.2005.032516
10. Agus ZS, Puschett JB, Senesky D et al (1971) Mode of action of parathyroid hormone and cyclic adenosine 3′,5′-monophosphate on renal tubular phosphate reabsorption in the dog. J Clin Invest 50(3):617–626. https://doi.org/10.1172/JCI106532
11. Traebert M, Völkl H, Biber J et al (2000) Luminal and contraluminal action of 1–34 and 3–34 PTH peptides on renal type IIa Na-Pi cotransporter. Am J Physiol Ren Physiol 278(5):F792–F798. https://doi.org/10.1152/ajprenal.2000.278.5.F792
12. Bacic D, Schulz N, Biber J et al (2003) Involvement of the MAPK-kinase pathway in the PTH-mediated regulation of the proximal tubule type IIa Na+/Pi cotransporter in mouse kidney. Pflugers Arch 446(1):52–60. https://doi.org/10.1007/s00424-002-0969-8
13. Kempson SA, Lotscher M, Kaissling B et al (1995) Parathyroid hormone action on phosphate transporter mRNA and protein in rat renal proximal tubules. Am J Phys 268(4):F784–F791. https://doi.org/10.1152/ajprenal.1995.268.4.F784
14. Keusch I, Traebert M, Lötscher M et al (1998) Parathyroid hormone and dietary phosphate provoke a lysosomal routing of the proximal tubular Na/Pi-cotransporter type II. Kidney Int 54(4):1224–1232. https://doi.org/10.1046/j.1523-1755.1998.00115.x
15. Kaufmann M, Muff R, Stieger B et al (1994) Apical and basolateral parathyroid hormone receptors in rat renal cortical membranes. Endocrinology 134(3): 1173–1178. https://doi.org/10.1210/endo.134.3.8119156
16. Lavi-Moshayoff V, Wasserman G, Meir T et al (2010) PTH increases FGF23 gene expression and mediates the high-FGF23 levels of experimental kidney failure: a bone parathyroid feedback loop. Am J Physiol Ren Physiol 299(4):F882–F889. https://doi.org/10.1152/ajprenal.00360.2010
17. Rubin MR, Bilezikian JP, McMahon DJ et al (2008) The natural history of primary hyperparathyroidism with or without parathyroid surgery after 15 years. J Clin Endocrinol Metab 93(9):3462–3470. https://doi.org/10.1210/jc.2007-1215
18. Dobnig H, Turner RT (1997) The effects of programmed administration of human parathyroid hormone fragment (1–34) on bone histomorphometry and serum chemistry in rats. Endocrinology 138(11): 4607–4612. https://doi.org/10.1210/endo.138.11.5505
19. Greenspan SL, Bone HG, Ettinger MP et al (2007) Effect of recombinant human parathyroid hormone (1-84) on vertebral fracture and bone mineral density in postmenopausal women with osteoporosis. Ann Intern Med 146(5):326–339. https://doi.org/10.7326/0003-4819-146-5-200703060-00005
20. Clarke BL, Kay Berg J, Fox J et al (2014) Pharmacokinetics and pharmacodynamics of subcutaneous recombinant parathyroid hormone (1–84) in patients with hypoparathyroidism: an open-label, single-dose, phase I study. Clin Ther 36(5):722–736. https://doi.org/10.1016/j.clinthera.2014.04.001
21. Rubin MR, Dempster DW, Zhou H et al (2008) Dynamic and structural properties of the skeleton in hypoparathyroidism. J Bone Miner Res 23(12): 2018–2024. https://doi.org/10.1359/jbmr.080803
22. Michigami T, Kawai M, Yamazaki M et al (2018) Phosphate as a signaling molecule and its sensing mechanism. Physiol Rev 98(4):2317–2348. https://doi.org/10.1152/physrev.00022.2017
23. Bon N, Couasnay G, Bourgine A et al (2018) Phosphate (Pi)-regulated heterodimerization of the high-affinity sodium-dependent Pi transporters PiT1/Slc20a1 and PiT2/Slc20a2 underlies extracellular Pi sensing independently of Pi uptake. J Biol Chem 293(6):2102–2114. https://doi.org/10.1074/jbc.M117.807339
24. Bon N, Frangi G, Sourice S et al (2018) Phosphate-dependent FGF23 secretion is modulated by PiT2/Slc20a2. Mol Metab 11:197–204. https://doi.org/10.1016/j.molmet.2018.02.007
25. Takashi Y, Kinoshita Y, Hori M et al (2017) Patients with FGF23-related hypophosphatemic rickets/osteomalacia do not present with left ventricular hypertrophy. Endocr Res 42(2):132–137. https://doi.org/10.1080/07435800.2016.1242604
26. Kato K, Jeanneau C, Tarp MA et al (2006) Polypeptide GalNAc-transferase T3 and familial tumoral calcinosis: secretion of fibroblast growth factor 23 requires O-glycosylation. J Biol Chem 281(27): 18370–18377. https://doi.org/10.1074/jbc.M602469200
27. Bergwitz C, Banerjee S, Abu-Zahra H et al (2009) Defective O-glycosylation due to a novel homozygous S129P mutation is associated with lack of fibroblast growth factor 23 secretion and tumoral calcinosis. J Clin Endocrinol Metab 94(11): 4267–4274. https://doi.org/10.1210/jc.2009-0961
28. Takashi Y, Kosako H, Sawatsubashi S et al (2019) Activation of unliganded FGF receptor by extracellular phosphate potentiates proteolytic protection of

FGF23 by its O-glycosylation. Proc Natl Acad Sci U S A 116(23):11418–11427. https://doi.org/10.1073/pnas.1815166116
29. Brown EM, Gamba G, Riccardi D et al (1993) Cloning and characterization of an extracellular Ca2+-sensing receptor from bovine parathyroid. Nature 366(6455):575–580. https://doi.org/10.1038/366575a0
30. Centeno PP, Herberger A, Mun H et al (2019) Phosphate acts directly on the calcium-sensing receptor to stimulate parathyroid hormone secretion. Nat Commun 10(1):4693. https://doi.org/10.1038/s41467-019-12399-9
31. Koumakis E, Cormier C, Roux C et al (2021) The causes of hypo-and hyperphosphatemia in humans. Calcif Tissue Int 108(1):41–73. https://doi.org/10.1007/s00223-020-00664-9
32. Ganesh SK, Stack AG, Levin NW et al (2001) Association of elevated serum PO4, Ca× PO4 product, and parathyroid hormone with cardiac mortality risk in chronic hemodialysis patients. J Am Soc Nephrol 12(10):2131–2138. https://doi.org/10.1681/ASN.V12102131
33. Blacher J, Guerin AP, Pannier B et al (2001) Arterial calcifications, arterial stiffness, and cardiovascular risk in end-stage renal disease. Hypertension 38(4):938–942. https://doi.org/10.1161/hy1001.096358
34. London GM, Guerin AP, Marchais SJ et al (2003) Arterial media calcification in end-stage renal disease: impact on all-cause and cardiovascular mortality. Nephrol Dial Transplant 18(9):1731–1740
35. Kestenbaum B, Sampson JN, Rudser KD et al (2005) Serum phosphate levels and mortality risk among people with chronic kidney disease. J Am Soc Nephrol 16(2):520–528
36. Russo D, Corrao S, Battaglia Y et al (2011) Progression of coronary artery calcification and cardiac events in patients with chronic renal disease not receiving dialysis. Kidney Int 80(1):112–118
37. Chang AR, Lazo M, Appel LJ et al (2014) High dietary phosphorus intake is associated with all-cause mortality: results from NHANES III. Am J Clin Nutr 99(2):320–327
38. Dhingra R, Sullivan LM, Fox CS et al (2007) Relations of serum phosphorus and calcium levels to the incidence of cardiovascular disease in the community. Arch Intern Med 167(9):879–885
39. McGovern AP, de Lusignan S, van Vlymen J et al (2013) Serum phosphate as a risk factor for cardiovascular events in people with and without chronic kidney disease: a large community-based cohort study. PLoS One 8(9):e74996
40. Foley RN, Collins AJ, Herzog CA et al (2009) Serum phosphorus levels associate with coronary atherosclerosis in young adults. J Am Soc Nephrol 20(2):397–404
41. Ix JH, De Boer IH, Peralta CA et al (2009) Serum phosphorus concentrations and arterial stiffness among individuals with normal kidney function to moderate kidney disease in MESA. Clin J Am Soc Nephrol 4(3):609–615
42. Kwak SM, Kim JS, Choi Y et al (2014) Dietary intake of calcium and phosphorus and serum concentration in relation to the risk of coronary artery calcification in asymptomatic adults. Arterioscler Thromb Vasc Biol 34(8):1763–1769
43. Park KS, Park J, Choi SH et al (2016) Serum phosphorus concentration and coronary artery calcification in subjects without renal dysfunction. PLoS One 11(3):e0151007
44. Dhingra R, Gona P, Benjamin EJ et al (2010) Relations of serum phosphorus levels to echocardiographic left ventricular mass and incidence of heart failure in the community. Eur J Heart Fail 12(8):812–818
45. Yamamoto KT, Robinson-Cohen C, De Oliveira MC et al (2013) Dietary phosphorus is associated with greater left ventricular mass. Kidney Int 83(4):707–714
46. Foley RN, Collins AJ, Herzog CA et al (2009) Serum phosphate and left ventricular hypertrophy in young adults: the coronary artery risk development in young adults study. Kidney Blood Press Res 32(1):37–44
47. Mohammad J, Scanni R, Bestmann L et al (2018) A controlled increase in dietary phosphate elevates BP in healthy human subjects. J Am Soc Nephrol 29(8):2089–2098
48. Giachelli CM (2009) The emerging role of phosphate in vascular calcification. Kidney Int 75(9):890–897
49. Voelkl J, Lang F, Eckardt K et al (2019) Signaling pathways involved in vascular smooth muscle cell calcification during hyperphosphatemia. Cell Mol Life Sci 76(11):2077–2091
50. Villa-Bellosta R, Bogaert YE, Levi M et al (2007) Characterization of phosphate transport in rat vascular smooth muscle cells: implications for vascular calcification. Arterioscler Thromb Vasc Biol 27(5):1030–1036
51. Chavkin NW, Chia JJ, Crouthamel MH et al (2015) Phosphate uptake-independent signaling functions of the type III sodium-dependent phosphate transporter, PiT-1, in vascular smooth muscle cells. Exp Cell Res 333(1):39–48
52. Li X, Yang H, Giachelli CM (2006) Role of the sodium-dependent phosphate cotransporter, Pit-1, in vascular smooth muscle cell calcification. Circ Res 98(7):905–912
53. Yamada S, Leaf EM, Chia JJ et al (2018) PiT-2, a type III sodium-dependent phosphate transporter, protects against vascular calcification in mice with chronic kidney disease fed a high-phosphate diet. Kidney Int 94(4):716–727
54. Zhang D, Bi X, Liu Y et al (2017) High phosphate-induced calcification of vascular smooth muscle cells is associated with the TLR4/NF-κb signaling pathway. Kidney Blood Press Res 42(6):1205–1215
55. Voelkl J, Luong TT, Tuffaha R et al (2018) SGK1 induces vascular smooth muscle cell calcification

through NF-κB signaling. J Clin Invest 128(7):3024–3040
56. Zhao M, Xu M, Cai Y et al (2011) Mitochondrial reactive oxygen species promote p65 nuclear translocation mediating high-phosphate-induced vascular calcification in vitro and in vivo. Kidney Int 79(10):1071–1079
57. Lee H, Woo KM, Ryoo H et al (2010) Tumor necrosis factor-α increases alkaline phosphatase expression in vascular smooth muscle cells via MSX2 induction. Biochem Biophys Res Commun 391(1):1087–1092
58. Sheen CR, Kuss P, Narisawa S et al (2015) Pathophysiological role of vascular smooth muscle alkaline phosphatase in medial artery calcification. J Bone Miner Res 30(5):824–836
59. Yoshida T, Yamashita M, Horimai C et al (2017) Smooth muscle–selective nuclear factor-κB inhibition reduces phosphate-induced arterial medial calcification in mice with chronic kidney disease. J Am Heart Assoc 6(11):e007248
60. Rong S, Zhao X, Jin X et al (2014) Vascular calcification in chronic kidney disease is induced by bone morphogenetic protein-2 via a mechanism involving the Wnt/β-catenin pathway. Cell Physiol Biochem 34(6):2049–2060
61. Yao L, Sun Y, Sun W et al (2015) High phosphorus level leads to aortic calcification via β-catenin in chronic kidney disease. Am J Nephrol 41(1):28–36
62. Freise C, Kretzschmar N, Querfeld U (2016) Wnt signaling contributes to vascular calcification by induction of matrix metalloproteinases. BMC Cardiovasc Disord 16(1):1–8
63. Liao R, Wang L, Li J et al (2020) Vascular calcification is associated with Wnt-signaling pathway and blood pressure variability in chronic kidney disease rats. Nephrology 25(3):264–272
64. Deng D, Diao Z, Han X et al (2016) Secreted frizzled-related protein 5 attenuates high phosphate-induced calcification in vascular smooth muscle cells by inhibiting the wnt/ss-catenin pathway. Calcif Tissue Int 99(1):66–75
65. Martínez-Moreno JM, Muñoz-Castañeda JR, Herencia C et al (2012) In vascular smooth muscle cells paricalcitol prevents phosphate-induced Wnt/β-catenin activation. Am J Physiol Ren Physiol 303(8):F1136–F1144
66. Tian B, Yao L, Sheng Z et al (2019) Specific knockdown of WNT8b expression protects against phosphate-induced calcification in vascular smooth muscle cells by inhibiting the Wnt–β-catenin signaling pathway. J Cell Physiol 234(4):3469–3477
67. Zhou P, Zhang X, Guo M et al (2019) Ginsenoside Rb1 ameliorates CKD-associated vascular calcification by inhibiting the Wnt/β-catenin pathway. J Cell Mol Med 23(10):7088–7098
68. Yuan Z, Li Q, Luo S et al (2016) PPARγ and Wnt signaling in adipogenic and osteogenic differentiation of mesenchymal stem cells. Curr Stem Cell Res Ther 11(3):216–225
69. Cheng L, Zhang L, Yang J et al (2017) Activation of peroxisome proliferator-activated receptor γ inhibits vascular calcification by upregulating Klotho. Exp Ther Med 13(2):467–474
70. Liu L, Liu Y, Zhang Y et al (2018) High phosphate-induced downregulation of PPARγ contributes to CKD-associated vascular calcification. J Mol Cell Cardiol 114:264–275
71. Gao M, Chen T, Wu L et al (2017) Effect of pioglitazone on the calcification of rat vascular smooth muscle cells through the downregulation of the Wnt/β-catenin signaling pathway. Mol Med Rep 16(5):6208–6213
72. Cui L, Bai Y, Zhang J et al (2016) Effects of extracellular acid stimulation on rat vascular smooth muscle cell in Gas6/Axl or PI3K/Akt signaling pathway. Clin Exp Hypertens 38(5):451–456
73. Ponnusamy A, Sinha S, Hyde GD et al (2018) FTI-277 inhibits smooth muscle cell calcification by up-regulating PI3K/Akt signaling and inhibiting apoptosis. PLoS One 13(4):e0196232
74. Frantz C, Stewart KM, Weaver VM (2010) The extracellular matrix at a glance. J Cell Sci 123(24):4195–4200
75. Basalyga DM, Simionescu DT, Xiong W et al (2004) Elastin degradation and calcification in an abdominal aorta injury model: role of matrix metalloproteinases. Circulation 110(22):3480–3487
76. Pai A, Leaf EM, El-Abbadi M et al (2011) Elastin degradation and vascular smooth muscle cell phenotype change precede cell loss and arterial medial calcification in a uremic mouse model of chronic kidney disease. Am J Pathol 178(2):764–773
77. Amann K (2008) Media calcification and intima calcification are distinct entities in chronic kidney disease. Clin J Am Soc Nephrol 3(6):1599–1605
78. El-Abbadi MM, Pai AS, Leaf EM et al (2009) Phosphate feeding induces arterial medial calcification in uremic mice: role of serum phosphorus, fibroblast growth factor-23, and osteopontin. Kidney Int 75(12):1297–1307
79. Pai AS, Giachelli CM (2010) Matrix remodeling in vascular calcification associated with chronic kidney disease. J Am Soc Nephrol 21(10):1637–1640
80. Aikawa E, Aikawa M, Libby P et al (2009) Clinical perspective. Circulation 119(13):1785–1794
81. Chen NX, O'Neill KD, Chen X et al (2011) Activation of arterial matrix metalloproteinases leads to vascular calcification in chronic kidney disease. Am J Nephrol 34(3):211–219
82. Hecht E, Freise C, Websky K et al (2016) The matrix metalloproteinases 2 and 9 initiate uraemic vascular calcifications. Nephrol Dial Transplant 31(5):789–797
83. Hosaka N, Mizobuchi M, Ogata H et al (2009) Elastin degradation accelerates phosphate-induced mineralization of vascular smooth muscle cells. Calcif Tissue Int 85(6):523–529
84. Chung AW, Yang HH, Kim JM et al (2009) Upregulation of matrix metalloproteinase-2 in the arterial vasculature contributes to stiffening and vasomotor dysfunction in patients with chronic kidney disease. Circulation 120(9):792–801

85. Bai Y, Zhang J, Xu J et al (2015) Alteration of type I collagen in the radial artery of patients with end-stage renal disease. Am J Med Sci 349(4):292–297
86. Jover E, Silvente A, Marin F et al (2018) Inhibition of enzymes involved in collagen cross-linking reduces vascular smooth muscle cell calcification. FASEB J 32(8):4459–4469
87. Huang A, Guo G, Yu Y et al (2021) The roles of collagen in chronic kidney disease and vascular calcification. J Mol Med (Berl) 99(1):75–92
88. Roszkowska M, Strzelecka-Kiliszek A, Bessueille L et al (2018) Collagen promotes matrix vesicle-mediated mineralization by vascular smooth muscle cells. J Inorg Biochem 186:1–9
89. Watson KE, Parhami F, Shin V et al (1998) Fibronectin and collagen I matrixes promote calcification of vascular cells in vitro, whereas collagen IV matrix is inhibitory. Arterioscler Thromb Vasc Biol 18(12): 1964–1971
90. Nikolov IG, Joki N, Galmiche A et al (2013) Farnesyltransferase inhibitor R115777 protects against vascular disease in uremic mice. Atherosclerosis 229(1):42–51
91. Shroff RC, McNair R, Figg N et al (2008) Dialysis accelerates medial vascular calcification in part by triggering smooth muscle cell apoptosis. Circulation 118(17):1748–1757
92. Proudfoot D, Skepper JN, Hegyi L et al (2000) Apoptosis regulates human vascular calcification in vitro: evidence for initiation of vascular calcification by apoptotic bodies. Circ Res 87(11):1055–1062
93. Clarke MC, Littlewood TD, Figg N et al (2008) Chronic apoptosis of vascular smooth muscle cells accelerates atherosclerosis and promotes calcification and medial degeneration. Circ Res 102(12): 1529–1538
94. Son B, Kozaki K, Iijima K et al (2006) Statins protect human aortic smooth muscle cells from inorganic phosphate-induced calcification by restoring Gas6-Axl survival pathway. Circ Res 98(8):1024–1031
95. Takemura A, Iijima K, Ota H et al (2011) Sirtuin 1 retards hyperphosphatemia-induced calcification of vascular smooth muscle cells. Arterioscler Thromb Vasc Biol 31(9):2054–2062
96. Yamada S, Tatsumoto N, Tokumoto M et al (2015) Phosphate binders prevent phosphate-induced cellular senescence of vascular smooth muscle cells and vascular calcification in a modified, adenine-based uremic rat model. Calcif Tissue Int 96(4):347–358
97. Xu D, Zeng F, Han L et al (2019) The synergistic action of phosphate and interleukin-6 enhances senescence-associated calcification in vascular smooth muscle cells depending on p53. Mech Ageing Dev 182:111124
98. Dai X, Zhao M, Cai Y et al (2013) Phosphate-induced autophagy counteracts vascular calcification by reducing matrix vesicle release. Kidney Int 83(6): 1042–1051
99. Liu D, Cui W, Liu B et al (2014) Atorvastatin protects vascular smooth muscle cells from TGF-β1-stimulated calcification by inducing autophagy via suppression of the β-catenin pathway. Cell Physiol Biochem 33(1):129–141
100. Frauscher B, Kirsch AH, Schabhüttl C et al (2018) Autophagy protects from uremic vascular media calcification. Front Immunol 9:1866
101. Xu T, Sheng Z, Li Y et al (2020) OGT knockdown counteracts high phosphate-induced vascular calcification in chronic kidney disease through autophagy activation by downregulating YAP. Life Sci 261: 118121
102. Lanzer P, Boehm M, Sorribas V et al (2014) Medial vascular calcification revisited: review and perspectives. Eur Heart J 35(23):1515–1525
103. Villa-Bellosta R (2020) New insights into endogenous mechanisms of protection against arterial calcification. Atherosclerosis 306:68–74
104. Villa-Bellosta R, González-Parra E, Egido J (2016) Alkalosis and dialytic clearance of phosphate increases phosphatase activity: a hidden consequence of hemodialysis. PLoS One 11(7):e0159858
105. Lomashvili KA, Khawandi W, O'Neill WC (2005) Reduced plasma pyrophosphate levels in hemodialysis patients. J Am Soc Nephrol 16(8):2495–2500
106. Azpiazu D, González-Parra E, Egido J et al (2018) Hydrolysis of extracellular pyrophosphate increases in post-hemodialysis plasma. Sci Rep 8(1):1–8
107. Lomashvili KA, Narisawa S, Millán JL et al (2014) Vascular calcification is dependent on plasma levels of pyrophosphate. Kidney Int 85(6):1351–1356
108. Villa-Bellosta R, Sorribas V (2011) Calcium phosphate deposition with normal phosphate concentration–role of pyrophosphate. Circ J 75(11):2705–2710
109. Riser BL, Barreto FC, Rezg R et al (2011) Daily peritoneal administration of sodium pyrophosphate in a dialysis solution prevents the development of vascular calcification in a mouse model of uraemia. Nephrol Dial Transplant 26(10):3349–3357
110. O'neill WC, Lomashvili KA, Malluche HH et al (2011) Treatment with pyrophosphate inhibits uremic vascular calcification. Kidney Int 79(5):512–517
111. Mutluay R, Değertekin CK, DERİCİ Ü et al (2019) Serum fetuin-A is associated with the components of MIAC (malnutrition, inflammation, atherosclerosis, calcification) syndrome in different stages. Turk J Med Sci 49(1):327–335
112. Yamada S, Tokumoto M, Tsuruya K et al (2015) Fetuin-A decrease induced by a low-protein diet enhances vascular calcification in uremic rats with hyperphosphatemia. Am J Physiol Ren Physiol 309(8):F744–F754
113. Westenfeld R, Schäfer C, Krüger T et al (2009) Fetuin-A protects against atherosclerotic calcification in CKD. J Am Soc Nephrol 20(6):1264–1274
114. Westenfeld R, Schäfer C, Smeets R et al (2007) Fetuin-A (AHSG) prevents extraosseous calcification

induced by uraemia and phosphate challenge in mice. Nephrol Dial Transplant 22(6):1537–1546
115. Moe SM, Reslerova M, Ketteler M et al (2005) Role of calcification inhibitors in the pathogenesis of vascular calcification in chronic kidney disease (CKD). Kidney Int 67(6):2295–2304
116. Cozzolino M, Galassi A, Biondi ML et al (2006) Serum fetuin-A levels link inflammation and cardiovascular calcification in hemodialysis patients. Am J Nephrol 26(5):423–429
117. Muzasti RA, Loesnihari R (2019) High fetuin-A level as a protective factor to abdominal aortic calcification in Indonesian regular hemodialysis patients. Open Access Maced J Med Sci 7(5):721
118. Ulutas O, Taskapan MC, Dogan A et al (2018) Vascular calcification is not related to serum fetuin-A and osteopontin levels in hemodialysis patients. Int Urol Nephrol 50(1):137–142
119. Luo G, Ducy P, McKee MD et al (1997) Spontaneous calcification of arteries and cartilage in mice lacking matrix GLA protein. Nature 386(6620):78–81
120. Julien M, Khoshniat S, Lacreusette A et al (2009) Phosphate-dependent regulation of MGP in osteoblasts: role of ERK1/2 and Fra-1. J Bone Miner Res 24(11):1856–1868
121. Julien M, Magne D, Masson M et al (2007) Phosphate stimulates matrix Gla protein expression in chondrocytes through the extracellular signal regulated kinase signaling pathway. Endocrinology 148(2):530–537
122. Louvet L, Büchel J, Steppan S et al (2013) Magnesium prevents phosphate-induced calcification in human aortic vascular smooth muscle cells. Nephrol Dial Transplant 28(4):869–878. https://doi.org/10.1093/ndt/gfs520
123. Aoun M, Makki M, Azar H et al (2017) High dephosphorylated-uncarboxylated MGP in hemodialysis patients: risk factors and response to vitamin K 2, a pre-post intervention clinical trial. BMC Nephrol 18(1):1–10
124. Delanaye P, Krzesinski J, Warling X et al (2014) Dephosphorylated-uncarboxylated Matrix Gla protein concentration is predictive of vitamin K status and is correlated with vascular calcification in a cohort of hemodialysis patients. BMC Nephrol 15(1):1–8
125. Thamratnopkoon S, Susantitaphong P, Tumkosit M et al (2017) Correlations of plasma desphosphorylated uncarboxylated matrix Gla protein with vascular calcification and vascular stiffness in chronic kidney disease. Nephron 135(3):167–172
126. Moe SM, O'Neill KD, Duan D et al (2002) Medial artery calcification in ESRD patients is associated with deposition of bone matrix proteins. Kidney Int 61(2):638–647
127. Beck GR, Zerler B, Moran E (2000) Phosphate is a specific signal for induction of osteopontin gene expression. Proc Natl Acad Sci U S A 97(15):8352–8357
128. Chen NX, O'Neill KD, Duan D et al (2002) Phosphorus and uremic serum up-regulate osteopontin expression in vascular smooth muscle cells. Kidney Int 62(5):1724–1731
129. Giachelli CM, Speer MY, Li X et al (2005) Regulation of vascular calcification: roles of phosphate and osteopontin. Circ Res 96(7):717–722
130. Wada T, McKee MD, Steitz S et al (1999) Calcification of vascular smooth muscle cell cultures: inhibition by osteopontin. Circ Res 84(2):166–178
131. Speer MY, Chien Y, Quan M et al (2005) Smooth muscle cells deficient in osteopontin have enhanced susceptibility to calcification in vitro. Cardiovasc Res 66(2):324–333
132. Paloian NJ, Leaf EM, Giachelli CM (2016) Osteopontin protects against high phosphate-induced nephrocalcinosis and vascular calcification. Kidney Int 89(5):1027–1036
133. Van Campenhout A, Golledge J (2009) Osteoprotegerin, vascular calcification and atherosclerosis. Atherosclerosis 204(2):321–329
134. Lee SJ, Lee I, Jeon J (2020) Vascular calcification—New insights into Its mechanism. Int J Mol Sci 21(8):2685
135. Orita Y, Yamamoto H, Kohno N et al (2007) Role of osteoprotegerin in arterial calcification: development of new animal model. Arterioscler Thromb Vasc Biol 27(9):2058–2064
136. Nitta K, Akiba T, Uchida K et al (2003) The progression of vascular calcification and serum osteoprotegerin levels in patients on long-term hemodialysis. Am J Kidney Dis 42(2):303–309
137. Sigrist MK, Levin A, Er L et al (2009) Elevated osteoprotegerin is associated with all-cause mortality in CKD stage 4 and 5 patients in addition to vascular calcification. Nephrol Dial Transplant 24(10):3157–3162
138. Morena M, Dupuy A, Jaussent I et al (2009) A cut-off value of plasma osteoprotegerin level may predict the presence of coronary artery calcifications in chronic kidney disease patients. Nephrol Dial Transplant 24(11):3389–3397
139. Mesquita M, Demulder A, Damry N et al (2009) Plasma osteoprotegerin is an independent risk factor for mortality and an early biomarker of coronary vascular calcification in chronic kidney disease. Clin Chem Lab Med 47(3):339–346
140. Desjardins M, Sidibé A, Fortier C et al (2018) Association of interleukin-6 with aortic stiffness in end-stage renal disease. J Am Soc Hypertens 12(1):5–13
141. Jean G, Terrat J, Vanel T et al (2009) High levels of serum fibroblast growth factor (FGF)-23 are associated with increased mortality in long haemodialysis patients. Nephrol Dial Transplant 24(9):2792–2796
142. Benz K, Varga I, Neureiter D et al (2017) Vascular inflammation and media calcification are already present in early stages of chronic kidney disease. Cardiovasc Pathol 27:57–67

143. Tintut Y, Patel J, Parhami F et al (2000) Tumor necrosis factor-α promotes in vitro calcification of vascular cells via the cAMP pathway. Circulation 102(21):2636–2642
144. Wen C, Yang X, Yan Z et al (2013) Nalp3 inflammasome is activated and required for vascular smooth muscle cell calcification. Int J Cardiol 168(3): 2242–2247
145. Sun M, Chang Q, Xin M et al (2017) Endogenous bone morphogenetic protein 2 plays a role in vascular smooth muscle cell calcification induced by interleukin 6 in vitro. Int J Immunopathol Pharmacol 30(3): 227–237
146. Yamada S, Tokumoto M, Tatsumoto N et al (2014) Phosphate overload directly induces systemic inflammation and malnutrition as well as vascular calcification in uremia. Am J Physiol Ren Physiol 306(12): F1418–F1428
147. Agharazii M, St-Louis R, Gautier-Bastien A et al (2015) Inflammatory cytokines and reactive oxygen species as mediators of chronic kidney disease-related vascular calcification. Am J Hypertens 28(6):746–755
148. Aghagolzadeh P, Bachtler M, Bijarnia R et al (2016) Calcification of vascular smooth muscle cells is induced by secondary calciprotein particles and enhanced by tumor necrosis factor-α. Atherosclerosis 251:404–414
149. Martínez-Moreno JM, Herencia C, de Oca AM et al (2017) High phosphate induces a pro-inflammatory response by vascular smooth muscle cells and modulation by vitamin D derivatives. Clin Sci 131(13): 1449–1463
150. Wang N, Wang X, Xing C et al (2010) Role of TGF-β1 in bone matrix production in vascular smooth muscle cells induced by a high-phosphate environment. Nephron Exp Nephrol 115(3):e60–e68
151. Zhang K, Zhang Y, Feng W et al (2017) Interleukin-18 enhances vascular calcification and osteogenic differentiation of vascular smooth muscle cells through TRPM7 activation. Arterioscler Thromb Vasc Biol 37(10):1933–1943
152. Al-Aly Z (2008) Arterial calcification: a tumor necrosis factor-alpha mediated vascular Wnt-opathy. Transl Res 151(5):233–239
153. Shao J, Cheng S, Pingsterhaus JM et al (2005) Msx2 promotes cardiovascular calcification by activating paracrine Wnt signals. J Clin Invest 115(5): 1210–1220
154. Zickler D, Luecht C, Willy K et al (2018) Tumour necrosis factor-alpha in uraemic serum promotes osteoblastic transition and calcification of vascular smooth muscle cells via extracellular signal-regulated kinases and activator protein 1/c-FOS-mediated induction of interleukin 6 expression. Nephrol Dial Transplant 33(4):574–585
155. Deuell KA, Callegari A, Giachelli CM et al (2012) RANKL enhances macrophage paracrine pro-calcific activity in high phosphate-treated smooth muscle cells: dependence on IL-6 and TNF-α. J Vasc Res 49(6):510–521
156. Callegari A, Coons ML, Ricks JL et al (2014) Increased calcification in osteoprotegerin-deficient smooth muscle cells: dependence on receptor activator of NF-κB ligand and interleukin 6. J Vasc Res 51(2):118–131
157. Leibrock CB, Alesutan I, Voelkl J et al (2015) NH4Cl treatment prevents tissue calcification in klotho deficiency. J Am Soc Nephrol 26(10): 2423–2433
158. Watson KE, Boström K, Ravindranath R et al (1994) TGF-beta 1 and 25-hydroxycholesterol stimulate osteoblast-like vascular cells to calcify. J Clin Invest 93(5):2106–2113
159. Wang N, Wang X, Sun B et al (2013) Role of TGF-β1 in production of fibronectin in vascular smooth muscle cells cultured under high-phosphate conditions. J Nephrol 26(1):213–218
160. Smith ER, Hewitson TD, Jahnen-Dechent W (2020) Calciprotein particles: mineral behaving badly? Curr Opin Nephrol Hypertens 29(4):378–386
161. Silaghi CN, Ilyés T, Van Ballegooijen AJ et al (2020) Calciprotein particles and serum calcification propensity: hallmarks of vascular calcifications in patients with chronic kidney disease. J Clin Med 9(5):1287
162. Smith ER, Hewitson TD, Cai MM et al (2017) A novel fluorescent probe-based flow cytometric assay for mineral-containing nanoparticles in serum. Sci Rep 7(1):1–17
163. Yamada H, Kuro-o M, Ishikawa S et al (2018) Daily variability in serum levels of calciprotein particles and their association with mineral metabolism parameters: a cross-sectional pilot study. Nephrology 23(3):226–230
164. Köppert S, Büscher A, Babler A et al (2018) Cellular clearance and biological activity of calciprotein particles depend on their maturation state and crystallinity. Front Immunol 9:1991
165. Herrmann M, Schäfer C, Heiss A et al (2012) Clearance of fetuin-A–containing calciprotein particles is mediated by scavenger receptor-A. Circ Res 111(5): 575–584
166. Smith ER, Ford ML, Tomlinson LA et al (2014) Serum calcification propensity predicts all-cause mortality in predialysis CKD. J Am Soc Nephrol 25(2):339–348
167. Nakazato J, Hoshide S, Wake M et al (2019) Association of calciprotein particles measured by a new method with coronary artery plaque in patients with coronary artery disease: a cross-sectional study. J Cardiol 74(5):428–435
168. Pasch A, Farese S, Gräber S et al (2012) Nanoparticle-based test measures overall propensity for calcification in serum. J Am Soc Nephrol 23(10): 1744–1752
169. Pasch A, Block GA, Bachtler M et al (2017) Blood calcification propensity, cardiovascular events, and survival in patients receiving hemodialysis in the

EVOLVE trial. Clin J Am Soc Nephrol 12(2): 315–322
170. Bundy JD, Cai X, Scialla JJ et al (2019) Serum calcification propensity and coronary artery calcification among patients with CKD: the CRIC (Chronic Renal Insufficiency Cohort) study. Am J Kidney Dis 73(6):806–814
171. Bundy JD, Cai X, Mehta RC et al (2019) Serum calcification propensity and clinical events in CKD. Clin J Am Soc Nephrol 14(11):1562–1571
172. Smith ER, Hanssen E, McMahon LP et al (2013) Fetuin-A-containing calciprotein particles reduce mineral stress in the macrophage. PLoS One 8(4): e60904
173. Chue CD, Edwards NC, Moody WE et al (2012) Serum phosphate is associated with left ventricular mass in patients with chronic kidney disease: a cardiac magnetic resonance study. Heart 98(3):219–224
174. Zou J, Yu Y, Wu P et al (2016) Serum phosphorus is related to left ventricular remodeling independent of renal function in hospitalized patients with chronic kidney disease. Int J Cardiol 221:134–140
175. Faul C, Amaral AP, Oskouei B et al (2011) FGF23 induces left ventricular hypertrophy. J Clin Invest 121(11)
176. Hsu HJ, Wu M (2009) Fibroblast growth factor 23: a possible cause of left ventricular hypertrophy in hemodialysis patients. Am J Med Sci 337(2): 116–122
177. Mirza MA, Larsson A, Melhus H et al (2009) Serum intact FGF23 associate with left ventricular mass, hypertrophy and geometry in an elderly population. Atherosclerosis 207(2):546–551
178. Kirkpantur A, Balci M, Gurbuz OA et al (2011) Serum fibroblast growth factor-23 (FGF-23) levels are independently associated with left ventricular mass and myocardial performance index in maintenance haemodialysis patients. Nephrol Dial Transplant 26(4):1346–1354
179. Shroff R (2013) Phosphate is a vascular toxin. Pediatr Nephrol 28(4):583–593
180. Grabner A, Amaral AP, Schramm K et al (2015) Activation of cardiac fibroblast growth factor receptor 4 causes left ventricular hypertrophy. Cell Metab 22(6):1020–1032
181. Hu MC, Shi M, Cho HJ et al (2015) Klotho and phosphate are modulators of pathologic uremic cardiac remodeling. J Am Soc Nephrol 26(6): 1290–1302
182. Hu MC, Scanni R, Ye J et al (2020) Dietary vitamin D interacts with high phosphate-induced cardiac remodeling in rats with normal renal function. Nephrol Dial Transplant 35(3):411–421
183. Peri-Okonny P, Baskin KK, Iwamoto G et al (2019) High-phosphate diet induces exercise intolerance and impairs fatty acid metabolism in mice. Circulation 139(11):1422–1434
184. Moench I, Aravindhan K, Kuziw J et al (2021) High FGF23 levels failed to predict cardiac hypertrophy in animal models of hyperphosphatemia and chronic renal failure. J Endocr Soc. https://doi.org/10.1210/jendso/bvab066
185. Amann K, Törnig J, Kugel B et al (2003) Hyperphosphatemia aggravates cardiac fibrosis and microvascular disease in experimental uremia. Kidney Int 63(4):1296–1301
186. Neves KR, Graciolli FG, Dos Reis LM et al (2004) Adverse effects of hyperphosphatemia on myocardial hypertrophy, renal function, and bone in rats with renal failure. Kidney Int 66(6):2237–2244
187. Huang Y, Wang S, Zhou J et al (2020) IRF1-mediated downregulation of PGC1α contributes to cardiorenal syndrome type 4. Nat Commun 11(1): 1–16
188. Calvo MS, Moshfegh AJ, Tucker KL (2014) Assessing the health impact of phosphorus in the food supply: issues and considerations. Adv Nutr 5(1):104–113
189. Kim H, Mizuno M, Vongpatanasin W (2019) Phosphate, the forgotten mineral in hypertension. Curr Opin Nephrol Hypertens 28(4):345
190. Patel RK, Jeemon P, Stevens KK et al (2015) Association between serum phosphate and calcium, long-term blood pressure, and mortality in treated hypertensive adults. J Hypertens 33(10):2046–2053
191. Mendes M, Resende L, Teixeira A et al (2017) Blood pressure and phosphate level in diabetic and non-diabetic kidney disease: results of the cross-sectional “Low Clearance Consultation” study. Porto Biomed J 2(6):301–305
192. Mizuno M, Mitchell JH, Crawford S et al (2016) High dietary phosphate intake induces hypertension and augments exercise pressor reflex function in rats. Am J Phys Regul Integr Comp Phys 311(1):R39–R48
193. Six I, Maizel J, Barreto FC et al (2012) Effects of phosphate on vascular function under normal conditions and influence of the uraemic state. Cardiovasc Res 96(1):130–139. https://doi.org/10.1093/cvr/cvs240
194. Bozic M, Panizo S, Sevilla MA et al (2014) High phosphate diet increases arterial blood pressure via a parathyroid hormone mediated increase of renin. J Hypertens 32(9):1822–1832
195. Gaasbeek A, Meinders AE (2005) Hypophosphatemia: an update on its etiology and treatment. Am J Med 118(10):1094–1101. https://doi.org/10.1016/j.amjmed.2005.02.014
196. Martins D, Wolf M, Pan D et al (2007) Prevalence of cardiovascular risk factors and the serum levels of 25-hydroxyvitamin D in the United States: data From the Third National Health and Nutrition Examination Survey. Arch Intern Med 167(11):1159–1165. https://doi.org/10.1001/archinte.167.11.1159
197. Xiang W, Kong J, Chen S et al (2005) Cardiac hypertrophy in vitamin D receptor knockout mice: role of the systemic and cardiac renin-angiotensin systems. Am J Physiol Endocrinol Metab 288(1):

E125–E132. https://doi.org/10.1152/ajpendo.00224.2004
198. Simpson RU, Hershey SH, Nibbelink KA (2007) Characterization of heart size and blood pressure in the vitamin D receptor knockout mouse. J Steroid Biochem Mol Biol 103(3):521–524
199. O'Keefe EL, DiNicolantonio JJ, O'Keefe JH et al (2018) Alcohol and CV health: jekyll and hyde J-curves. Prog Cardiovasc Dis 61(1):68–75
200. Wacker MJ, Touchberry CD, Silswal N et al (2016) Skeletal muscle, but not cardiovascular function, is altered in a mouse model of autosomal recessive hypophosphatemic rickets. Front Physiol 7:173. https://doi.org/10.3389/fphys.2016.00173
201. Milionis HJ, Alexandrides GE, Liberopoulos EN et al (2002) Hypomagnesemia and concurrent acid–base and electrolyte abnormalities in patients with congestive heart failure. Eur J Heart Fail 4(2): 167–173
202. Rozentryt P, Nowak J, Niedziela J et al (2014) Serum phosphorus level is related to degree of clinical response to up-titration of heart failure pharmacotherapy. Int J Cardiol 177(1):248–254. https://doi.org/10.1016/j.ijcard.2014.09.034
203. Schwartz A, Brotfain E, Koyfman L et al (2014) Association between hypophosphatemia and cardiac arrhythmias in the early stage of sepsis: could phosphorus replacement treatment reduce the incidence of arrhythmias? Electrolyte Blood Press 12(1):19–25. https://doi.org/10.5049/EBP.2014.12.1.19
204. Cariem AK, Lemmer ER, Adams MG et al (1994) Severe hypophosphataemia in anorexia nervosa. Postgrad Med J 70(829):825–827. https://doi.org/10.1136/pgmj.70.829.825
205. Hu C, Lee B, Cheng H et al (2015) Acetazolamide-related life-threatening hypophosphatemia in a glaucoma patient. J Glaucoma 24(4):e31–e33. https://doi.org/10.1097/IJG.0000000000000185
206. Osuka A, Matsuoka T, Idoguchi K (2009) Is this the worst outcome of metabolic syndrome? Hypophosphatemia and resulting cardiac arrest during the treatment of diabetic ketoacidosis with hypertriglyceridemia. Intern Med 48(16): 1391–1395. https://doi.org/10.2169/internalmedicine.48.2236
207. O'Connor LR, Wheeler WS, Bethune JE (1977) Effect of hypophosphatemia on myocardial performance in man. N Engl J Med 297(17):901–903. https://doi.org/10.1056/NEJM197710272971702
208. Zazzo J, Troché G, Ruel P et al (1995) High incidence of hypophosphatemia in surgical intensive care patients: efficacy of phosphorus therapy on myocardial function. Intensive Care Med 21(10): 826–831. https://doi.org/10.1007/BF01700966
209. Rasmussen A, Buus S, Hessov I (1985) Postoperative myocardial performance during glucose-induced hypophosphatemia. Acta Chir Scand 151(1):13–15
210. Davis SV, Olichwier KK, Chakko SC (1988) Reversible depression of myocardial performance in hypophosphatemia. Am J Med Sci 295(3):183–187
211. Francis F, Hennig S, Korn B et al (1995) A gene (PEX) with homologies to endopeptidases is mutated in patients with X–linked hypophosphatemic rickets. Nat Genet 11(2):130–136. https://doi.org/10.1038/ng1095-130
212. Yamazaki Y, Okazaki R, Shibata M et al (2002) Increased circulatory level of biologically active full-length FGF-23 in patients with hypophosphatemic rickets/osteomalacia. J Clin Endocrinol Metab 87(11):4957–4960. https://doi.org/10.1210/jc.2002-021105
213. Alon US, Monzavi R, Lilien M et al (2003) Hypertension in hypophosphatemic rickets—role of secondary hyperparathyroidism. Pediatr Nephrol 18(2): 155–158. https://doi.org/10.1007/s00467-002-1044-6
214. Nehgme R, Fahey JT, Smith C et al (1997) Cardiovascular abnormalities in patients with X-linked hypophosphatemia. J Clin Endocrinol Metab 82(8): 2450–2454. https://doi.org/10.1210/jcem.82.8.4181
215. Hernández-Frías O, Gil-Peña H, Pérez-Roldán JM et al (2019) Risk of cardiovascular involvement in pediatric patients with X-linked hypophosphatemia. Pediatr Nephrol 34(6):1077–1086. https://doi.org/10.1007/s00467-018-4180-3
216. Nakamura Y, Takagi M, Takeda R et al (2017) Hypertension is a characteristic complication of X-linked hypophosphatemia. Endocr J 64(3): 283–289. https://doi.org/10.1507/endocrj.EJ16-0199
217. Vered Z, Battler A, Motro M et al (1984) Left ventricular function in patients with chronic hypophosphatemia. Am Heart J 107(4):796–798
218. Vered I, Whyte MP, Vered Z et al (1990) Normal left ventricular performance in children with x-linked hypophosphatemic rickets: a doppler echocardiography study. J Bone Miner Res 5(5):469–474
219. Leifheit-Nestler M, Richter B, Basaran M et al (2018) Impact of altered mineral metabolism on pathological cardiac remodeling in elevated fibroblast growth factor 23. Front Endocrinol 9:333. https://doi.org/10.3389/fendo.2018.00333
220. Liu ES, Thoonen R, Petit E et al (2018) Increased circulating FGF23 does not lead to cardiac hypertrophy in the male hyp mouse model of XLH. Endocrinology 159(5):2165–2172. https://doi.org/10.1210/en.2018-00174
221. Andrukhova O, Slavic S, Smorodchenko A et al (2014) FGF23 regulates renal sodium handling and blood pressure. EMBO Mol Med 6(6):744–759
222. Han X, Ross J, Kolumam G et al (2018) Cardiovascular effects of renal distal tubule deletion of the FGF receptor 1 gene. J Am Soc Nephrol 29(1):69–80. https://doi.org/10.1681/ASN.2017040412
223. Pastor-Arroyo E, Gehring N, Krudewig C et al (2018) The elevation of circulating fibroblast growth factor 23 without kidney disease does not increase cardiovascular disease risk. Kidney Int 94(1):49–59. https://doi.org/10.1016/j.kint.2018.02.017
224. Lorenz-Depiereux B, Bastepe M, Benet-Pagès A et al (2006) DMP1 mutations in autosomal recessive

hypophosphatemia implicate a bone matrix protein in the regulation of phosphate homeostasis. Nat Genet 38(11):1248–1250. https://doi.org/10.1038/ng1868
225. Feng JQ, Ward LM, Liu S et al (2006) Loss of DMP1 causes rickets and osteomalacia and identifies a role for osteocytes in mineral metabolism. Nat Genet 38(11):1310–1315. https://doi.org/10.1038/ng1905
226. Liu S, Zhou J, Tang W et al (2008) Pathogenic role of Fgf23 in Dmp1-null mice. Am J Physiol Endocrinol Metab 295(2):E254–E261. https://doi.org/10.1152/ajpendo.90201.2008
227. Fuller TJ, Nichols WW, Brenner BJ et al (1978) Reversible depression in myocardial performance in dogs with experimental phosphorus deficiency. J Clin Invest 62(6):1194–1200. https://doi.org/10.1172/JCI109239
228. Capasso JM, Aronson RS, Strobeck JE et al (1982) Effects of experimental phosphate deficiency on action potential characteristics and contractile performance of rat myocardium. Cardiovasc Res 16(2): 71–79. https://doi.org/10.1093/cvr/16.2.71
229. Schwartz A, Gurman GM, Cohen G et al (2002) Association between hypophosphatemia and cardiac arrhythmias in the early stages of sepsis. Eur J Intern Med 13(7):434–438
230. Cohen J, Kogan A, Sahar G et al (2004) Hypophosphatemia following open heart surgery: incidence and consequences. Eur J Cardiothorac Surg 26(2):306–310. https://doi.org/10.1016/j.ejcts.2004.03.004
231. Heames RM, Cope RA (2006) Hypophosphataemia causing profound cardiac failure after cardiac surgery. Anaesthesia 61(12):1211–1213
232. Frustaci A, Pennestrì, Scoppetta C (1984) Myocardial damage due to hypokalaemia and hypophosphataemia. Postgrad Med J 60(708):679–681. https://doi.org/10.1136/pgmj.60.708.679
233. Foley RN, Collins AJ, Ishani A et al (2008) Calcium-phosphate levels and cardiovascular disease in community-dwelling adults: the Atherosclerosis Risk in Communities (ARIC) study. Am Heart J 156(3):556–563
234. Larsson TE, Olauson H, Hagström E et al (2010) Conjoint effects of serum calcium and phosphate on risk of total, cardiovascular, and noncardiovascular mortality in the community. Arterioscler Thromb Vasc Biol 30(2):333–339. https://doi.org/10.1161/ATVBAHA.109.196675
235. Ketteler M, Block GA, Evenepoel P et al (2017) Executive summary of the 2017 KDIGO Chronic Kidney Disease–Mineral and Bone Disorder (CKD-MBD) guideline update: what's changed and why it matters. Kidney Int 92(1):26–36
236. Finch JL, Lee DH, Liapis H et al (2013) Phosphate restriction significantly reduces mortality in uremic rats with established vascular calcification. Kidney Int 84(6):1145–1153
237. Di Iorio B, Di Micco L, Torraca S et al (2012) Acute effects of very-low-protein diet on FGF23 levels: a randomized study. Clin J Am Soc Nephrol 7(4): 581–587. https://doi.org/10.2215/CJN.07640711
238. Oliveira RB, Cancela ALE, Graciolli FG et al (2010) Early control of PTH and FGF23 in normophosphatemic CKD patients: a new target in CKD-MBD therapy? Clin J Am Soc Nephrol 5(2): 286–291. https://doi.org/10.2215/CJN.05420709
239. Block GA, Wheeler DC, Persky MS et al (2012) Effects of phosphate binders in moderate CKD. J Am Soc Nephrol 23(8):1407–1415. https://doi.org/10.1681/ASN.2012030223
240. Russo D, Miranda I, Ruocco C et al (2007) The progression of coronary artery calcification in predialysis patients on calcium carbonate or sevelamer. Kidney Int 72(10):1255–1261
241. Yokoyama K, Hirakata H, Akiba T et al (2014) Ferric citrate hydrate for the treatment of hyperphosphatemia in nondialysis-dependent CKD. Clin J Am Soc Nephrol 9(3):543–552. https://doi.org/10.2215/CJN.05170513
242. Block GA, Block MS, Smits G et al (2019) A pilot randomized trial of ferric citrate coordination complex for the treatment of advanced CKD. J Am Soc Nephrol 30(8):1495–1504. https://doi.org/10.1681/ASN.2018101016
243. Isakova T, Barchi-Chung A, Enfield G et al (2013) Effects of dietary phosphate restriction and phosphate binders on FGF23 levels in CKD. Clin J Am Soc Nephrol 8(6):1009–1018. https://doi.org/10.2215/CJN.09250912
244. Maizel J, Six I, Dupont S et al (2013) Effects of sevelamer treatment on cardiovascular abnormalities in mice with chronic renal failure. Kidney Int 84(3): 491–500
245. Block GA, Pergola PE, Fishbane S et al (2019) Effect of ferric citrate on serum phosphate and fibroblast growth factor 23 among patients with nondialysis-dependent chronic kidney disease: path analyses. Nephrol Dial Transplant 34(7):1115–1124. https://doi.org/10.1093/ndt/gfy318
246. David V, Martin A, Isakova T et al (2016) Inflammation and functional iron deficiency regulate fibroblast growth factor 23 production. Kidney Int 89(1): 135–146
247. Hanudel MR, Chua K, Rappaport M et al (2016) Effects of dietary iron intake and chronic kidney disease on fibroblast growth factor 23 metabolism in wild-type and hepcidin knockout mice. Am J Physiol Ren Physiol 311(6):F1369–F1377. https://doi.org/10.1152/ajprenal.00281.2016
248. Francis C, Courbon G, Gerber C et al (2019) Ferric citrate reduces fibroblast growth factor 23 levels and improves renal and cardiac function in a mouse model of chronic kidney disease. Kidney Int 96(6): 1346–1358
249. Spencer AG, Labonte ED, Rosenbaum DP et al (2014) Intestinal inhibition of the Na+/H+ exchanger 3 prevents cardiorenal damage in rats and inhibits Na + uptake in humans. Sci Transl Med 6(227). https://doi.org/10.1126/scitranslmed.3007790
250. Labonté ED, Carreras CW, Leadbetter MR et al (2015) Gastrointestinal inhibition of sodium-hydrogen exchanger 3 reduces phosphorus

absorption and protects against vascular calcification in CKD. J Am Soc Nephrol 26(5):1138–1149. https://doi.org/10.1681/ASN.2014030317

251. Beck L, Karaplis AC, Amizuka N et al (1998) Targeted inactivation of Npt2 in mice leads to severe renal phosphate wasting, hypercalciuria, and skeletal abnormalities. Proc Natl Acad Sci U S A 95(9): 5372–5377. https://doi.org/10.1073/pnas.95.9.5372
252. Klar J, Anja G, Ehrmann A et al (2020) MO053 (ABSTRACT) inhibition of sodium phoshate transporter NPT2A improves experimental vascular calcificatio and phosphate homeostasis in rats. Nephrol Dial Transplant 35. https://doi.org/10.1093/ndt/gfaa140.MO053
253. Bergwitz C, Roslin NM, Tieder M et al (2006) SLC34A3 mutations in patients with hereditary hypophosphatemic rickets with hypercalciuria predict a key role for the sodium-phosphate cotransporter NaPi-IIc in maintaining phosphate homeostasis. Am J Hum Genet 78(2):179–192
254. Tenenhouse HS, Martel J, Gauthier C et al (2003) Differential effects of Npt2a gene ablation and X-linked Hyp mutation on renal expression of Npt2c. Am J Phys Renal Phys 285(6):F1271–F1278. https://doi.org/10.1152/ajprenal.00252.2003
255. Montes de Oca A, Guerrero F, Martinez-Moreno J et al (2014) Magnesium inhibits Wnt/β-catenin activity and reverses the osteogenic transformation of vascular smooth muscle cells. PLoS One 9(2). https://doi.org/10.1371/journal.pone.0089525
256. ter Braake AD, Tinnemans PT, Shanahan CM et al (2018) Magnesium prevents vascular calcification in vitro by inhibition of hydroxyapatite crystal formation. Sci Rep 8(1):2069. https://doi.org/10.1038/s41598-018-20241-3
257. Sakaguchi Y, Fujii N, Shoji T et al (2015) Magnesium modifies the cardiovascular mortality risk associated with hyperphosphatemia in patients undergoing hemodialysis: a cohort study. PLoS One 9(12). https://doi.org/10.1371/journal.pone.0116273
258. Yao Z, Xu Y, Ma W et al (2018) Magnesium citrate protects against vascular calcification in an adenine-induced chronic renal failure rat model. J Cardiovasc Pharmacol 72(6):270–276. https://doi.org/10.1097/FJC.0000000000000590
259. Diaz-Tocados J, Peralta-Ramirez A, Rodríguez-Ortiz ME et al (2017) Dietary magnesium supplementation prevents and reverses vascular and soft tissue calcifications in uremic rats. Kidney Int 92(5): 1084–1099
260. Kaesler N, Goettsch C, Weis D et al (2020) Magnesium but not nicotinamide prevents vascular calcification in experimental uraemia. Nephrol Dial Transplant 35(1):65–73. https://doi.org/10.1093/ndt/gfy410
261. Sakaguchi Y, Hamano T, Obi Y et al (2019) A randomized trial of magnesium oxide and oral carbon adsorbent for coronary artery calcification in predialysis CKD. J Am Soc Nephrol 30(6): 1073–1085. https://doi.org/10.1681/ASN.2018111150
262. Bressendorff I, Hansen D, Schou M et al (2018) The effect of increasing dialysate magnesium on serum calcification propensity in subjects with end stage kidney disease. Clin J Am Soc Nephrol 13(9): 1373–1380. https://doi.org/10.2215/CJN.13921217

Coordination of Phosphate and Magnesium Metabolism in Bacteria

12

Roberto E. Bruna, Christopher G. Kendra, and Mauricio H. Pontes

Abstract

The majority of cellular phosphate (PO_4^{-3}; Pi) exists as nucleoside triphosphates, mainly adenosine triphosphate (ATP), and ribosomal RNA (rRNA). ATP and rRNA are also the largest cytoplasmic reservoirs of magnesium (Mg^{2+}), the most abundant divalent cation in living cells. The co-occurrence of these ionic species in the cytoplasm is not coincidental. Decades of work in the Pi and Mg^{2+} starvation responses of two model enteric bacteria, *Escherichia coli* and *Salmonella enterica*, have led to the realization that the metabolisms of Pi and Mg^{2+} are interconnected. Bacteria must acquire these nutrients in a coordinated manner to achieve balanced growth and avoid loss of viability. In this chapter, we will review how bacteria sense and respond to fluctuations in environmental and intracellular Pi and Mg^{2+} levels. We will also discuss how these two compounds are functionally linked, and how cells elicit physiological responses to maintain their homeostasis.

R. E. Bruna · C. G. Kendra
Department of Pathology and Laboratory Medicine, Penn State College of Medicine, Hershey, PA, USA
e-mail: rbruna@pennstatehealth.psu.edu; ckendra@pennstatehealth.psu.edu

M. H. Pontes (✉)
Department of Pathology and Laboratory Medicine, Department of Microbiology and Immunology, Penn State College of Medicine, Hershey, PA, USA
e-mail: mpontes@pennstatehealth.psu.edu

Keywords

Phosphorus · Magnesium · Bacteria · Cytotoxicity · Transport · *Salmonella*

12.1 Phosphorus Acquisition in Bacteria

Phosphorus (P) is an intrinsic component of all living cells. The importance of this chemical element is illustrated by its various structural and biochemical functions. P is a structural constituent of lipids and complex carbohydrates, including glycolipids, lipopolysaccharides (LPS) and (lipo)teichoic acids. These molecules are major structural components of biological membranes and cell walls, and are required for the formation of cellular boundaries and membrane-bound organelles. P in phosphoryl groups enables the implementation of phosphorelays and the establishment of signal transduction cascades [41]. P is also a component of nucleotides. In addition to their role as biosynthetic precursors for polymers, nucleotides can act as second messengers, promoting the transfer of information in signaling networks [14, 46, 53, 136], and store and release chemical energy to be used in energy-dependent processes. Finally, the structural role of P as a component of nucleic acids enables the storage, expression and transmission of genetic information. How do bacterial cells acquire P from their environment?

M. S. Razzaque (ed.), *Phosphate Metabolism*, Advances in Experimental Medicine and Biology 1362,
https://doi.org/10.1007/978-3-030-91623-7_12

Although bacteria are able to acquire and metabolize P from organic molecules such as organophosphates (phosphate esters, C-O-P bonds) or phosphonates (C-P bonds) [75, 115, 119, 120, 130], these organisms typically prefer to utilize inorganic orthophosphate (PO_4^{3-}, Pi, phosphate) as their P source (Fig. 12.1). In this context, the acquisition of P relies primarily on inorganic Pi transporters present at the cytoplasmic membrane. Three biochemically distinct classes of Pi transporters have been identified in bacteria: NptA, Pit, and Pst (Fig. 12.2a). NptA (Na-dependent phosphate transport) belongs to the type II sodium/phosphate (Na/Pi) cotransporters family. Homologs of this transport family are common in eukaryotic organisms, where they have been extensively studied [7, 47, 125]. Studies of bacterial NptA homologs are scarce [57, 61, 137]. Nonetheless, the biochemical characterization of the NptA homolog from *Vibrio cholerae* indicates that like its eukaryotic counterparts, this transporter has a low affinity for Pi (Km of ~300 μM) and uses a sodium (Na^+) gradient to facilitate the movement of Pi into the cell at a stoichiometry of 3:1 (Na^+:Pi), and a rate of ~9 pmol Pi/min/mg of protein [61].

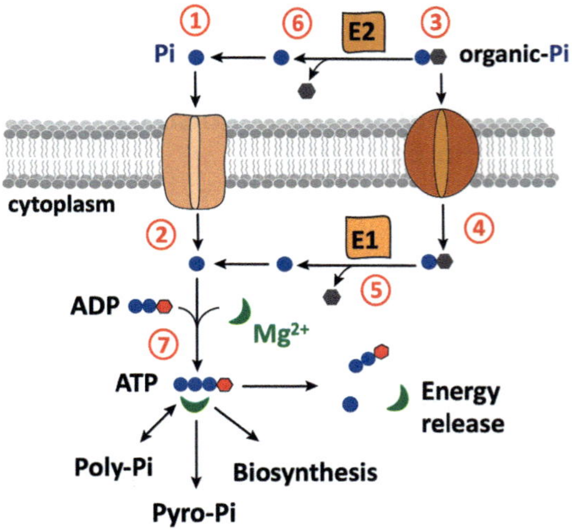

Fig. 12.1 Schematic representation of bacterial Pi metabolism. Inorganic Pi is the preferred P-source **(1)**. Pi is translocated into the cytoplasm by specialized membrane-bound transporters **(2)**. Alternatively, bacteria can use organic-P sources, such as organophosphate and phosphonates **(3)**, which are transported into the cytoplasm by a variety of dedicated membrane transporters **(4)**. Enzymatic Pi extraction from organic molecules can occur either in the cytoplasm **(5)** by intracellular enzymes **(E1)**, or extracytoplasmically **(6)** by specialized enzymes **(E2)**. Pi is assimilated by the cell during the synthesis of ATP **(7)**. Intracellular ATP functions as the P-donor molecule, serving as the source of Pi for the biosynthesis of inorganic (e.g., pyro-Pi and poly-Pi) and organic (e.g., DNA and lipids) molecules, and powering energy dependent processes **(7)**

The Pit (phosphate inorganic transport) family is widespread among all domains of life [76, 85, 113]. Pit systems use the proton motive force to transport soluble, neutral divalent metal-phosphate complexes (Me:Pi) across biological membranes. Bacterial Pit transporters have low affinity for Pi (Km ≈ 1.9–25 μM) and can transport Pi at a rate of 19–58 nmol Pi /min/mg of dry cell [45, 77, 114, 127]. This system can mediate both the efflux and homologous exchange of Me: Pi complexes [114], and displays some promiscuity towards non-Pi oxyanions such as arsenate [128] (Fig. 12.2a). In *Escherichia coli*, the Pit transporter PitA is often thought to function as a constitutive housekeeping Pi transporter. PitA expression is, nonetheless, regulated: *pitA* transcription is stimulated by exposure of cells to zinc-phosphate ($Zn_3(PO_4)_2$) salt [54]. Furthermore, PitA expression is post-transcriptionally repressed during Mg^{2+} starvation [134].

In addition to NptA and Pit, bacteria often encode inducible, ATP-dependent Pi transporters. The Pst system is a complex of four proteins (PstSCAB) that can transport Pi with high affinity (Km 0.18–0.40 μM Pi) and at a rate of ~16 nmol Pi/min/mg of dry cell [127] (Fig. 12.2a). PstS is an extracytoplasmic substrate-binding protein. PstC and PstA are transmembrane components that comprise the Pi channel, and PstB harbors the ATP-binding domain needed for coupling energy hydrolysis with the release of Pi into the cytoplasm [38, 92]. The Pst system is highly specific for Pi, efficiently discriminating it from other inorganic oxyanions such as arsenate and

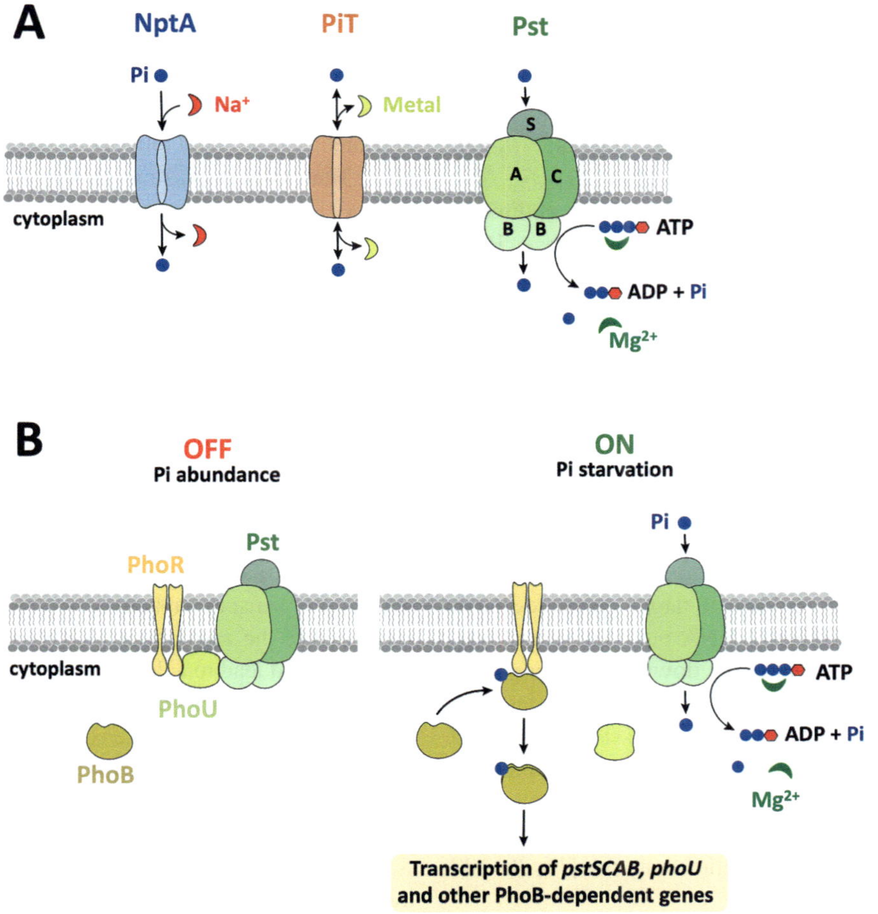

Fig. 12.2 **(a) Cartoon depicting the known classes of Pi transport systems found in bacteria:** NptA, Pit and Pst. NptA is a sodium/phosphate symporter. Pit functions as a metal:phosphate (M:Pi) proton symporter (proton not shown for simplicity). Pst is a multi-component, high affinity, ATP-dependent Pi transporter. PstS is the extracytoplasmic Pi-binding protein, which binds and delivers Pi to the transmembrane channel comprised of the PstC and PstA components. PstB is responsible for the ATP binding, hydrolysis, and release of ADP that powers Pi transport. **(b) Schematics depicting the multicomponent signal complex controlling Pi-responsive PhoB/PhoR two-component system activity.** (Left-hand side panel) During growth in Pi abundant conditions, the PhoU regulatory protein interacts with and represses the activity of the Pst transport system and PhoR kinase activity. (Right-hand side panel) A decrease in cytoplasmic Pi levels alters or disrupts the PhoU interaction with Pst and PhoR. Pst begins to import extracellular Pi while PhoR phosphorylates its cognate response regulator, PhoB. PhoB-P activates the transcription of its target genes, including the *pstSCAB-phoU* operon, *phoBR*, and other genes required for Pi acquisition

sulfate [6, 77, 94, 127]. Noteworthy, Pst-coding genes are usually arranged in an operon (*pstSCAB-phoU*) along with a regulatory accessory gene *phoU*, which codes for a peripheral membrane protein [110]. The *pstSCAB-phoU* operon is transcriptionally activated by the PhoB/PhoR two-component system (see below).

12.2 Sensing and Responding to Pi Starvation

When Pi is plentiful, bacteria can maintain adequate supplies of cytoplasmic Pi using their housekeeping, low affinity Pi transporter systems. However, when Pi is limited and these transporters can no longer keep up with cellular demand, bacteria activate a Pi starvation response. In *E. coli* and most bacteria, this response is governed by the PhoB/PhoR two-component signal transduction system [27, 33, 38, 40, 52, 80, 119–121, 129].

PhoR is a membrane-bound homodimeric, bifunctional histidine kinase/phosphatase. PhoB is its cognate transcriptional response regulator, comprised of an N-terminal receiver domain and a C-terminal DNA-binding domain. Scarce environmental Pi promotes the kinase state of PhoR. PhoR autophosphorylates at a conserved histidine residue and subsequently transfers this phosphoryl group to the receiver domain of PhoB [30, 73, 121]. Phosphorylated PhoB (PhoB-P) binds to DNA sequences known as Pho boxes [3, 35, 74]. Binding of PhoB-P to Pho boxes located at the promoter regions of target genes results in either recruitment of RNA polymerase and transcriptional activation of downstream gene(s), or interference with RNA polymerase and downregulation of gene expression [9, 13, 29, 56, 72, 79, 132, 135]. Abundant environmental Pi promotes the phosphatase state of PhoR, leading to the dephosphorylation of PhoB-P and inactivation of the Pi starvation response [16, 38, 100, 119–121].

While the phosphorylation events controlling the activation state of PhoB/PhoR are well understood, how this two-component system senses environmental Pi is still unclear. Unlike other prototypical bifunctional histidine kinases/phosphatases, PhoR does not contain a sizable extracytoplasmic domain, which is typically involved in the recognition of extracellular ligands [97]. Hence, PhoR is unlikely to directly detect Pi via extracytoplasmic sensory domain(s). Rather, the ability of PhoB/PhoR to respond to Pi requires the Pst transport system and the PhoU protein. PhoU mediates a physical interaction between PhoR and the PstB component of the Pst system, creating a multicomponent signaling complex capable of modulating PhoB/PhoR activity in response to environmental Pi [37, 38, 107, 110, 120, 121] (Fig. 12.2b). Pi limitation is presumed to disrupt (or alter) this signaling complex, promoting the kinase activity of PhoR, phosphorylation of PhoB and activation of Pi starvation response (Fig. 12.2b).

It is also unclear whether Pi starvation is sensed by the extracytoplasmic or cytoplasmic portions of this multicomponent signaling complex. The extracytoplasmic sensing model relies on two lines of evidence. First, intracellular Pi levels have been measured by ^{31}P NMR and shown only a modest decrease under conditions in which PhoB is activated [91]. However, given the lack of positive and negative controls in this study, it is difficult to determine whether the observed differences are biologically significant or if larger discrepancies in intracellular Pi concentrations would be revealed by the experimental set up. Second, certain mutations in the PstC and PstA components of the Pst system have been shown to abolish Pi transport activity while retaining Pi sensing capacity [24, 25]. This argument makes the assumption that an inactive Pst system would lead to a decrease in intracellular Pi levels—that is, it does not consider that cells can compensate for a decrease in intracellular Pi, resulting from a defective Pst system, by upregulating other transporters such as PitA [15, 49].

The cytoplasmic sensing model has been evoked to account for several experimental results that can only be explained if Pi starvation is being sensed by the cytoplasmic portion of the PhoR-PhoU-Pst signaling complex. For example, when metabolic mutants of *E. coli* are grown in medium containing high Pi, PhoB/PhoR is activated under conditions that promote an expansion of cytoplasmic pools of ATP [126]. Similarly, a decrease in cytoplasmic concentrations of Mg^{2+} activates PhoB/PhoR in both *E. coli* and *Salmonella enterica* (*Salmonella*) grown in high Pi media [15, 87, 90]. Notably, cytoplasmic Mg^{2+} starvation destabilizes the bacterial ribosomal subunits [90]. Decreased ATP consumption by functionally compromised ribosomes causes a rise in cytoplasmic ATP concentrations and, presumably, a concomitant decrease in levels of free Pi that is proposed to activate the PhoB/PhoR two-component system. This model is independently supported by the demonstration that antibiotics that inhibit translation cause a rise in intracellular ATP levels, resulting in PhoB/PhoR activation in both *E. coli* and *Salmonella* grown in high Pi medium. Importantly, activation of PhoB/PhoR by translation inhibitors is not observed if the rise in ATP levels is prevented by inducing the expression of an ATPase prior to antibiotic treatment [87]. Finally, in *Salmonella*, inactivation of the *pitA* transporter causes an increase in the basal level of *pstSCAB* transcription, presumably because this mutation lowers the intracellular concentration of Pi [15].

Which genes are expressed in response to Pi starvation? In *E. coli*, the PhoB regulon contains at least 38 genes including: *phoBR*, for the positive autoregulation of the system; *phoA*, an extracytoplasmic alkaline phosphatase; *phoE*, an outer membrane porin channel for anions (including Pi); *pstSCAB-phoU*, coding the high affinity Pst system and the chaperon like inhibitory protein PhoU (see above); *phnCDEFGHIJKLMNOP*, phosphonate uptake and breakdown genes; *ugpBAECQ*, genes for glyceraldehyde-3-phosphate uptake plus a phosphodiesterase that hydrolyzes glycerophosphoryl diesters (deacylated phospholipids) (For reviews see [38, 92, 119, 120]). Notably, this response goes beyond the mere acquisition of environmental Pi and alternative P sources. Some PhoB-activated genes allow *E. coli* to scavenge Pi from its own cellular components. For example, the PhoB activated *waaH* gene encodes a glucuronic acid transferase. WaaH modifies the core region of the *E. coli* LPS, replacing a Pi residue, which can be re-purposed by the cell, with a glucuronic acid [4, 60]. Pi-starvation driven changes in cell composition have also been documented in the distantly related species *Bacillus subtilis*, and may be a general strategy adopted by bacteria to reallocate assimilated Pi [10, 11, 68, 96].

Importantly, the genes within the PhoB regulon are neither expressed at the same time, nor at the same levels. An elegant analysis of the PhoB-dependent temporal patterns of gene expression revealed the existence of *early* genes, encoding proteins required for PhoBR autofeedback regulation and Pi transport (i.e. *phoBR* and *pstSCAB-phoU*); and *late* genes, encoding proteins required for the acquisition and utilization of alternative, typically organic phosphorus sources (i.e. *phoA*, and *ugp* and *phn* operons) [35]. This temporal expression pattern can be explained, in large part, by variations in the DNA sequence of Pho boxes, which produce different binding affinities to PhoB-P. While the promoter regions of genes needed at the initial stages of starvation usually have high affinity Pho Boxes, requiring lower cytoplasmic concentrations of PhoB-P for transcriptional activation, the promoter regions of genes needed at the later stages of the response have low affinity Pho Boxes, requiring higher concentrations of PhoB-P [35].

12.3 Phosphate Cytotoxicity via Pst

While P is essential, excessive transport of Pi into the cytoplasm is toxic. In some cases, toxicity is mediated by the co-transported counterions, rather than Pi itself. For instance, some Pi transporters import Pi:Me salts (Fig. 12.2a). Excessive transport of Pi:Me (Pi:Co^{2+}; Pi:Cu^{2+}; Pi:Mn^{2+}; Pi:Zn^{2+}) salts by the Pho84 transporter of *Saccharomyces cerevisiae* results in metal toxicity [55, 81, 95]. An analogous phenomenon occurs in the PitA transporter of *E. coli*, which can render cells liable to Zn^{2+} poisoning [5]. Toxicity mediated by these transport systems likely reflect the fact that relatively small disturbances in the intracellular concentrations of these metals are sufficient to promote unproductive mismetallation events that hinder enzymatic reactions [34].

In bacteria, Pi toxicity *per se* has been observed in cells experiencing uncontrolled Pst-mediated Pi uptake resulting from either (1) missense mutations in Pst components that increase Pi influx [122], (2) null mutations in *phoU* that derepresses the PhoB/PhoR regulon [93, 107], or (3) over-expression of the Pst system [15]. The PstA and PstC proteins form the Pst Pi channel that spans through the cytoplasmic membrane [92, 122] (Fig. 12.2a). In *E. coli*, certain amino acid substitutions within PstC transmembrane helices cause a severe growth defect when the Pi concentration in the medium is raised to levels approaching that of the cytoplasm. These substitutions are thought to lock the transport systems into an "open" conformation that allows unrestricted Pi uptake. In agreement with this notion, the Pi sensitivity of cells harboring these PstC variants is suppressed by additional amino acid substitutions in PstC that abolish Pi transport, presumably by locking the channel into a "closed" conformation [122].

The involvement of Pst in potentiating Pi toxicity is also inferred from genetic studies of the *phoU* regulatory gene (Fig. 12.2b). Pioneer work in *E. coli* established that inactivation of *phoU* results in a severe growth defect due to increased Pst expression and transport activity. Specifically, the growth impairment of *phoU* mutants is suppressed by secondary mutations that inactivate components of the Pst transporter (*pstB* or *pstSCAB*), or prevent its expression (*phoB* or *phoR*) [107], and it is improved by reducing the concentration of Pi in the medium [93]. Importantly, this relationship is not restricted to *E. coli*: *phoU* mutations also lead to growth impairment in species where PhoU participates in the repression of PhoB/PhoR and Pst transport activity. These include *Pseudomonas aeruginosa* [2], *Synechocystis* sp. [78], *Streptococcus pneumoniae* [137], *Staphylococcus epidermidis* [118], *Mycobacterium marinum* [117], *Sinorhizobium meliloti* [28] and *Caulobacter crescentus* [69]. In *C. crescentus*, *phoU* inactivating mutations are conditionally lethal. These genetic lesions can only be introduced in strains already containing mutations in *phoBR* or *pstSCAB* [69]. Similarly, in *Sinorhizobium meliloti*, *phoU* can only be inactivated if cells are grown in medium containing low concentrations of Pi, or if the Pst transporter is also inactivated [28].

The dysregulation of Pst-mediated Pi uptake observed in *phoU* mutants has pleiotropic effects on cellular physiology, reflecting the pervasive role of Pi in the cell. In addition to hindering growth, mutations in *phoU* are frequently linked to increased sensitivity to antibiotics and other environmental stresses [2, 117, 137]. Furthermore, these mutations lead to elevated intracellular ATP [117, 118] and over accumulation of polyphosphate (poly-Pi) [2, 28, 48, 69, 78, 117]. Notably, secondary mutations that hinder poly-Pi synthesis do not suppress the lethality of *phoU* mutations in *C. crescentus* [69], indicating that toxicity does not originate from poly-Pi accumulation.

In *Salmonella*, overexpression of the Pst transporter from an inducible promoter is sufficient to hinder growth and induce a rise in cytoplasmic ATP levels [15]. These phenotypes are also accompanied by a global decrease in translation and the transcription of the *mgtCBRU* operon, encoding genes that respond to cytoplasmic Mg^{2+} starvation (see below). Surprisingly, the effect of *pstSCAB* overexpression on growth can be drastically improved by enzymatically decreasing intracellular ATP concentration through the controlled expression of soluble components of the F1Fo ATPase [15]. Excessive Pi, therefore, becomes toxic following its incorporation into ATP, and this toxicity can be remediated by providing cells with Mg^{2+} [15].

12.4 The Role of Mg^{2+} on the Utilization of Assimilated Pi

Pi imported into the cytoplasm is incorporated into biomass primarily through the synthesis of ATP. ATP functions as the main P-carrier molecule in the cell, mediating the transfer of Pi in many biosynthetic reactions, and serving as the main source of phosphoanhydride bonds used to power energy-dependent processes (Fig. 12.1) [70, 120]. In bacteria and other living organisms, most of the assimilated Pi is found in ATP, and ribosomal ribonucleic acids (rRNA), the most abundant RNA species in cells [12, 31, 39, 89].

Following assimilation, the negative charges from Pi residues in biomolecules often undergo charge neutralization by cytoplasmic cationic species. Mg^{2+} plays a pivotal role in this process due to its high charge density (the highest among all biologically relevant cations), and its presence as the most abundant divalent cation in cells [116] (Fig. 12.1). The Mg^{2+} concentrations in bacteria are estimated to be around 75–100 mM, of which two-thirds (50–65 mM) are intracellular and one-third (25–35 mM) are bound to cell envelop components such as the LPS of Gram-negative species [89]. Half of the intracellular Mg^{2+} is complexed to nucleoside-triphosphates (NTPs), mainly ATP [89]. Importantly, it is this ATP: Mg^{2+} salt, rather than the ATP anion, which serve as the substrate for most ATP-dependent enzymatic reactions [71, 108].

The assembly of rRNAs into functional ribosomes also requires the association of Mg^{2+} ions to reduce the electrostatic repulsion among Pi residues present in the sugar-phosphate backbone, enabling the compaction of the rRNA into stable structures and the association of ribosomal proteins [59]. A prototypical bacterial ribosome binds approximately 170 Mg^{2+} ions [98]. During rapid exponential growth, ribosomes pools are at their highest and are estimated to chelate the equivalent of 25% of the total intracellular Mg^{2+} [89]. Importantly, Mg^{2+} associates with ribosomes in a specific manner [58, 123, 124]. Compared to other common biological cations such as Na^+, K^+, Ca^{2+}, or polyamines [32], Mg^{2+} has a greater affinity for oxygen atoms in phosphodiester bonds. Mg^{2+} binds oxygen atoms with a well-defined geometry, bringing them into its first hydration shell. Consequently, rRNA:Mg^{2+} complexes are also stabilized by non-electrostatic components of the binding such as polarization, charge transfer and exchange interactions [86].

Given the central role of Mg^{2+} on the function of assimilated Pi, it is not fortuitous that ATP and rRNA, the largest cytoplasmic reservoirs of Pi, are also the largest cytoplasmic repositories of Mg^{2+} [89]. How do bacteria maintain adequate concentrations of Mg^{2+} in their cytoplasm? Do they alter their Pi metabolism in response to insufficient Mg^{2+}?

12.5 Sensing and Responding to Mg^{2+} Starvation

Bacterial cells have also evolved means to sense and respond to decreases in the concentration of extra- and intracellular Mg^{2+}. In many bacterial species, the Mg^{2+} starvation response is controlled by the PhoP/PhoQ two-component signal transduction system. This regulatory system has been extensively studied in *Salmonella*, from which most of our knowledge is derived (for extensive reviews see [42, 43]. PhoQ is a membrane-bound, homodimeric, bifunctional histidine kinase/phosphatase. PhoQ senses extracytoplasmic Mg^{2+} through a patch of acidic amino acid residues located at its extracytoplasmic domain [18, 20, 36]. Binding of Mg^{2+} to this acidic patch forms salt bridges that reduce electrostatic repulsion against the negatively charged cytoplasmic membrane, stabilizing PhoQ in a rigid conformation that favors its phosphatase activity. In this conformation, PhoQ maintains its cognate DNA-binding transcriptional regulator PhoP in an inactive, unphosphorylated state [17, 20, 36, 66, 102] (Fig. 12.3).

As the concentration of extracellular Mg^{2+} decreases, the Mg^{2+} bridges between the acidic amino acid patch of PhoQ and the outer leaflet of cytoplasmic membrane are destabilized, promoting conformational changes favoring a kinase state [20, 36]. PhoQ undergoes autophosphorylation and, subsequently, transfers its phosphoryl residue to PhoP. Phosphorylated PhoP (PhoP-P) binds with high-affinity to target DNA sites, named PhoP boxes [65, 66, 102, 131]. The majority of PhoP boxes are located at promoters, where binding of PhoP-P stimulates the transcription of downstream genes through physical interactions with RNA polymerase [22, 84, 102, 138, 139] (Fig. 12.3).

The response to Mg^{2+} starvation orchestrated by PhoP/PhoQ occurs in an orderly manner, with transcription of individual genes being induced at specific stages of the stress, in a fashion resembling the PhoB/PhoR response to Pi starvation (see above). However, unlike PhoB [35], the time and order of transcriptional activation of genes within the regulon is not dictated solely by the affinity of PhoP-P to PhoP boxes. In *Salmonella*, the expression order of genes is influenced by additional regulatory mechanisms, including the existence of Mg^{2+} sensing RNAs and other regulatory elements at the 5′-end of transcripts [26, 50, 62, 63, 82, 99, 106], transcriptional silencing by nucleoid associated proteins

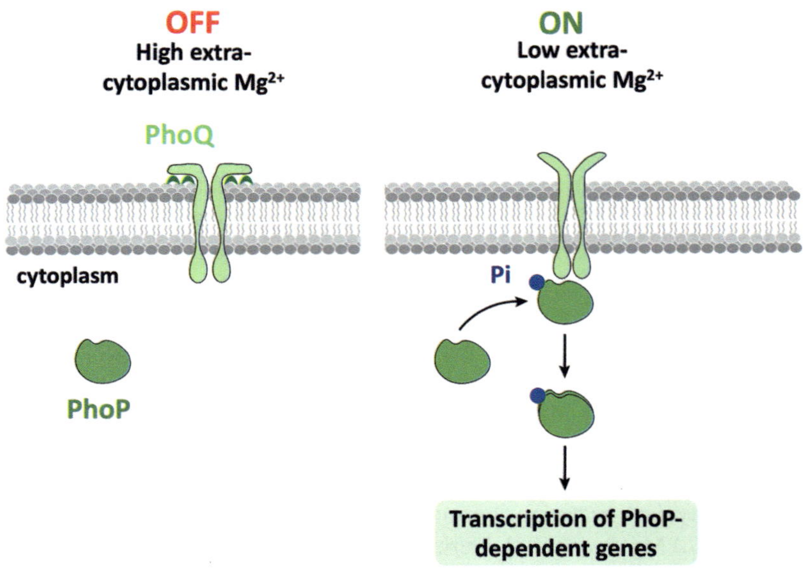

Fig. 12.3 Representation of the Mg^{2+}-sensing PhoP/PhoQ two-component system. (Left-hand side panel) High extracellular Mg^{2+} levels stabilize the sensor protein PhoQ in a rigid conformation. This state promotes PhoQ phosphatase activity, and maintains its cognate response regulator, PhoP, in an inactive, unphosphorylated state. (Right-hand side panel) Low extracytoplasmic Mg^{2+} activates PhoQ kinase activity, increasing the intracellular concentration of PhoP-P, which, in turn, promotes transcription initiation at PhoP-dependent promoters, such as those involved in adaptation to low Mg^{2+} environments (e.g., cell envelope remodeling, Mg^{2+} scavenging, etc.)

[23, 140], and modulation of PhoQ activity by members of the regulon [67, 83]. Nonetheless, the timely expression of genes in the regulon is linked to their physiological functions. For instance, during the early stages of Mg^{2+} starvation, PhoP activates the expression of several enzymes involved in the remodeling of the cell envelope. These proteins carry out covalent modifications that replace negatively charged Pi-containing residues in the LPS. The result is a reduction in the net negative charge of the outer membrane, which facilitates charge neutralization when Mg^{2+} is limited, and frees Mg^{2+} ions bound to the LPS to be imported into the cytoplasm [8, 19, 44, 51, 101, 105, 112].

The transcription of genes responding to physiological disturbances caused by prolonged Mg^{2+} starvation is often conditioned by additional events taking place in the cytoplasm. For instance, while phosphorylation of PhoP resulting from low extracytoplasmic Mg^{2+} sensing by PhoQ causes transcription to initiate at the promoters of the *mgtA* gene and the *mgtCBRU* operon, transcription elongation into their corresponding coding regions takes place only when cells also experience a decrease in free cytoplasmic Mg^{2+} [26, 106]. Therefore, MgtA and the proteins within the *mgtCBRU* operon are expressed following a substantial period of starvation, when the housekeeping transporter, CorA, is no longer able to maintain an adequate supply of Mg^{2+} to the cytoplasm. Whereas *mgtA* and *mgtB* encode high affinity, ATP-dependent Mg^{2+} transporters [103, 104, 111], *mgtC* encodes a protein that reduces ATP levels by inhibiting Pi uptake via an unidentified Pi transporter [15, 63, 64, 88]. (MgtR and MgtU encode regulatory proteins that control the degradation of MgtA, MgtB, and MgtC [1, 21, 133]). Hence, MgtA and MgtB import extracellular Mg^{2+} into the cytoplasm and MgtC prevents the sequestration of this scarce ion by ATP molecules

[63, 64]. Notably, the transcription of a subset of PhoP-activated genes requires the activity of the MgtA transporter. The importation of extracellular Mg^{2+} by this transporter increases the levels of PhoP-P, presumably by removing inhibitory Mg^{2+} from the vicinity of PhoQ [83].

12.6 Coordination of Pi and Mg^{2+} Homeostasis in Bacteria

Pi toxicity resulting from increased Pst transporter activity explains why this transporter is not constitutive, being only expressed in low Pi environments. However, this does not reveal a biologically relevant context where Pi toxicity would occur. Do natural populations of bacteria experience Pi toxicity? If so, under what circumstances? In *Salmonella*, prolonged Mg^{2+} starvation causes a decrease in cytoplasmic Mg^{2+}. At the initial stages of cytoplasmic Mg^{2+} starvation, the assembly of ribosomal subunits is compromised while Mg^{2+}-dependent enzymatic reactions remain unaffected (Fig. 12.4a). This phenomenon likely reflects differences between the Mg^{2+} requirements of the ribosome and enzymes. Whereas the ribosome has a high, continuous requirement for Mg^{2+} (>170 ions for their structural stabilization), enzymes have a low and transient need for this cation (one or a few ions during each catalytic cycle). Consequently, cells experience a decrease in translation efficiency, but can still carry out other biosynthetic reactions such as ATP synthesis and transcription [90]. Because translation is the most expensive anabolic activity in the cell, consuming over 70% of the ATP that is utilized in biosynthetic processes [89, 109], this ribosome assembly defect leads to a reduction in ATP consumption and a concomitant rise in cytoplasmic ATP concentration [15, 87, 90] (Fig. 12.4a). This sets in motion two chains of events.

First, as a substantial portion of intracellular Pi comprise phosphoryl residues in ATP, a decrease in ATP hydrolysis from translation reactions reduces the intracellular recycling of Pi. This presumably decreases cytoplasmic Pi concentrations, which triggers PhoB/PhoR activation [87]. PhoB promotes Pst expression, causing an increase in Pi uptake, and ATP synthesis [15]. Increased ATP concentrations stimulate rRNA synthesis, the rate-limiting step in ribosome biosynthesis. However, newly synthesized rRNA cannot assemble into functional ribosomes due to insufficient Mg^{2+} [90] (Fig. 12.4a). Consequently, cells are subjected to a vicious cycle, whereby the transport of Pi by Pst increases cytoplasmic ATP levels and rRNA synthesis, promoting additional chelation of Mg^{2+} that further destabilizes the ribosomes and stimulates Pst expression [15, 90].

Second, this reduction in translation efficiency also triggers the expression of genes responding to cytoplasmic Mg^{2+} starvation. Inefficient translation promotes transcription elongation into the coding regions of *mgtA* and *mgtCBRU* through a number of regulatory mechanisms acting at their 5′-mRNA leaders [62, 63, 82, 99, 106]. MgtA and MgtB import Mg^{2+}, promoting the stabilization of ribosomes [90] (Fig. 12.4b). MgtC reduces ATP levels [63] by hindering the activity of a yet unidentified Pi transport system [15] (Fig. 12.4b). By inhibiting Pi uptake, MgtC hampers all ATP generating reactions in the cell. On the one hand, a reduction in ATP quenches rRNA synthesis, preventing further entrapment of Mg^{2+} by nascent rRNA (Fig. 12.4b). On the other hand, as new steady-state levels of ribosomes are reached, the degradation of excess rRNA and ribosomes liberates additional Mg^{2+} ions into the cytoplasm. Overall, the concerted activities of MgtA, MgtB and MgtC restore ribosomal function, effectively silencing PhoB/PhoR and inhibiting Pst expression [15, 87, 90].

12.7 Concluding Remarks

The utilization of assimilated Pi into ATP and ribosomes requires Mg^{2+}. Because cytoplasmic Pi is rapidly assimilated into ATP [15], cells coordinate Pi uptake with Mg^{2+} availability. In *Salmonella*, the response to insufficient cytoplasmic Mg^{2+} entails the inhibition of Pi uptake as a general strategy to prevent sequestration of

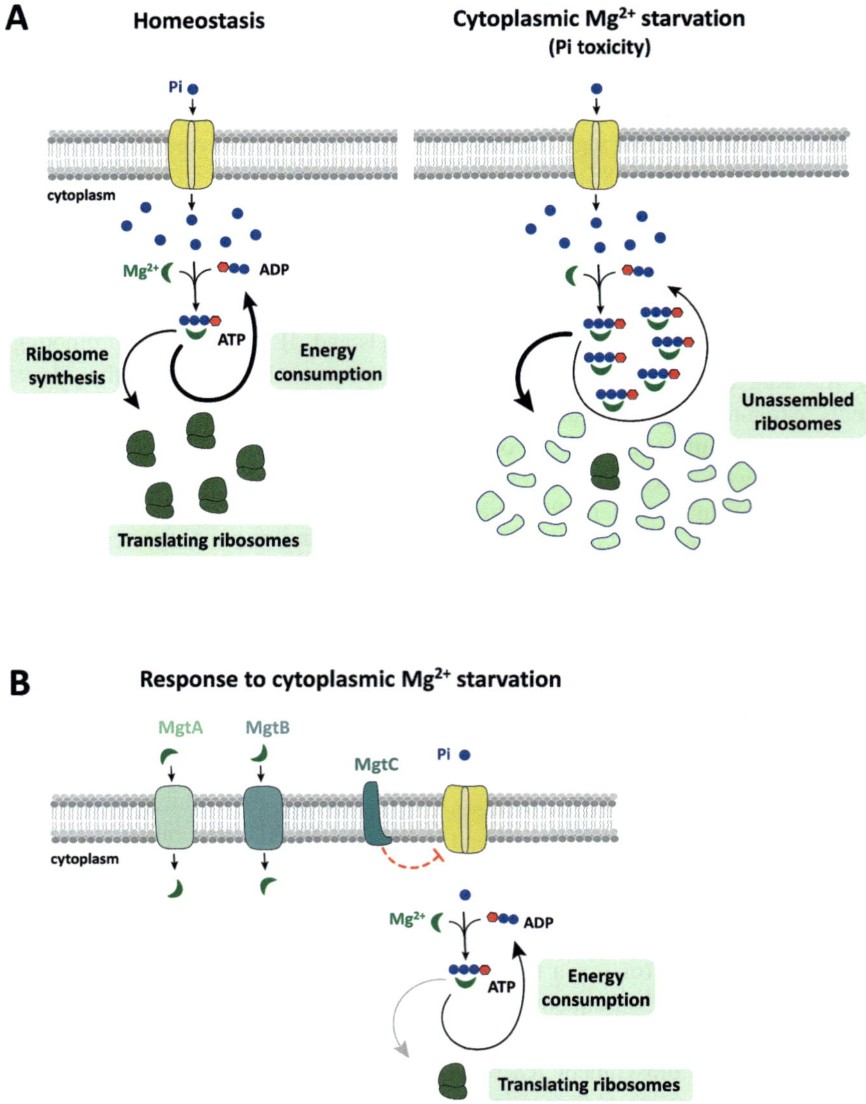

Fig. 12.4 (a) **Model illustrating the interplay between Pi and Mg^{2+} metabolisms in bacteria.** (Left-hand side panel) During homeostasis, Pi is translocated into the cytoplasm by dedicated membrane-bound transporters. Cells assimilate imported Pi through the synthesis of ATP, which is primarily neutralized by Mg^{2+} cations, yielding ATP:Mg^{2+}, the main physiological form of ATP. In enteric bacteria, ATP:Mg^{2+} stimulates rRNA synthesis, thus promoting ribosome biogenesis. Large amounts of Mg^{2+} ions are required to support assembled, translating ribosomes which consume the majority of intracellular ATP pools, thereby recycling Pi and Mg^{2+}. (Right-hand side panel) After exhausting the environmental Mg^{2+}, cells eventually experience a shortage in cytoplasmic Mg^{2+} levels. Insufficient cytoplasmic Mg^{2+} impairs ribosomal subunit assembly, lowering translation efficiency and ATP consumption. Yet, the combined activity of all Pi-assimilating reactions continues to promote ATP synthesis, which furthers rRNA transcription. This results in the futile accumulation of unassembled ribosomal subunits, which are now unable to hydrolyze ATP:Mg^{2+} to recycle Pi and Mg^{2+}. **(b) Cellular response to cytoplasmic Mg^{2+} starvation in *Salmonella*.** The decrease in cytoplasmic Mg^{2+} levels promotes transcription elongation into the coding regions of the PhoP-dependent *mgtA*, *mgtB*, and *mgtC* genes. MgtA and MgtB proteins import Mg^{2+} into the cytoplasm, and the MgtC protein reduces the uptake of Pi, thus, preventing its assimilation into Mg^{2+}-chelating ATP. The increase in free cytoplasmic Mg^{2+} in combination with a diminished pool of ribosomes restores translation, normalizing ATP hydrolysis and recycling of intracellular Pi and Mg^{2+}

already scarce Mg^{2+} by assimilated Pi. Interestingly, artificially increasing Pi intake through Pst expression induces the transcription of genes that respond to cytoplasmic Mg^{2+} starvation (see above). This occurs even when cells are grown in conditions where Mg^{2+} is plentiful [15]. Therefore, excessive cytoplasmic Pi mimics conditions resulting from insufficient cytoplasmic Mg^{2+}, and vice-versa. Given that Mg^{2+}, ATP and rRNA are indispensable constituents of every living cell, we anticipate that the inherent interplay among them hereby described is widespread in nature.

References

1. Alix E, Blanc-Potard AB (2008) Peptide-assisted degradation of the *Salmonella* MgtC virulence factor. EMBO J 27(3):546–557
2. Almeida LG De, Ortiz H, Schneider RP, Spira B (2015) *phoU* inactivation in *Pseudomonas aeruginosa* enhances accumulation of ppGpp and polyphosphate. Appl Environ Microbiol 81(9):3006–3015
3. Bachhawat P, Swapna GVT, Montelione GT, Stock AM (2005) Mechanism of activation for transcription factor PhoB suggested by different modes of dimerization in the inactive and active states. Structure 13(9):1353–1363
4. Baek JH, Lee SY (2006) Novel gene members in the Pho regulon of *Escherichia coli*. FEMS Microbiol Lett 264(1):104–109
5. Beard SJ, Hashim R, Wu G, Binet MR, Hughes MN, Poole RK (2000) Evidence for the transport of zinc (II) ions via the Pit inorganic phosphate transport system in *Escherichia coli*. FEMS Microbiol Lett 184(2):231–235
6. Bennett RL, Malamy MH (1970) Arsenate resistant mutants of *Escherichia coli* and phosphate transport. Biochem Biophys Res Commun 40(2):496–503
7. Biber J, Custer M, Magagnin S, Hayes G, Werner A, Lötscher M et al (1996) Renal Na/Pi-cotransporters. Kidney Int 49(4):981–985
8. Bishop RE, Gibbons HS, Guina T et al (2000) Transfer of palmitate from phospholipids to lipid A in outer membranes of Gram-negative bacteria. EMBO J 19(19):5071–5080
9. Blanco AG, Canals A, Bernués J, Solà M, Coll M (2011) The structure of a transcription activation subcomplex reveals how σ70 is recruited to PhoB promoters. EMBO J 30(18):3776–3785
10. Botella E, Devine SK, Hubner S, Salzberg LI, Gale RT, Brown ED et al (2014) PhoR autokinase activity is controlled by an intermediate in wall teichoic acid metabolism that is sensed by the intracellular PAS domain during the PhoPR-mediated phosphate limitation response of *Bacillus subtilis*. Mol Microbiol 94(6):1242–1259
11. Botella E, Hübner S, Hokamp K, Hansen A, Bisicchia P, Noone D et al (2011) Cell envelope gene expression in phosphate-limited *Bacillus subtilis* cells. Microbiology 157(Pt 9):2470–2484
12. Bremer H, Dennis PP (2008) Modulation of chemical composition and other parameters of the cell at different exponential growth rates. EcoSal Plus 3(1)
13. Browning DF, Busby SJW (2004) The regulation of bacterial transcription initiation. Nat Rev Microbiol 2(1):57–65
14. Bruckner R, Titgemeyer F (2002) Carbon catabolite repression in bacteria: choice of the carbon source and autoregulatory limitation of sugar utilization. FEMS Microbiol Lett 209(2):141–148
15. Bruna RE, Kendra CG, Groisman EA, Pontes MH (2021) Limitation of phosphate assimilation maintains cytoplasmic magnesium homeostasis. Proc Natl Acad Sci U S A 118(11):e2021370118
16. Carmany DO, Hollingsworth K, McCleary WR (2003) Genetic and biochemical studies of phosphatase activity of PhoR. J Bacteriol 185(3):1112–1115
17. Castelli ME, García Véscovi E, Soncini FC (2000) The phosphatase activity is the target for Mg^{2+} regulation of the sensor protein PhoQ in *Salmonella*. J Biol Chem 275(30):22948–22954
18. Chamnongpol S, Cromie M, Groisman EA (2003) Mg^{2+} sensing by the Mg^{2+} sensor PhoQ of *Salmonella enterica*. J Mol Biol 325(4):795–807
19. Chen HD, Groisman EA (2013) The biology of the PmrA/PmrB two-component system: the major regulator of lipopolysaccharide modifications. Annu Rev Microbiol 67:83–112
20. Cho US, Bader MW, Amaya MF, Daley ME, Klevit RE, Miller SI et al (2006) Metal bridges between the PhoQ sensor domain and the membrane regulate transmembrane signaling. J Mol Biol 356(5): 1193–1206
21. Choi E, Lee KY, Shin D (2012) The MgtR regulatory peptide negatively controls expression of the MgtA Mg^{2+} transporter in *Salmonella enterica* serovar Typhimurium. Biochem Biophys Res Commun 417(1):318–323
22. Christian Perez J, Groisman EA (2009) Transcription factor function and promoter architecture govern the evolution of bacterial regulons. Proc Natl Acad Sci U S A 106(11):4319–4324
23. Christian Perez J, Latifi T, Groisman EA (2008) Overcoming H-NS-mediated transcriptional silencing of horizontally acquired genes by the PhoP and SlyA proteins in *Salmonella enterica*. J Biol Chem 283(16):10773–10783
24. Cox GB, Webb D, Godovac-Zimmermann J, Rosenberg H (1988) Arg-220 of the PstA protein is required for phosphate transport through the

phosphate-specific transport system in *Escherichia coli* but not for alkaline phosphatase repression. J Bacteriol 170(5):2283–2286
25. Cox GB, Webb D, Rosenberg H (1989) Specific amino acid residues in both the PstB and PstC proteins are required for phosphate transport by the *Escherichia coli* Pst system. J Bacteriol 171(3): 1531–1534
26. Cromie MJ, Shi Y, Latifi T, Groisman EA (2006) An RNA sensor for intracellular Mg^{2+}. Cell 125(1): 71–84
27. Danhorn T, Hentzer M, Givskov M, Parsek MR, Fuqua C (2004) Phosphorus limitation enhances biofilm formation of the plant pathogen *Agrobacterium tumefaciens* through the PhoR-PhoB regulatory system. J Bacteriol 186(14):4492–4501
28. diCenzo GC, Sharthiya H, Nanda A, Zamani M, Finan TM (2017) PhoU allows rapid adaptation to high phosphate concentrations by modulating PstSCAB transport rate in *Sinorhizobium meliloti*. J Bacteriol 199(18):e00143–e00117
29. Egan SE, Fliege R, Tong S, Shibata A, Wolf RE Jr, Conway T (1992) Molecular characterization of the Entner-Doudoroff pathway in *Escherichia coli*: sequence analysis and localization of promoters for the *edd-eda* operon. J Bacteriol 174(14):4638–4646
30. Ellison DW, McCleary WR (2000) The unphosphorylated receiver domain of PhoB silences the activity of its output domain. J Bacteriol 182(23): 6592–6597
31. Elser JJ, Acharya K, Kyle M, Cotner J, Makino W, Markow T et al (2003) Growth rate-stoichiometry couplings in diverse biota. Ecol Lett 6(10):936–943
32. Feuerstein BG, Williams LD, Basu HS, Marton LJ (1991) Implications and concepts of polyamine-nucleic acid interactions. J Cell Biochem 46(1): 37–47
33. Filloux A, Bally M, Soscia C, Murgier M, Lazdunski A (1988) Phosphate regulation in *Pseudomonas aeruginosa*: cloning of the alkaline phosphatase gene and identification of *phoB*- and *phoR*-like genes. Mol Gen Genet 212(3):510–513
34. Foster AW, Osman D, Robinson NJ (2014) Metal preferences and metallation. J Biol Chem 289(41): 28095–28103
35. Gao R, Stock AM (2015) Temporal hierarchy of gene expression mediated by transcription factor binding affinity and activation dynamics. MBio 6(3): e00686-15
36. García Véscovi E, Soncini FC, Groisman EA (1996) Mg^{2+} as an extracellular signal: environmental regulation of *Salmonella* virulence. Cell 84(1):165–174
37. Gardner SG, Johns KD, Tanner R, McCleary WR (2014) The PhoU protein from *Escherichia coli* interacts with PhoR, PstB, and metals to form a phosphate-signaling complex at the membrane. J Bacteriol 196(9):1741–1752
38. Gardner SG, McCleary WR (2019) Control of the *phoBR* regulon in *Escherichia coli*. EcoSal Plus 8(2)
39. Gillooly JF, Allen AP, Brown JH, Elser JJ, Martinez del Rio C, Savage VM et al (2005) The metabolic basis of whole-organism RNA and phosphorus content. Proc Natl Acad Sci U S A 102(33):11923–11927
40. Glover RT, Kriakov J, Garforth SJ, Baughn AD, Jacobs WR Jr (2007) The two-component regulatory system *senX3-regX3* regulates phosphate-dependent gene expression in *Mycobacterium smegmatis*. J Bacteriol 189(15):5495–5503
41. Groisman EA (2016) Feedback control of two-component regulatory systems. Annu Rev Microbiol 70:103–124
42. Groisman EA, Hollands K, Kriner MA, Lee EJ, Park SY, Pontes MH (2013) Bacterial Mg^{2+} homeostasis, transport, and virulence. Annu Rev Genet 47:625–646
43. Groisman EA, Kato A (2008) The PhoQ/PhoP regulatory network of *Salmonella enterica*. Adv Exp Med Biol 631:7–21
44. Guo L, Lim KB, Poduje CM, Daniel M, Gunn JS, Hackett M et al (1998) Lipid A acylation and bacterial resistance against vertebrate antimicrobial peptides. Cell 95(2):189–198
45. Harris RM, Webb DC, Howitt SM, Cox GB (2001) Characterization of PitA and PitB from *Escherichia coli*. J Bacteriol 183(17):5008–5014
46. Hengge R (2021) High-specificity local and global c-di-GMP signaling. Trends Microbiol:S0966-842X (21)00037-8
47. Hernando N, Gagnon K, Lederer E (2021) Phosphate transport in epithelial and nonepithelial tissue. Physiol Rev 101(1):1–35
48. Hirota R, Motomura K, Nakai S, Handa T, Ikeda T, Kuroda A (2013) Stable polyphosphate accumulation by a pseudo-revertant of an *Escherichia coli phoU* mutant. Biotechnol Lett 35(5):695–701
49. Hoffer SM, Tommassen J (2001) The phosphate-binding protein of *Escherichia coli* is not essential for Pi-regulated expression of the Pho regulon. J Bacteriol 183(19):5768–5771
50. Hollands K, Proshkin S, Sklyarova S, Epshtein V, Mironov A, Nudler E et al (2012) Riboswitch control of Rho-dependent transcription termination. Proc Natl Acad Sci U S A 109(14):5376–5381
51. Hong X, Chen HD, Groisman EA (2018) Gene expression kinetics governs stimulus-specific decoration of the *Salmonella* outer membrane. Sci Signal 11(529):eaar7921
52. Hulett FM, Lee J, Shi L, Sun G, Chesnut R, Sharkova E et al (1994) Sequential action of two-component genetic switches regulates the PHO regulon in *Bacillus subtilis*. J Bacteriol 176(5):1348–1358
53. Irving SE, Choudhury NR, Corrigan RM (2021) The stringent response and physiological roles of (pp)-pGpp in bacteria. Nat Rev Microbiol 19(4):256–271

54. Jackson RJ, Binet MR, Lee LJ, Ma R, Graham AI, McLeod CW et al (2008) Expression of the PitA phosphate/metal transporter of *Escherichia coli* is responsive to zinc and inorganic phosphate levels. FEMS Microbiol Lett 289(2):219–224
55. Jensen LT, Ajua-Alemanji M, Culotta VC (2003) The *Saccharomyces cerevisiae* high affinity phosphate transporter encoded by PHO84 also functions in manganese homeostasis. J Biol Chem 278(43): 42036–42040
56. Jiang J, Yu K, Qi L, Liu Y, Cheng S, Wu M et al (2018) A proteomic view of *Salmonella* Typhimurium in response to phosphate limitation. Proteomes 6(2):19
57. Kelliher JL, Radin JN, Grim KP, Párraga Solórzano PK, Degnan PH, Kehl-Fie TE (2018) Acquisition of the phosphate transporter NptA enhances *Staphylococcus aureus* pathogenesis by improving phosphate uptake in divergent environments. Infect Immun 86(1):e00631–e00617
58. Kimes BW, Morris DR (1973) Cations and ribosome structure. II. Effects on the 50S subunit of substituting polyamines for magnesium ion. Biochemistry 12(3):442–449
59. Klein DJ, Moore PB, Steitz TA (2004) The contribution of metal ions to the structural stability of the large ribosomal subunit. RNA 10(9):1366–1379
60. Klein G, Müller-Loennies S, Lindner B, Kobylak N, Brade H, Raina S (2013) Molecular and structural basis of inner core lipopolysaccharide alterations in *Escherichia coli*: incorporation of glucuronic acid and phosphoethanolamine in the heptose region. J Biol Chem 288(12):8111–8127
61. Lebens M, Lundquist P, Söderlund L, Todorovic M, Carlin NI (2002) The *nptA* gene of *Vibrio cholerae* encodes a functional sodium-dependent phosphate cotransporter homologous to the type II cotransporters of eukaryotes. J Bacteriol 184(16): 4466–4474
62. Lee EJ, Groisman EA (2012a) Tandem attenuators control expression of the *Salmonella mgtCBR* virulence operon. Mol Microbiol 86(1):212–224
63. Lee EJ, Groisman EA (2012b) Control of a *Salmonella* virulence locus by an ATP-sensing leader messenger RNA. Nature 486(7402):271–275
64. Lee EJ, Pontes MH, Groisman EA (2013) A bacterial virulence protein promotes pathogenicity by inhibiting the bacterium' s own F1Fo ATP synthase. Cell 154(1):146–156
65. Lejona S, Aguirre A, Cabeza ML, García Véscovi E, Soncini FC (2003) Molecular characterization of the Mg^{2+}-responsive PhoP-PhoQ regulon in *Salmonella enterica*. J Bacteriol 185(21):6287–6294
66. Lejona S, Castelli ME, Cabeza ML, Kenney LJ, García Véscovi E, Soncini FC (2004) PhoP can activate its target genes in a PhoQ-independent manner. J Bacteriol 186(8):2476–2480
67. Lippa AM, Goulian M (2009) Feedback inhibition in the PhoQ/PhoP signaling system by a membrane peptide. PLoS Genet 5(12):e1000788
68. Liu W, Eder S, Hulett FM (1998) Analysis of *Bacillus subtilis tagAB* and *tagDEF* expression during phosphate starvation identifies a repressor role for PhoP~P. J Bacteriol 180(3):753–758
69. Lubin EA, Henry JT, Fiebig A, Crosson S, Laub MT (2016) Identification of the PhoB regulon and role of PhoU in the phosphate starvation response of *Caulobacter crescentus*. J Bacteriol 198(1):187–200
70. Madigan MT, Martinko JM, Bender KS, Buckley DH (2015) Microbial metabolism. In: Madigan MT, Martinko JM, Bender KS et al (eds) Brock biology of microorganisms, 14th edn. Pearson, London, pp 73–105
71. Maguire ME, Cowan JA (2002) Magnesium chemistry and biochemistry. Biometals 15(3):203–210
72. Makino K, Amemura M, Kawamoto T, Kimura S, Shinagawa H, Nakata A et al (1996) DNA binding of PhoB and its interaction with RNA polymerase. J Mol Biol 259(1):15–26
73. Makino K, Shinagawa H, Amemura M, Kawamoto T, Yamada M, Nakata A (1989) Signal transduction in the phosphate regulon of *Escherichia coli* involves phosphotransfer between PhoR and PhoB proteins. J Mol Biol 210(3):551–559
74. Makino K, Shinagawa H, Amemura M, Kimura S, Nakata A, Ishihama A (1988) Regulation of the phosphate regulon of *Escherichia coli*. Activation of *pstS* transcription by PhoB protein *in vitro*. J Mol Biol 203(1):85–95
75. Masotti F, Garavaglia BS, Piazza A, Burdisso P, Altabe S, Gottig N et al (2021) Bacterial isolates from Argentine Pampas and their ability to degrade glyphosate. Sci Total Environ 774:145761
76. McCarthy S, Ai C, Wheaton G, Tevatia R, Eckrich V, Kelly R et al (2014) Role of an archaeal PitA transporter in the copper and arsenic resistance of *Metallosphaera sedula*, an extreme thermoacidophile. J Bacteriol 196(20):3562–3570
77. Medveczky N, Rosenberg H (1971) Phosphate transport in *Escherichia coli*. Biochim Biophys Acta 241(2):494–506
78. Morohoshi T, Maruo T, Shirai Y, Kato J, Ikeda T, Takiguchi N et al (2002) Accumulation of inorganic polyphosphate in *phoU* mutants of *Escherichia coli* and *Synechocystis* sp. strain PCC6803. Appl Environ Microbiol 68(8):4107–4110
79. Murray EL, Conway T (2005) Multiple regulators control expression of the Entner-Doudoroff aldolase (Eda) of *Escherichia coli*. J Bacteriol 187(3): 991–1000
80. Novak R, Cauwels A, Charpentier E, Tuomanen E (1999) Identification of a *Streptococcus pneumoniae* gene locus encoding proteins of an ABC phosphate transporter and a two-component regulatory system. J Bacteriol 181(4):1126–1133

81. Ofiteru AM, Ruta LL, Rotaru C, Dumitru I, Ene CD, Neagoe A et al (2012) Overexpression of the PHO84 gene causes heavy metal accumulation and induces Ire1p-dependent unfolded protein response in *Saccharomyces cerevisiae* cells. Appl Microbiol Biotechnol 94(2):425–435
82. Park SY, Cromie MJ, Lee EJ, Groisman EA (2010) A bacterial mRNA leader that employs different mechanisms to sense disparate intracellular signals. Cell 142(5):737–748
83. Park SY, Groisman EA (2014) Signal-specific temporal response by the *Salmonella* PhoP/PhoQ regulatory system. Mol Microbiol 91(1):135–144
84. Perez JC, Shin D, Zwir I, Latifi T, Hadley TJ, Groisman EA (2009) Evolution of a bacterial regulon controlling virulence and Mg^{2+} homeostasis. PLoS Genet 5(3):e1000428
85. Persson BL, Petersson J, Fristedt U, Weinander R, Berhe A, Pattison J (1999) Phosphate permeases of *Saccharomyces cerevisiae*: structure, function and regulation. Biochim Biophys Acta 1422(3):255–272
86. Petrov AS, Bowman JC, Harvey SC, Williams LD (2011) Bidentate RNA-magnesium clamps: on the origin of the special role of magnesium in RNA folding. RNA 17(2):291–297
87. Pontes MH, Groisman EA (2018) Protein synthesis controls phosphate homeostasis. Genes Dev 32(1): 79–92
88. Pontes MH, Lee E-J, Choi J, Groisman EA (2015a) *Salmonella* promotes virulence by repressing cellulose production. Proc Natl Acad Sci U S A 112(16): 5183–5188
89. Pontes MH, Sevostyanova A, Groisman EA (2015b) When too much ATP is bad for protein synthesis. J Mol Biol 427(16):2586–2594
90. Pontes MH, Yeom J, Groisman EA (2016) Reducing ribosome biosynthesis promotes translation during low Mg^{2+}stress. Mol Cell 64(3):480–492
91. Rao NN, Roberts MF, Torriani A, Yashphe J (1993) Effect of *glpT* and *glpD* mutations on expression of the *phoA* gene in *Escherichia coli*. J Bacteriol 175(1): 74–79
92. Rao NN, Torriani A (1990) Molecular aspects of phosphate transport in *Escherichia coli*. Mol Microbiol 4(7):1083–1090
93. Rice CD, Pollard JE, Lewis ZT, McCleary WR (2009) Employment of a promoter-swapping technique shows that PhoU modulates the activity of the $PstSCAB_2$ ABC transporter in *Escherichia coli*. Appl Environ Microbiol 75(3):573–582
94. Rosenberg H, Gerdes RG, Chegwidden K (1977) Two systems for the uptake of phosphate in *Escherichia coli*. J Bacteriol 131(2):505–511
95. Rosenfeld L, Culotta VC (2012) Phosphate disruption and metal toxicity in *Saccharomyces cerevisiae*: effects of RAD23 and the histone chaperone HPC2. Biochem Biophys Res Commun 418(2):414–419
96. Salzberg LI, Botella E, Hokamp K, Antelmann H, Maaß S, Becher D et al (2015) Genome-wide analysis of phosphorylated PhoP binding to chromosomal DNA reveals several novel features of the PhoPR-mediated phosphate limitation response in *Bacillus subtilis*. J Bacteriol 197(8):1492–1506
97. Scholten M, Tommassen J (1993) Topology of the PhoR protein of *Escherichia coli* and functional analysis of internal deletion mutants. Mol Microbiol 8(2): 269–275
98. Schuwirth BS, Borovinskaya MA, Hau CW, Zhang W, Vila-Sanjurjo A, Holton JM et al (2005) Structures of the bacterial ribosome at 3.5 A resolution. Science 310(5749):827–834
99. Sevostyanova A, Groisman EA (2015) An RNA motif advances transcription by preventing Rho-dependent termination. Proc Natl Acad Sci U S A 112(50):E6835–E6843
100. Shi L, Liu W, Hulett FM (1999) Decay of activated *Bacillus subtilis* Pho response regulator, PhoP~P, involves the PhoR~P intermediate. Biochemistry 38(31):10119–10125
101. Shi Y, Cromie MJ, Hsu FF, Turk J, Groisman EA (2004) PhoP-regulated *Salmonella* resistance to the antimicrobial peptides magainin 2 and polymyxin B. Mol Microbiol 53(1):229–241
102. Shin D, Groisman EA (2005) Signal-dependent binding of the response regulators PhoP and PmrA to their target promoters *in vivo*. J Biol Chem 280(6): 4089–4094
103. Snavely MD, Florer JB, Miller CG, Maguire ME (1989) Magnesium transport in *Salmonella typhimurium*: $^{28}Mg^{2+}$ transport by the CorA, MgtA, and MgtB systems. J Bacteriol 171(9):4761–4766
104. Snavely MD, Miller CG, Maguire ME (1991) The *mgtB* Mg^{2+} transport locus of *Salmonella typhimurium* encodes a P-type ATPase. J Biol Chem 266(2): 815–823
105. Soncini FC, García Véscovi E, Solomon F, Groisman EA (1996) Molecular basis of the magnesium deprivation response in *Salmonella typhimurium*: identification of PhoP-regulated genes. J Bacteriol 178(17): 5092–5099
106. Spinelli SV, Pontel LB, García Véscovi E, Soncini FC (2008) Regulation of magnesium homeostasis in *Salmonella*: Mg^{2+} targets the *mgtA* transcript for degradation by RNase E. FEMS Microbiol Lett 280(2): 226–234
107. Steed PM, Wanner BL (1993) Use of the *rep* technique for allele replacement to construct mutants with deletions of the *pstSCAB-phoU* operon: evidence of a new role for the PhoU protein in the phosphate regulon. J Bacteriol 175(21):6797–6809
108. Storer AC, Cornish-Bowden A (1976) Concentration of $MgATP^{2-}$ and other ions in solution. Calculation of the true concentrations of species present in mixtures of associating ions. Biochem J 159(1):1–5
109. Stouthamer AH (1973) A theoretical study on the amount of ATP required for synthesis of microbial cell material. Antonie Van Leeuwenhoek 39(3): 545–565

110. Surin BP, Rosenberg H, Cox GB (1985) Phosphate-specific transport system of *Escherichia coli*: Nucleotide sequence and gene-polypeptide relationships. J Bacteriol 161(1):189–198
111. Tao T, Snavely MD, Farr SG, Maguire ME (1995) Magnesium transport in *Salmonella typhimurium*: *mgtA* encodes a P-type ATPase and is regulated by Mg^{2+} in a manner similar to that of the *mgtB* P-type ATPase. J Bacteriol 177(1):2654–2662
112. Trent MS, Pabich W, Raetz CRH, Miller SI (2001) A PhoP/PhoQ-induced Lipase (PagL) that catalyzes 3-O-deacylation of lipid A precursors in membranes of *Salmonella typhimurium*. J Biol Chem 276(12): 9083–9092
113. van Veen HW (1997) Phosphate transport in prokaryotes: molecules, mediators and mechanisms. Antonie van Leeuwenhoek Int J Gen Mol Microbiol 72(4):299–315
114. van Veen HW, Abee T, Kortstee GJ, Konings WN, Zehnder AJ (1994) Translocation of metal phosphate via the phosphate inorganic transport system of *Escherichia coli*. Biochemistry 33(7):1766–1770
115. Vila-Costa M, Sebastián M, Pizarro M, Cerro-Gálvez E, Lundin D, Gasol JM et al (2019) Microbial consumption of organophosphate esters in seawater under phosphorus limited conditions. Sci Rep 9(1): 233
116. Wacker WE (1969) The biochemistry of magnesium. Ann N Y Acad Sci 162(2):717–726
117. Wang C, Mao Y, Yu J, Zhu L, Li M, Wang D et al (2013) PhoY2 of mycobacteria is required for metabolic homeostasis and stress response. J Bacteriol 195(2):243–252
118. Wang X, Han H, Lv Z, Lin Z, Shang Y, Xu T et al (2017) PhoU2 but not PhoU1 as an important regulator of biofilm formation and tolerance to multiple stresses by participating in various fundamental metabolic processes in *Staphylococcus epidermidis*. J Bacteriol 199(24):e00219–e00217
119. Wanner BL (1993) Gene regulation by phosphate in enteric bacteria. J Cell Biochem 51(1):47–54
120. Wanner BL (1996) Phosphorus assimilation and control of the phosphate regulon. In: Neidhardt FC, Curtiss R, Ingraham JL et al (eds) *Escherichia coli* and *Salmonella*: cellular and molecular biology, 2nd edn. American Society for Microbiology, Washington, DC, pp 1357–1381
121. Wanner BL, Hsieh Y-J (2010) Global regulation by the seven-component Pi signaling system. Curr Opin Microbiol 13(2):198–203
122. Webb DC, Rosenberg H, Cox GB (1992) Mutational analysis of the *Escherichia coli* phosphate-specific transport system, a member of the traffic ATPase (or ABC) family of membrane transporters. A role for proline residues in transmembrane helices. J Biol Chem 267(34):24661–24668
123. Weiss RL, Kimes BW, Morris DR (1973) Cations and ribosome structure. 3. Effects on the 30S and 50S subunits of replacing bound Mg^{2+} by inorganic cations. Biochemistry 12(3):450–456
124. Weiss RL, Morris DR (1973) Cations and ribosome structure. I. Effects on the 30S subunit of substituting polyamines for magnesium ion. Biochemistry 12(3): 435–441
125. Werner A, Kinne RKH (2001) Evolution of the Na-Pi cotransport systems. Am J Phys Regul Integr Comp Phys 280(2):R301–R312
126. Wilkins AS (1972) Physiological factors in the regulation of alkaline phosphatase synthesis in *Escherichia coli*. J Bacteriol 110(2):616–623
127. Willsky GR, Malamy MH (1980a) Characterization of two genetically separable inorganic phosphate transport systems in *Escherichia coli*. J Bacteriol 144(1):356–365
128. Willsky GR, Malamy MH (1980b) Effect of arsenate on inorganic phosphate transport in *Escherichia coli*. J Bacteriol 144(1):366–374
129. Wösten MM, Parker CT, van Mourik A, Guilhabert MR, van Dijk L, van Putten JP (2006) The *Campylobacter jejuni* PhosS/PhosR operon represents a non-classical phosphate-sensitive two-component system. Mol Microbiol 62(1):278–291
130. Wuttge S, Bommer M, Jäger F, Martins BM, Jacob S, Licht A et al (2012) Determinants of substrate specificity and biochemical properties of the sn-glycerol-3-phosphate ATP binding cassette transporter (UgpB-AEC2) of *Escherichia coli*. Mol Microbiol 86(4):908–920
131. Yamamoto K, Ogasawara H, Fujita N, Utsumi R, Ishihama A (2002) Novel mode of transcription regulation of divergently overlapping promoters by PhoP, the regulator of two-component system sensing external magnesium availability. Mol Microbiol 45(2):423–438
132. Yang C, Huang TW, Wen SY, Chang CY, Tsai SF, Wu WF et al (2012) Genome-wide PhoB binding and gene expression profiles reveal the hierarchical gene regulatory network of phosphate starvation in *Escherichia coli*. PLoS One 7(10):e47314
133. Yeom J, Shao Y, Groisman EA (2020) Small proteins regulate *Salmonella* survival inside macrophages by controlling degradation of a magnesium transporter. Proc Natl Acad Sci U S A 117(33):20235–20243
134. Yin X, Wu Orr M, Wang H, Hobbs EC, Shabalina SA, Storz G (2019) The small protein MgtS and small RNA MgrR modulate the PitA phosphate symporter to boost intracellular magnesium levels. Mol Microbiol 111(1):131-144.
135. Yoshida Y, Sugiyama S, Oyamada T, Yokoyama K, Makino K (2012) Novel members of the phosphate regulon in *Escherichia coli* O157: H7 identified using a whole-genome shotgun approach. Gene 502(1): 27–35
136. Zarrella TM, Bai G (2020) The many roles of the bacterial second messenger cyclic di-AMP in adapting to stress cues. J Bacteriol 203(1):e00348–e00320

137. Zheng JJ, Sinha D, Wayne KJ, Winkler ME (2016) Physiological roles of the dual phosphate transporter systems in low and high phosphate conditions and in capsule maintenance of *Streptococcus pneumoniae* D39. Front Cell Infect Microbiol 6:63
138. Zwir I, Latifi T, Perez JC, Huang H, Groisman EA (2012) The promoter architectural landscape of the *Salmonella* PhoP regulon. Mol Microbiol 84(3): 463–485
139. Zwir I, Shin D, Kato A, Nishino K, Latifi T, Solomon F et al (2005) Dissecting the PhoP regulatory network of *Escherichia coli* and *Salmonella enterica*. Proc Natl Acad Sci U S A 102(8):2862–2867
140. Zwir I, Yeo WS, Shin D, Latifi T, Huang H, Groisman EA (2014) Bacterial nucleoid-associated protein uncouples transcription levels from transcription timing. MBio 5(5):e01485-14

Phosphate Dysregulation and Neurocognitive Sequelae

13

John Acquaviva, Hosam G. Abdelhady, and Mohammed S. Razzaque

Abstract

The endocrine regulator proteins, fibroblast growth factor 23 (FGF23) and Klotho have been well studied as mediators of phosphate metabolism. FGF23 has been implicated in the renal excretion of phosphate by limiting the docking of sodium-dependent phosphate transporters, Npt2a and Npt2c, into the luminal side of renal proximal tubular epithelial cells. By limiting Npt2a/c activity in the renal tubular epithelial cells, phosphate is reabsorbed at lower rates and is excreted at higher rates. The action of Klotho is relatively less understood but has been implicated as an FGF23 cofactor in receptor binding. Klotho is mostly synthesized in the distal tubules of the nephron relative to FGF23's activity in proximal renal tubules. The neurological sequelae due to alterations in the FGF23-Klotho axis may be explained by the direct effects of these phosphate-regulating proteins on neuronal tissues or by the roles of these proteins in phosphate metabolism. Hyperphosphatemia has been associated with vascular wall stiffness that may alter blood flow and weakenvessels in the brain. In contrast, hypophosphatemia may alter ATP usage and metabolism in the central nervous system (CNS), leading to neurological compromise. Altered levels of FGF23 and Klotho have both been associated with neurocognitive decline, clinical dementia, memory loss, and poor executive function in humans. Furthermore, FGF23 and Klotho dysregulation has been linked to structural and functional changes of the cardiovascular system with an increased risk of stroke. Subsequent research should focus on characterizing the neuropathology associated with alterations in the FGF23-Klotho system and dysregulated phosphate metabolism.

Keywords

Phosphate · Brain · Cognitive function · FGF23 · Klotho

13.1 Introduction

The metabolic regulation of phosphate relies on an interplay between hydroxypatite deposition and breakdown, intestinal absorption, and renal excretion. Within the intestines, phosphate is reabsorbed by a paracellular route through tight junctions and an active transport route through

J. Acquaviva · M. S. Razzaque (✉)
Department of Pathology, Lake Erie College of Osteopathic Medicine, Erie, PA, USA
e-mail: JAcquaviva82648@med.lecom.edu; mrazzaque@lecom.edu; msr.nagasaki@gmail.com

H. G. Abdelhady
Department of Physiology and Pharmacology, College of Osteopathic Medicine, Sam Huston State University, Conroe, TX, USA
e-mail: Hosam.Abdelhady@shsu.edu

M. S. Razzaque (ed.), *Phosphate Metabolism*, Advances in Experimental Medicine and Biology 1362,
https://doi.org/10.1007/978-3-030-91623-7_13

sodium-dependent phosphate transporters (Npt2b) [13, 17, 39]. Calcitriol and a low phosphate diet induce the expression of Npt2b on the apical membranes of intestine mucosal cells and therefore control active transport of phosphate [13, 40]. How the human body senses extracellular phosphate status is not completely elucidated, as a phosphate sensor is not yet identified but a likely mechanism could be mediated by parathyroid hormone (PTH) and its effects on bone and renal tubules. For example, high extracellular phosphate levels have been associated with elevated PTH, and in response, PTH suppresses Npt2a and Npt2c activities in proximal renal tubules [44]. Decreased Npt2a/c activity in the proximal renal tubules causes increased urinary excretion of phosphate. Similarly, the bone-derived fibroblast growth factor 23 (FGF23) and the membrane-bound protein, Klotho have augmentative actions with PTH on phosphate excretion within the renal tubules. FGF23 causes the suppression of proximal renal tubular Npt2a/c expression through direct inhibition or modulation of the parathyroid axis. This suppression causes decreased cotransport activity of Npt2a/c, which leads to decreased reabsorption of phosphate from the renal tubules. Murine and subsequent human genetic studies have linked hypophosphatemia with overexpression of FGF23, further characterizing the role of FGF23 in phosphate excretion [3, 7]. The actions of Klotho are relatively less understood, however, Klotho has been implicated as an FGF23 obligatory cofactor [47]. Conversely, Klotho is mostly expressed in the distal tubules while FGF23 is functional in the proximal tubules [16]. Klotho expression in the distal tubules may implicate this protein in an isolated action on phosphate metabolism that has yet to be fully explained. Additionally, Klotho knockout mice exhibited similar features as FGF23 knockout mice, both causing hyperphosphatemia and manifestations such as short lifespan, infertility, arteriosclerosis, skin atrophy, osteoporosis, and emphysema [21, 32–34, 38]. The similar manifestations of FGF23 knockout and Klotho knockout imply a connection between the two molecules that control phosphate metabolism [28]. This article is shedding the light on the effects of hyperphosphatemia versus hypophosphatemia on the neuronal dysfunction and neurovascular complications, and discuss the effect of FGF23-Klotho axis on the neuronal functions.

13.2 Hyperphosphatemia and Neuronal Dysfunction

The effects of hyperphosphatemia on the central nervous system (CNS) have been documented in humans, particularly in individuals with chronic kidney disease (CKD) and/or those on hemodialysis but have also been independently linked to acute/subacute neurological dysfunction and chronic cognitive decline. Other common causes of hyperphosphatemia are tumor lysis syndrome, increased exogenous phosphate ingestion (laxatives), vitamin D intoxication, pseudohypoparathyroidism, hypoparathyroidism, and genetic disorders such as Albright hereditary osteodystrophy [11]. The acute/subacute neurological symptoms of hyperphosphatemia are varied but can include paresthesia, seizures, reflex hyperexcitability, tetany, headaches, dizziness, delirium, and coma [11]. Elevated levels of phosphate in individuals with CKD have also been associated with cognitive impairment and poor performance on the Mini-Mental Status Examination (MMSE) [27]. Furthermore, higher phosphate levels have been associated with specific cognitive dysfunction, such as poor executive function, in older men when assessed with the Modified Mini-Mental Status Examination (3MS) and Trailmaking Test B (Trails B) cognitive function measures [45]. Higher serum phosphate levels have also been documented as a risk factor for incident dementia in those under 60 years old [24]. While cognitive decline is a frequently documented complication of elevated phosphate levels, vascular changes that cause neurological emergencies have also been documented in individuals with higher serum phosphate levels. In a study that measured the phosphate levels of 3437 individuals on hemodialysis, it was the individuals with the highest phosphate levels who were documented as

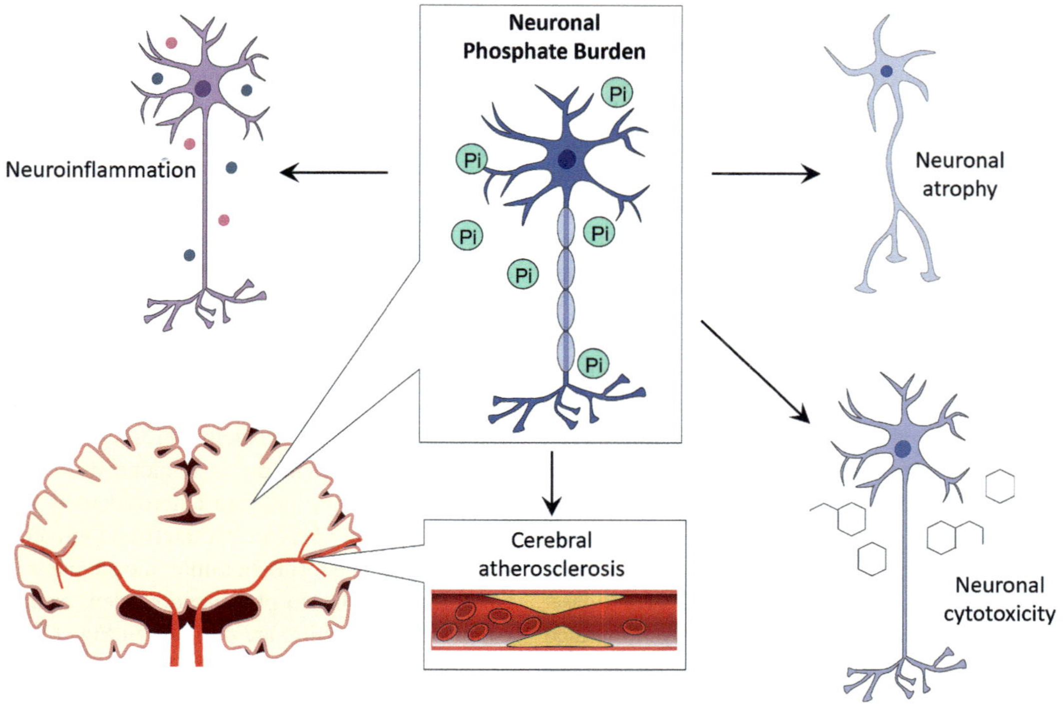

Fig. 13.1 Possible neuropathological events, initiated by neuronal phosphate burden, which can eventually contribute to overall declining brain function and cognitive impairment

experiencing future hemorrhagic brain infarctions [52]. Some of the possible effects of neuronal phosphate burden are outlined in Fig. 13.1.

13.3 Hyperphosphatemia-Induced Neurovascular Complications

It has been well-documented that chronically elevated levels of phosphate have a deleterious effect on brain tissue, but possible mechanisms of this damage have not been fully elucidated. The majority of the documented cases of hyperphosphatemia-induced vascular changes have been described in individuals with CKD as they are at the highest risk of retaining substantial levels of phosphate [5, 6, 35]. Systemically, hyperphosphatemia seems to have a calcific response in peripheral vasculature that may extend to the neurovasculature or act as a risk factor for cerebrovascular damage. Hyperphosphatemia causes vascular calcification by inducing vessel wall inflammation and converting vascular smooth muscle cells (VSMCs) into osteoblast-like cells [15, 48, 51]. The conversion of VSMCs to osteoblast-like cells is mediated by the effect of hyperphosphatemia on the sodium-dependent channels, PiT-1 and PiT-2 [6]. Hyperphosphatemia and hypercalcemia directly stimulate PiT-1 and PiT-2, which upregulates gene transformations that are associated with matrix deposition [6]. Once the VSMCs develop osteoblast-like properties, they deposit hydroxyapatite crystals into the vessel walls [15]. The calcific changes established by the phenotype conversion of VSMCs are identical to atherosclerosis and compromise the tunica intima and the tunica media [48]. Furthermore, entry of phosphate through PiT-1 and PiT-2 into the VSMCs can cause damage to these cells, eventual apoptosis, and compromise of vessel functions [51]. The high phosphate microenvironment can also induce endothelial cell damage that further exacerbates vascular dysfunction.

Vessel wall inflammation may also have a prominent role in vascular changes associated with hyperphosphatemia. When rat models with induced CKD were fed with high-phosphate-diets, elevated levels of TNF-alpha were detected in serum and tissue, representing increased inflammatory responses in these rats relative to the control rats [51]. Furthermore, rats exposed to a high phosphate diet also had increased markers of general stress responses such as decreased body weight and lower serum albumin levels [51]. Hyperphosphatemic inflammation within vasculature may also be mediated by oxidant formation by mitochondria in VSMCs. When murine aortas were exposed to even low levels of phosphate, levels of the NADPH oxidase, Nox4 were elevated and subsequently stimulated oxidant formation, inflammation, and fibrosis in vitro [2]. The increases in inflammatory markers, oxidants, and stress responses may augment the atherosclerotic response associated with hyperphosphatemia. The increased inflammation associated with hyperphosphatemia may also directly affect nervous tissue and have a role in CNS impairment due to neuroinflammatory responses [26]. The neuroinflammatory responses associated with phosphate burden may be mediated by the production of oxidants, such as superoxide, by the mitochondria in neurons; as discussed above in the context of vasculature inflammation [2, 26].

While vessel wall compromise has been described as the leading cause of vascular damage due to hyperphosphatemia, endothelial dysfunction has also been associated with hyperphosphatemia in vitro. When bovine aortic endothelial cells were exposed to high levels of extracellular phosphate, there was an induction of reactive oxygen species (ROS) measured by tetrazolium blue assay and decreased production of nitric oxide by these cells [2, 43]. Due to the reduced levels of nitric oxide by the endothelial cells, there was also a decreased capacity of endothelial-dependent aortic rings to vasodilate [43].

Conclusively, vascular compromise is a well-documented complication of hyperphosphatemia and may play a significant role in central nervous system damage directly and indirectly. When the hyperphosphatemic-induced atherosclerosis of vasculature extends to the CNS, it may lead to vessel stenosis and chronically low oxygenation, vessel wall weakness, and hemorrhage, or acute downstream embolic episodes associated with plaque rupture. It has been established that individuals with peripheral artery disease and large-vessel disease have a higher risk of stroke, and this may also partly apply to artery disease induced by hyperphosphatemia [18, 49]. Furthermore, hypertension induced by hyperphosphatemic-induced atherosclerosis may increase the risk of lacunar infarcts and intracerebral hemorrhage. The possible mechanisms of CNS damage, due to phosphate burden, causing cognitive dysfunction are outlined in Fig. 13.2.

13.4 Hypophosphatemia and Neuronal Dysfunction

Hypophosphatemia has also been linked to changes in the CNS that have acute and chronic effects on neurological functions. Hypophosphatemia can occur due to poor dietary sources of phosphate, malabsorption associated with diarrhea and medication binding, kidney damage, primary or secondary hyperparathyroidism, and can occur in settings that cause intracellular shifting of phosphate, such as refeeding syndrome [10, 42]. While mild-moderate hypophosphatemia is often asymptomatic, the acute/subacute neurological symptoms of severe hypophosphatemia are altered mental status, numbness, and weak reflexes [42]. In a study enrolling patients with mild cognitive impairment and autosomal dominant dementia, the patients who were identified as having Aβ protein on PET scan had lower levels of serum phosphate relative to patients who did not express Aβ protein [30]. This finding may link low phosphate levels with the onset of more severe forms of dementia and cognitive impairment. Furthermore, low levels of phosphate have been linked with

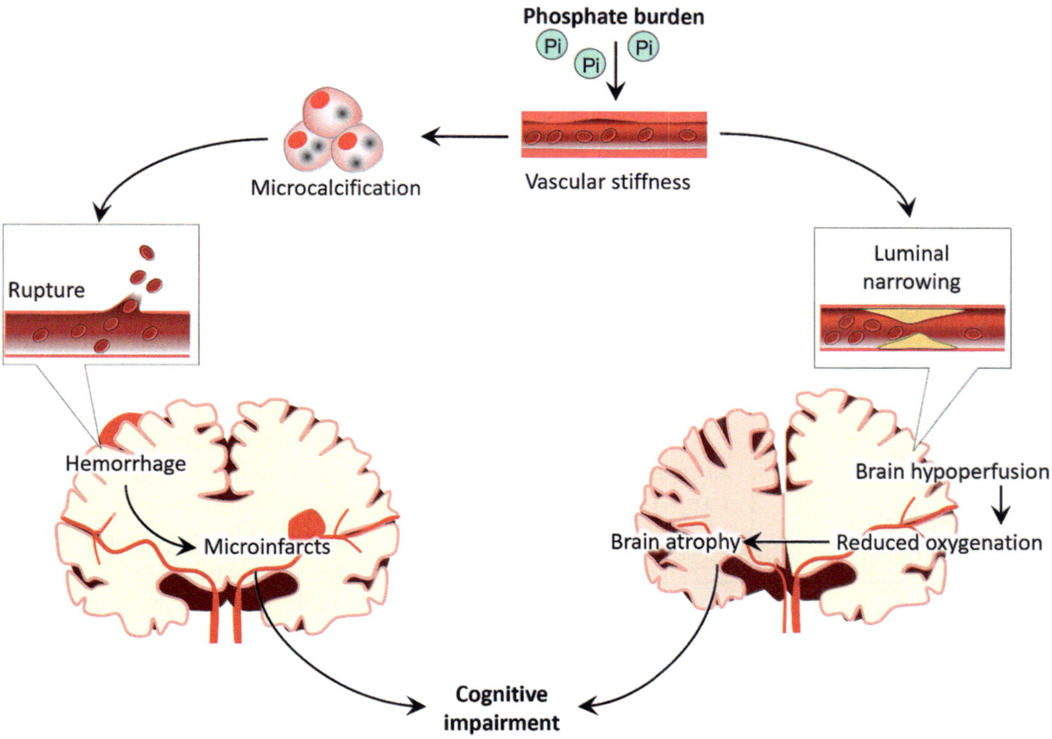

Fig. 13.2 Possible events in the brain, induced by phosphate burden, that could contribute to cognitive impairment

increased mortality in patients who have experienced an acute ischemic stroke and therefore may have a small role in neuronal energy use or vascular oxygen delivery [46].

13.5 Hypophosphatemia-Induced Neurovascular Complications

The mechanism by which hypophosphatemia causes neurological damage has not explicitly been elucidated yet, but it may be linked to decreased ability of nervous tissue to produce and use ATP, and thus nutrition of nervous tissue could be affected. In a study using induced hypophosphatemic mice, when insulin-stimulated rates of muscle ATP were measured between hypophosphatemic mice and controls, the hypophosphatemic mice had a 50% decrease in ATP synthesis [31]. Interestingly, resistant lactic acidosis is documented in multiple human case studies as a complication of hypophosphatemia which could represent altered energy metabolism as well [9]. Furthermore, lack of high-energy phosphate can decrease erythrocyte 2,3-DPG [41]. Reduced levels of 2,3-DPG cause a leftward shift of the hemoglobin dissociation curve that may lead to chronic decreases in oxygen delivery to nervous tissue and eventual neurological compromise [41]. It is plausible that the neurological manifestations associated with hypophosphatemia are the cumulative effects of chronic tissue deoxygenation and altered ATP production, although further research on this topic is necessary to understand the underlying pathology. The possible mechanisms of hypophosphatemia-induced cognitive impairment are outlined in Fig. 13.3.

13.6 FGF23-Klotho Axis and Neuronal Functions

While the FGF23-Klotho axis has been implicated in systemic phosphate metabolism,

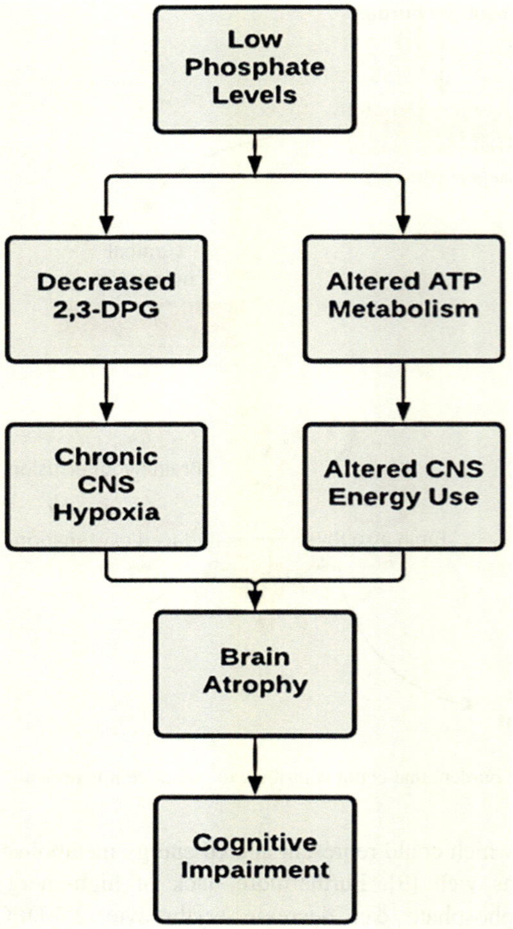

Fig. 13.3 Possible mechanisms of hypophosphatemic cognitive impairment

there has not been an explicit function of this axis uncovered in the CNS of humans. FGF23 was first found in the murine thalamic nuclei, which led to further studies addressing other locations within the CNS in which FGF23 can be found [53]. More recently, the presence of Klotho and FGF23 has been detected in the cerebrospinal fluid (CSF) of human children implicating the FGF23-Klotho axis in neurological activity outside of murine models [19]. Not only were FGF23 and Klotho identified in the CSF of human ventricles and sinuses, but a production locus for these proteins was found to be in the CNS as well, specifically from ependymal cells in the lateral ventricles [19]. Furthermore, Klotho and FGF23 have been uncovered in synapses of mice and within the hippocampus, implicating these proteins in possible neurotransmission and memory function [22, 23]. The finding of FGF23 and Klotho in neurotransmission and memory centers of the brain in murine models provides reason to study the possibility of the FGF23 and Klotho interaction further in the human nervous system.

13.7 FGF23 and Neuronal Functions

The possible functions of FGF23 in the CNS can be implied based on manifestations of the protein expression or suppression in murine models and human observations. When FGF23 deficiency was induced in murine models, mice showed decreased success in open-field velocity trials, and decreases in location memory tests after 5 weeks [22]. Furthermore, brain tissue analysis of the FGF23 deficient mice showed that hippocampal differentiation was impaired based on decreases in progenitor cell density in the dentate nucleus and subgranular zone [22, 29]. While FGF23 deficiency causes reduced hippocampal differentiation in murine models, FGF23 may also preserve hippocampal neuronal morphology and synaptic density [12]. When cultures of murine hippocampi are exposed to FGF23, synaptic density is increased, and enhancement and lengthening of dendrites occur [12]. The cognitive and neuro-morphological changes associated with deficiencies in FGF23 may be superimposed with hyperphosphatemia, as hyperphosphatemia has been correlated with multiple neurocognitive changes as mentioned above [24, 29, 45, 52]. However, FGF23 overexpression has been linked to CNS abnormalities and cognitive deficits in murine models, possibly due to effects of hypophosphatemia stimulated by the excretory actions of FGF23 [25]. Unlike the direct and beneficial effect of FGF23 on the murine hippocampus, FGF23 found in human plasma at elevated levels has been associated with poorer neurological outcomes. On MRI neuroimaging, 1170 individuals who were less than 40 years old and who had never been diagnosed with a stroke were more likely to have white matter hyperintensity if they had elevated levels of FGF23 in the plasma [50]. Furthermore, an increased level of FGF23 in human plasma has been associated with poorer performance on several individual tests primarily focused on short delay and delayed recall memory in a population of 263 human participants who were pre-tested for eventual dialysis [8]. Regarding cerebrovascular emergencies in humans, elevated levels of FGF23 have been correlated with increased risk of stroke and intracranial hemorrhage, however, the population analyzed in this association also had reduced glomerular filtration rate (GFR), which may be a moderator or confounder of this finding [8]. It can be hypothesized that an elevated level of FGF23 is also associated with kidney disease, and the neurological effects of elevated plasma FGF23 cannot be separated from the neurological deficits associated with CKD and related uremia. Phosphate is often elevated in CKD, and since FGF23 is associated with phosphate excretion, its elevation may be compensatory. The CNS abnormalities associated with elevated FGF23 may also be explained by hypophosphatemia as rescue phosphate ingestion in mice with elevated FGF23 ameliorates the cognitive deficits that were previously expressed [25].

13.8 Klotho and Neuronal Functions

The coreceptor Klotho has been associated with changes in neuronal architecture and cognitive function in human and mouse models. Comparatively to FGF23 overexpression and the resultant white matter hyperintensities, deficiency in the Klotho protein has shown to be a significant risk factor for diffuse white matter lesions [20, 50]. Furthermore, elevated levels of Klotho have been associated with higher MMSE scores, possibly due to induction of phosphate clearance [20]. Low levels of Klotho also have a significant association with vascular dementia onset and progression of dementia to more severe forms in humans [4]. The association of low Klotho and dementia may also be explained by the effects of Klotho on the hippocampus in murine models. In Klotho knockout mice, synapses in the hippocampus and synaptophysin levels that represent vesicular transport are both reduced [23]. Furthermore, Klotho deficient mice showed reduced nerve terminals within the stratum lucidum in the hippocampus [23]. While Klotho and FGF23 have been associated with hippocampal structural and functional changes, it is important to note that Klotho has been associated with changes in hippocampal proliferation and differentiation, while FGF23 has only been associated with changes in proliferation, implying a possible augmentative effect on the hippocampus between the two proteins in murine models [22]. Moreover, using Xenopus oocyte models, injection of Klotho alone showed no increase in glutamate current but adding Klotho with EAAT3/4 amino acid transporters caused a significant increase in glutaminergic current across neuronal membranes and the blood-brain-barrier [1]. Therefore, the protective activity of Klotho may be associated with excitatory glutamate action only if the necessary proteins are available for the transport of amino acids. Klotho has also been described as an "immunological gatekeeper" at the interface between the immune system and the choroid plexus in mice [54]. This finding implies that Klotho depletion with age or genetic Klotho deficiencies in mice may be caused by immune-mediated neuropathogenesis beginning in the choroid plexus. Klotho-deficient mice also showed increased activation of microglia, further identifying immune mechanisms of hippocampal degeneration due to Klotho deficiency [54].

13.9 Summary

While the mechanisms by which FGF23 and Klotho cause neurological sequelae and morphological changes are not completely understood, it appears likely that part of these changes is superimposed with dysregulation of phosphate homeostasis. FGF23 has a direct and/or an indirect role on Npt2a/c expression in the kidney tubules and may work with PTH to cause phosphate excretion. Klotho acts as a cofactor for FGF23 in the proximal tubules and is produced in the distal tubules, where it may have an isolated function in phosphate metabolism. FGF23 and Klotho have been isolated in brain tissue within the hippocampus, CSF, ventricles, and the cerebral neuron synapses of murine models. The locations of FGF23 and Klotho in the CNS may implicate these proteins in memory function, synaptic transmission, and immunoregulatory actions. Predictably, the cognitive decline and cerebrovascular changes associated with hyperphosphatemia and hypophosphatemia generally match the sequelae related to alterations in levels of FGF23 and Klotho. Hyperphosphatemia is associated with an increased risk of dementia, hemorrhagic stroke, isolated memory loss, and advancement of dementia [24, 29, 45, 52]. Hyperphosphatemia may also be associated with decreased brain growth, as observed in young mouse models [14]. Hypophosphatemia is associated with specific Aβ plaque deposition rates seen in individuals with Alzheimer disease, along with general dementia and ischemic stroke [30, 46, 52].

Future research should focus on quantifying and characterizing the relationships between FGF23, Klotho, and phosphate to uncover how altered phosphate concentrations affect neuronal functions. It is also important to further document the manifestations of Klotho and FGF23 deficiencies or excesses on the CNS, mainly in relation to phosphate levels. Finally, determining phosphate-independent effects of FGF23 and Klotho on neuronal growth, development, and maintenance will enhance our understanding of the pathophysiology of cognitive functions in general and beyond [36, 37].

Acknowledgments We want to express our sincere gratitude to Dr. Nuraly Akimbekov (Al-Farabi Kazakh National University, Kazakhstan) for his help in drawing the illustrations. John Acquaviva is an Osteopathic Medical Student (OMS III) at the Lake Erie College of Osteopathic Medicine, Erie (USA).

References

1. Almilaji A, Munoz C, Pakladok T, Alesutan I, Feger M et al (2013) Klotho sensitivity of the neuronal excitatory amino acid transporters EAAT3 and EAAT4. PLoS One 8:e70988
2. Asenjo-Bueno A, Alcalde-Estevez E, El Assar M, Olmos G, Plaza P et al (2021) Hyperphosphatemia-induced oxidant/antioxidant imbalance impairs vascular relaxation and induces inflammation and fibrosis in Old Mice. Antioxidants (Basel) 10:1308
3. Bai X, Miao D, Li J, Goltzman D, Karaplis AC (2004) Transgenic mice overexpressing human fibroblast growth factor 23 (R176Q) delineate a putative role for parathyroid hormone in renal phosphate wasting disorders. Endocrinology 145:5269–5279
4. Brombo G, Bonetti F, Ortolani B, Morieri ML, Bosi C et al (2018) Lower plasma Klotho concentrations are associated with vascular dementia but not late-onset Alzheimer's disease. Gerontology 64:414–421
5. Brown RB, Haq A, Stanford CF, Razzaque MS (2015) Vitamin D, phosphate, and vasculotoxicity. Can J Physiol Pharmacol 93:1077–1082
6. Cozzolino M, Ciceri P, Galassi A, Mangano M, Carugo S et al (2019) The key role of phosphate on vascular calcification. Toxins (Basel) 11:213
7. DeLuca S, Sitara D, Kang K, Marsell R, Jonsson K et al (2008) Amelioration of the premature ageing-like features of Fgf-23 knockout mice by genetically restoring the systemic actions of FGF-23. J Pathol 216:345–355
8. Drew DA, Tighiouart H, Scott TM, Lou KV, Fan L et al (2014) FGF-23 and cognitive performance in hemodialysis patients. Hemodial Int 18:78–86
9. Essa A, Kousa O, Awad D, Stevenson M, DeVrieze B, Moore D (2020) The Association of Hypophosphatemia with Resistant Lactic Acidosis in critical care illness. J Investig Med High Impact Case Rep 8:2324709620934963
10. Garcia Martin A, Varsavsky M, Cortes Berdonces M, Avila Rubio V, Alhambra Exposito MR et al (2020) Phosphate disorders and clinical management of hypophosphatemia and hyperphosphatemia. Endocrinol Diabetes Nutr (Engl Ed) 67:205–215
11. Goyal R, Jialal I (2021) Hyperphosphatemia. In: StatPearls. StatPearls Publishing LLC, Treasure Island
12. Hensel N, Schon A, Konen T, Lubben V, Forthmann B et al (2016) Fibroblast growth factor 23 signaling in hippocampal cells: impact on neuronal morphology and synaptic density. J Neurochem 137:756–769
13. Hernando N, Pastor-Arroyo EM, Marks J, Schnitzbauer U, Knopfel T et al (2021) 1,25(OH) 2 vitamin D3 stimulates active phosphate transport but not paracellular phosphate absorption in mouse intestine. J Physiol 599:1131–1150
14. Jin H, Hwang SK, Yu K, Anderson HK, Lee YS et al (2006) A high inorganic phosphate diet perturbs brain growth, alters Akt-ERK signaling, and results in changes in cap-dependent translation. Toxicol Sci 90: 221–229
15. Jono S, McKee MD, Murry CE, Shioi A, Nishizawa Y et al (2000) Phosphate regulation of vascular smooth muscle cell calcification. Circ Res 87:E10–E17
16. Kato Y, Arakawa E, Kinoshita S, Shirai A, Furuya A et al (2000) Establishment of the anti-klotho monoclonal antibodies and detection of klotho protein in

kidneys. Biochem Biophys Res Commun 267:597–602
17. Knopfel T, Himmerkus N, Gunzel D, Bleich M, Hernando N, Wagner CA (2019) Paracellular transport of phosphate along the intestine. Am J Physiol Gastrointest Liver Physiol 317:G233–GG41
18. Kolls BJ, Sapp S, Rockhold FW, Jordan JD, Dombrowski KE et al (2019) Stroke in patients with peripheral artery disease. Stroke 50:1356–1363
19. Kunert SK, Hartmann H, Haffner D, Leifheit-Nestler M (2017) Klotho and fibroblast growth factor 23 in cerebrospinal fluid in children. J Bone Miner Metab 35:215–226
20. Kuriyama N, Ozaki E, Mizuno T, Ihara M, Mizuno S et al (2018) Association between alpha-Klotho and deep white matter lesions in the brain: a pilot case control study using brain MRI. J Alzheimers Dis 61: 145–155
21. Kuro-o M, Matsumura Y, Aizawa H, Kawaguchi H, Suga T et al (1997) Mutation of the mouse klotho gene leads to a syndrome resembling ageing. Nature 390: 45–51
22. Laszczyk AM, Nettles D, Pollock TA, Fox S, Garcia ML, Wang J, Quarles LD, King GD (2019) FGF-23 deficiency impairs hippocampal-dependent cognitive function. eNeuro 6(2):ENEURO.0469-18.2019. https://doi.org/10.1523/ENEURO.0469-18.2019. PMID: 30911673; PMCID: PMC6430630
23. Li SA, Watanabe M, Yamada H, Nagai A, Kinuta M, Takei K (2004) Immunohistochemical localization of Klotho protein in brain, kidney, and reproductive organs of mice. Cell Struct Funct 29:91–99
24. Li T, Xie Y, Bowe B, Xian H, Al-Aly Z (2017) Serum phosphorus levels and risk of incident dementia. PLoS One 12:e0171377
25. Liu P, Chen L, Bai X, Karaplis A, Miao D, Gu N (2011) Impairment of spatial learning and memory in transgenic mice overexpressing human fibroblast growth factor-23. Brain Res 1412:9–17
26. McCarty MF, Lerner A, DiNicolantonio JJ, Iloki-Assanga SB (2021) High intakes of bioavailable phosphate may promote systemic oxidative stress and vascular calcification by boosting mitochondrial membrane potential-is good magnesium status an antidote? Cell 10:1744
27. Murray AM (2008) Cognitive impairment in the aging dialysis and chronic kidney disease populations: an occult burden. Adv Chronic Kidney Dis 15:123–132
28. Nakatani T, Sarraj B, Ohnishi M, Densmore MJ, Taguchi T et al (2009) In vivo genetic evidence for klotho-dependent, fibroblast growth factor 23 (Fgf23) -mediated regulation of systemic phosphate homeostasis. FASEB J 23:433–441
29. Otobe Y, Hiraki K, Izawa KP, Sakurada T, Shibagaki Y (2020) Relationship between serum inorganic phosphorus levels and cognitive decline over 2 years in older adults with pre-dialysis chronic kidney disease. Clin Exp Nephrol 24:286–287
30. Park JC, Han SH, Byun MS, Yi D, Lee JH et al (2017) Low serum phosphorus correlates with cerebral Abeta deposition in cognitively impaired subjects: results from the KBASE study. Front Aging Neurosci 9:362
31. Pesta DH, Tsirigotis DN, Befroy DE, Caballero D, Jurczak MJ et al (2016) Hypophosphatemia promotes lower rates of muscle ATP synthesis. FASEB J 30: 3378–3387
32. Razzaque MS (2009a) The FGF23-Klotho axis: endocrine regulation of phosphate homeostasis. Nat Rev Endocrinol 5:611–619
33. Razzaque MS (2009b) FGF23-mediated regulation of systemic phosphate homeostasis: is Klotho an essential player? Am J Physiol Renal Physiol 296:F470–F476
34. Razzaque MS (2011) Osteo-renal regulation of systemic phosphate metabolism. IUBMB Life 63:240–247
35. Razzaque MS (2013) Phosphate toxicity and vascular mineralization. Contrib Nephrol 180:74–85
36. Razzaque MS (2022) Salivary phosphate as a biomarker for human diseases. FASEB Bioadv 4: 102–108. https://doi.org/10.1096/fba.2021-00104
37. Razzaque MS (2022) Phosphate metabolism: from physiology to toxicity. Adv Exp Med Biol 1362:1–6. https://doi.org/10.1007/978-3-030-91623-7_1
38. Razzaque MS, Lanske B (2006) Hypervitaminosis D and premature aging: lessons learned from Fgf23 and Klotho mutant mice. Trends Mol Med 12:298–305
39. Sabbagh Y, Schiavi SC (2014) Role of NPT2b in health and chronic kidney disease. Curr Opin Nephrol Hypertens 23:377–384
40. Shao Y, Wen Q, Zhang S, Lu L, Zhang L et al (2019) Dietary supplemental vitamin D3 enhances phosphorus absorption and utilisation by regulating gene expression of related phosphate transporters in the small intestine of broilers. Br J Nutr 121:9–21
41. Sharma S, Brugnara C, Betensky RA, Waikar SS (2015) Reductions in red blood cell 2,3-diphosphoglycerate concentration during continuous renal replacment therapy. Clin J Am Soc Nephrol 10:74–79
42. Sharma S, Hashmi MF, Castro D (2021) Hypophosphatemia. In: StatPearls. StatPearls Publishing LLC, Treasure Island
43. Shuto E, Taketani Y, Tanaka R, Harada N, Isshiki M et al (2009) Dietary phosphorus acutely impairs endothelial function. J Am Soc Nephrol 20:1504–1512
44. Slatopolsky E, Finch J, Denda M, Ritter C, Zhong M et al (1996) Phosphorus restriction prevents parathyroid gland growth. High phosphorus directly stimulates PTH secretion in vitro. J Clin Invest 97: 2534–2540
45. Slinin Y, Vo T, Taylor BC, Murray AM, Schousboe J et al (2018) Serum phosphate and cognitive function in older men. Int J Geriatr Psychiatry 33:159–166
46. Sönmezler A, Keskek SO (2019) Low serum phosphorus levels and acute ischemic stroke. J Intensive Crit Care 5:3

47. Urakawa I, Yamazaki Y, Shimada T, Iijima K, Hasegawa H et al (2006) Klotho converts canonical FGF receptor into a specific receptor for FGF23. Nature 444:770–774
48. Vervloet M, Cozzolino M (2017) Vascular calcification in chronic kidney disease: different bricks in the wall? Kidney Int 91:808–817
49. Virani SS, Alonso A, Aparicio HJ, Benjamin EJ, Bittencourt MS et al (2021) Heart disease and stroke statistics-2021 update: a report from the American Heart Association. Circulation 143:e254–e743
50. Wright CB, Shah NH, Mendez AJ, DeRosa JT, Yoshita M et al (2016) Fibroblast growth factor 23 is associated with subclinical cerebrovascular damage: the northern Manhattan study. Stroke 47:923–928
51. Yamada S, Tokumoto M, Tatsumoto N, Taniguchi M, Noguchi H et al (2014) Phosphate overload directly induces systemic inflammation and malnutrition as well as vascular calcification in uremia. Am J Physiol Renal Physiol 306:F1418–F1428
52. Yamada S, Tsuruya K, Taniguchi M, Tokumoto M, Fujisaki K et al (2016) Association between serum phosphate levels and stroke risk in patients undergoing hemodialysis: the Q-Cohort study. Stroke 47:2189–2196
53. Yamashita T, Yoshioka M, Itoh N (2000) Identification of a novel fibroblast growth factor, FGF-23, preferentially expressed in the ventrolateral thalamic nucleus of the brain. Biochem Biophys Res Commun 277:494–498
54. Zhu L, Stein LR, Kim D, Ho K, Yu GQ et al (2018) Klotho controls the brain-immune system interface in the choroid plexus. Proc Natl Acad Sci U S A 115: E11388–E11E96

Index

M. S. Razzaque (ed.), *Phosphate Metabolism*, Advances in Experimental Medicine and Biology 1362,
https://doi.org/10.1007/978-3-030-91623-7

GPSR Compliance

The European Union's (EU) General Product Safety Regulation (GPSR) is a set of rules that requires consumer products to be safe and our obligations to ensure this.

If you have any concerns about our products, you can contact us on ProductSafety@springernature.com

In case Publisher is established outside the EU, the EU authorized representative is:

Springer Nature Customer Service Center GmbH
Europaplatz 3
69115 Heidelberg, Germany

Batch number: 10732809

Printed by Printforce, the Netherlands